Structural Reliability, Resilience and Design of Buildings against Multi-hazards

Structural Reliability, Resilience and Design of Buildings against Multi-hazards

Guest Editors

Xiaowei Zheng
Qinglin Wang
Yao Li

MDPI

Basel • Beijing • Wuhan • Barcelona • Belgrade • Novi Sad • Cluj • Manchester

Guest Editors
Xiaowei Zheng
School of Mechanics and Civil Engineering
China University of Mining and Technology
Xuzhou
China

Qinglin Wang
School of Mechanics and Civil Engineering
China University of Mining and Technology
Xuzhou
China

Yao Li
School of Mechanics and Civil Engineering
China University of Mining and Technology
Xuzhou
China

Editorial Office
MDPI AG
Grosspeteranlage 5
4052 Basel, Switzerland

This is a reprint of the Special Issue, published open access by the journal *Buildings* (ISSN 2075-5309), freely accessible at: https://www.mdpi.com/journal/buildings/special_issues/6VWCM7086V.

For citation purposes, cite each article independently as indicated on the article page online and as indicated below:

Lastname, A.A.; Lastname, B.B. Article Title. *Journal Name* **Year**, *Volume Number*, Page Range.

ISBN 978-3-7258-5209-3 (Hbk)
ISBN 978-3-7258-5210-9 (PDF)
https://doi.org/10.3390/books978-3-7258-5210-9

Contents

About the Editors

Xiaowei Zheng

Xiaowei Zheng received his Ph.D. degree from the Dalian University of Technology and worked as a Visiting Scholar at the Department of Civil and Environmental Engineering, University of Illinois at Urbana-Champaign from 2019 to 2020. He now works as an Associate Professor at the School of Mechanics and Civil Engineering, China University of Mining and Technology. His research interests include resilience assessment of engineering structures subjected to multiple hazards, especially that of tall buildings and bridges against multi-hazards such as earthquakes and winds. He has presided over five projects from the Provincial and National Foundations, including one project from the National Natural Science Foundation and one sub-project of the National Key Research and Development Program. Based on these projects, he has published more than 20 papers in peer-reviewed journals.

Qinglin Wang

Qinglin Wang earned his Ph.D. degree from Xi'an University of Architecture and Technology and is currently conducting postdoctoral research at the China University of Mining and Technology. His research focuses on seismic performance enhancement and resilience improvement of novel steel structural systems, as well as cold-formed steel structures. Currently, he leads two research projects: the Jiangsu Provincial Excellent Postdoctoral Program and the Fundamental Research Funds for the Central Universities (Youth Science and Technology Fund). He has published multiple papers as a first author or corresponding author in renowned journals including *Engineering Structures*, *Thin-Walled Structures*, the *Journal of Building Structures*, and *Engineering Mechanics*, and holds two granted invention patents.

Yao Li

Yao Li, Ph.D., is an Associate Professor in the School of Mechanics and Civil Engineering, China University of Mining and Technology. He joined CUMT in 2022. His research focuses on structural status evaluation and strengthening, high-performance civil engineering composite materials, and seismic resistance and durability for concrete structures. He receives financial support from the National Natural Science Foundation of China, China Postdoctoral Science Foundation, and Fundamental Research Funds for the Central Universities. He has also been selected by the Jiangsu Provincial Association for Science and Technology (JAST) Young Scientists Support Program (2024) and Excellent Postdoctoral Program of Jiangsu Province (2023), and has published more than 20 technical papers in refereed journals. Moreover, he serves as a Young Editorial Board Member for the *Journal of Central South University (Science and Technology)*. Dr. Li is also an active reviewer for several leading international journals, including *Engineering Structures*, the *Journal of Bridge Engineering*—ASCE, *Journal of Structural Engineering*—ASCE, and *Soil Dynamics and Earthquake Engineering*.

buildings

MDPI

Editorial

Editorial on the Structural Reliability, Resilience, and Design of Buildings Against Multi-Hazards

Xiaowei Zheng *, Qinglin Wang and Yao Li

School of Mechanics and Civil Engineering, China University of Mining and Technology, Xuzhou 221116, China
* Correspondence: xwzheng@cumt.edu.cn

1. Introduction

Against the backdrop of increasing seismic threats to cultural heritage globally, the vulnerability of historical masonry structures demands urgent attention. The nonlinear material behavior, structural heterogeneity, and time-sensitive conservation challenges necessitate innovative approaches to address these issues. This Special Issue compiles 14 pioneering studies that integrate simplified numerical modeling, smart monitoring, and advanced retrofitting to bridge the gap between engineering efficiency and heritage preservation. Collectively, these works establish a framework for rapid post-disaster decision-making, driving civil engineering toward the resilient and intelligent conservation of cultural assets in seismically active regions.

2. Overview of the Contributions

Zheng et al. (Contribution 1) developed an XGBoost-based surrogate model for the rapid seismic response assessment of building clusters (reducing the computation time by >90% with $R^2 > 0.99$) and a Bayesian probabilistic model for earthquake-induced debris width prediction, which reduced the prediction uncertainty by 60% and was validated via a campus case study showing 45% road blockage.

Demircan (Contribution 2) developed a simplified finite element (FE) approach to rapidly assess seismic vulnerability in the partially collapsed historic Habibi Neccar Mosque after the 2023 earthquakes. Using linear dynamic analysis, the study identified critical tensile stresses at dome–arch joints and vaults as primary failure triggers, exacerbated by soil amplification and out-of-plane wall deformations. The method prioritized computational efficiency, validated against field damage patterns, and offers a practical tool for emergency stabilization decisions in cultural heritage conservation.

Zhong et al. (Contribution 3) demonstrated, using shaking-table tests, that epoxy-bonded carbon-fiber cloth enhances RC Diaojiaolou seismic resilience, increasing the collapse PGA by 49.5% and redefining the damage into three phases: DS, YS, and PS. The natural periods and acceleration/displacement increments followed exponential laws, with carbon fiber minimizing the acceleration impact at the stilt layer (I) and optimizing the displacement resistance at the base for the unreinforced and stilt layer (I) for reinforced structures, providing a low-cost solution for mountain buildings in China's minority ethnic regions.

Jin et al. (Contribution 4) developed a nonlinear analytical framework using asymptotic methods to investigate the in-plane stability of pin-ended circular arches under uniform vertical loads. The parameter perturbation method effectively analyzed the snap-through buckling of shallow arches, with fifth-order solutions proving sufficiently accurate. The results demonstrated that the axial load components significantly influenced arches with large modified slenderness ratios or large rise-to-span ratios. For anti-symmetric buckling, the second-order WKB solution yielded lower critical loads than the third-order solution, providing a conservative estimate. The study confirmed the classical solutions' reliability for shallow arches with small rise-to-span ratios but highlighted the necessity of accounting for axial load effects in deep arches. The nonlinear bending moments consistently exceeded the linear predictions across all geometries.

Li et al. (Contribution 5) developed an innovative four-footstool GFRP connector for precast recycled concrete thermal insulation panels and experimentally evaluated its mechanical performance against traditional Thermomass MS connectors. The pull-out, shear, and flexural tests compared both connector types in conventional and recycled concrete. The results showed the GFRP connector increased the pull-out capacity by 14.8% and the shear resistance by 20.6% in recycled concrete, while the flexural strength improved by 16.5%, with more uniform crack distribution. A reduction factor $\gamma = 0.68$ was proposed to simplify the flexural design calculations, validated by close alignment with the experimental values. Recycled concrete demonstrated comparable performance to conventional concrete when paired with GFRP connectors, supporting sustainable construction applications.

Based on the cyclic loading tests of four prefabricated HDRPC joints, one cast-in-place HDRPC joint, and one normal concrete joint, Li et al. (Contribution 6) demonstrated that HDRPC significantly enhanced the seismic performance. The HDRPC increased the joint bearing capacity by 17.8% and the displacement ductility by 33.3%, reduced the crack width and shear deformation in the joint core, and allowed wider stirrup spacing, while outperforming normal concrete joints with denser reinforcement. The grouted sleeve connection ensured reliable force transfer, and a proposed crack-resistance equation aligned closely with the test results, supporting HDRPC's application in sustainable prefabricated structures.

Based on a seismic reliability framework, Li et al. (Contribution 7) proposed a collapse fragility analysis method for RC frame structures, considering the capacity uncertainty, utilizing incremental dynamic analysis to identify collapse points and the fractional exponential moments-based maximum entropy method to fit distributions of maximum inter-story drift ratios. They defined collapse via the stiffness degradation or MIDR thresholds, calculated the failure probabilities using a combined performance index, and fitted the fragility curves with a shifted generalized lognormal distribution. The results showed that accounting for the capacity uncertainty reduced the collapse probability, increasing the median collapse resistance by 13.2% and 87.3%, compared to the deterministic MIDR limits of 1/25 and 1/50, respectively, which demonstrated SGLD's superior accuracy over traditional lognormal models.

An et al. (Contribution 8) proposed a novel graded-yielding metal self-centering brace to minimize residual deformation in structures under strong earthquakes and achieve graded seismic protection. The brace utilized X-shaped steel plates and U-shaped steel plates for staged energy dissipation, combined with pre-compressed disc springs for self-centering. They derived a theoretical restoring force model and validated it using ABAQUS finite element simulations under low-cycle loading. The numerical results closely matched the theoretical model, demonstrating flag-shaped hysteresis curves, excellent energy dissipation, staged yielding, and effective self-centering. Parametric studies showed that increased disc spring pre-compression improved self-centering, while thicker steel plates enhanced energy dissipation but increased residual deformation.

Zhang et al. (Contribution 9) proposed and tested a composite damper with yielding reserve stiffness to address large structural deformations under seismic limit states. The YRSD combined a friction unit and a metal yield unit in series to provide multi-stage energy dissipation and yielding reserve stiffness. Low cyclic loading tests on three specimens with varying X-plate thicknesses and bolt preloads demonstrated full hysteresis curves, controllable multi-stage energy dissipation, and hardening post-yield stiffness. Doubling the bolt preload increased the friction energy dissipation by 53.8% and the envelop hysteresis area by 56.1%. Increasing the

X-plate thickness by 2 mm raised the resistance by 26.2% and the post-yield stiffness by 37.9%. The numerical simulations validated the experimental results with errors under 10%.

Ye et al. (Contribution 10) developed a multiscale damage identification method for beam-type structures using node curvature. They derived a damage index based on bending moment–curvature relationships and the assumption that micro-damage minimally affects stress redistribution. Wavelet analysis denoises signals before solving linear matrix equations via singular value decomposition. Laboratory tests on a simply supported beam verified the sensitivity to a 1% stiffness reduction. Field validation using a continuous beam bridge's multiscale FE model confirmed precise damage localization in small-scale elements and noise resistance with under 10% interference. The method effectively identified damage location/severity in both numerical and real-world scenarios, demonstrating robustness for structural health monitoring.

Bai et al. (Contribution 11) established seven 3D finite element models to investigate the influence of shallow soil reinforcement width and depth on the horizontal bearing capacity of a single pile in soft soil. The models, validated against field tests, analyzed reinforcement widths of 2*D*, 3*D*, and 4*D* and depths of 0.1 *L*, 0.2 *L*, and 0.3 *L*. The results showed that reinforcement significantly increased the horizontal bearing capacity by 83.0%, 104.3%, and 224.4% for increased widths and by 224.4%, 361.3%, and 456.8% for increased depths, respectively. The bending moments and deformations decreased with greater width, while the capacity gains diminished nonlinearly with a depth beyond 1.5*D*, indicating an optimal depth and advising against unevaluated deep reinforcement.

Chen et al. (Contribution 12) investigated the seismic performance of high-strength steel frames with Y-eccentric braces using a variable replaceable link. Parametric variations included the link's energy-dissipating region length, steel grade, span, and total link length. The results showed that reducing the span and using higher-grade steel enhanced the ductility and ultimate bearing capacity by up to 20.7%, while shorter energy-dissipating regions increased the shear force by 8.3%. An optimal link length improved the stiffness and energy dissipation, whereas an excessive length reduced the capacity and complicated the post-earthquake replacement. The plastic strain concentrated in the designed energy-dissipating zone, validating the replaceable link's effectiveness for targeted damage and easier repair.

Mao et al. (Contribution 13) proposed a kinetic energy distribution model for a rocking rigid body within a mass point system to refine the collision restitution coefficient calculation. Their model divided the collision into three stages: incorporating energy dissipation in the first two stages to determine total kinetic energy and redistributing the remaining kinetic energy in the third stage. This yielded an analytical solution for the coefficient. They validated the model using experimental and statistical data, finding significantly reduced errors and a correlation coefficient exceeding 0.9 compared to the classical Housner model, particularly for larger masses. The model provided a more reliable approach for the seismic analysis of rocking structures.

Ma et al. (Contribution 14) investigated the response modification factor (R) for high-strength steel frames with D-eccentric braces using the incremental dynamic analysis method. They designed 18 structural models varying in stories (4, 8, 12), link lengths, and steel strengths. The IDA method determined the R, displacement amplification factor (C_d), and overstrength factor (R_Ω) for each model. The results showed R values of approximately between 8.2 and 10.0, with effects of the story number, link length, and steel strength on R and C_d being complex and sometimes inconsistent. The study provided essential coefficients for the seismic design applications of D-HSS-EBFs.

3. Conclusions

These studies provide critical design coefficients, refine computational methodologies, and establish new benchmarks for earthquake-resistant infrastructure. The findings significantly advance the frontiers of structural engineering, offering actionable insights for resilient and sustainable construction in seismic-prone regions.

Acknowledgments: We extend our gratitude to the authors who contributed their research to this Special Issue and the reviewers for their meticulous evaluations. We also appreciate the editors for their dedication and perseverance in ensuring the success of this Special Issue.

Conflicts of Interest: The authors declare that they have no known competing financial interests or personal relationships that could have appeared to influence the work reported in this paper.

List of Contributions:

1. Zheng, X.; Hou, Y.; Cheng, J.; Xu, S.; Wang, W. Rapid Damage Assessment and Bayesian-Based Debris Prediction for Building Clusters Against Earthquakes. *Buildings* **2025**, *15*, 1481. https://doi.org/10.3390/buildings15091481.
2. Kılıç Demircan, R. Simplified FE-Based Post-Earthquake Vulnerability Assessment of a Partially Collapsed Historic Mosque. *Buildings* **2025**, *15*, 1849. https://doi.org/10.3390/buildings15111849.
3. Zhong, W.; Shu, Z.; Feng, W.; Zhang, X.; Ma, X.; Fei, Z. The Case Study of the Characteristic Analysis and Reinforcement Measures of RC Diaojiaolou Structures Under Different Seismic Intensities. *Buildings* **2025**, *15*, 1795. https://doi.org/10.3390/buildings15111795.
4. Jin, J.; Su, M. In-Plane Stability of Circular Arch Under Uniform Vertical Load Based on the Asymptotic Method. *Buildings* **2025**, *15*, 1149. https://doi.org/10.3390/buildings15071149.
5. Li, X.; Sun, H.; Zhang, T.; Bu, T.; Yu, H.; Sun, J.; Feng, H. Mechanical Properties of Precast Recycled Concrete Thermal Insulation Panels with GFRP Connectors. *Buildings* **2025**, *15*, 891. https://doi.org/10.3390/buildings15060891.
6. Li, X.; Sun, H.; Chen, K.; Yuan, T.; Wen, L.; Zheng, X.; Bu, T. Cyclic Behavior of Joints Assembled Using Prefabricated Beams and Columns with High-Ductility Recycled Powder Concrete. *Buildings* **2025**, *15*, 838. https://doi.org/10.3390/buildings15050838.
7. Zeng, T.; Li, Y. Collapse Fragility Analysis of RC Frame Structures Considering Capacity Uncertainty. *Buildings* **2025**, *15*, 694. https://doi.org/10.3390/buildings15050694.
8. An, Z.; Wang, W.; Wang, H.; Li, Z.; Wang, D.; Xie, G. Seismic Performance Research on a Graded-Yielding Metal Brace with Self-Centering Functions. *Buildings* **2024**, *14*, 3940. https://doi.org/10.3390/buildings14123940.
9. Zhang, X.; Zhu, L.; Liu, L.; Li, J. Hysteretic Performance of Composite Damper with Yielding Reserve Stiffness. *Buildings* **2024**, *14*, 3931. https://doi.org/10.3390/buildings14123931.
10. Ye, K.; Zhang, S.; Zhang, Q.; Zhong, R.; Wang, W. Multiscale Damage Identification Method of Beam-Type Structures Based on Node Curvature. *Buildings* **2024**, *14*, 3336. https://doi.org/10.3390/buildings14113336.
11. Bai, G.; Zhang, H.; Wang, B.; Chen, F.; Zhao, J.; Shu, Q. Investigation on the Bearing Performance of a Single Pile in Shallow Reinforced Soft Soil Foundation under Horizontal Load. *Buildings* **2024**, *14*, 3166. https://doi.org/10.3390/buildings14103166.
12. Chen, X.; Li, S.; Liang, G.; He, M. Numerical Parameter Analysis of High-Strength Steel Frame with Y-Eccentric Brace Using Variable Replaceable Link. *Buildings* **2024**, *14*, 2149. https://doi.org/10.3390/buildings14072149.
13. Mao, Q.; Deng, T.; Shen, B.; Wang, Y. Research on Collision Restitution Coefficient Based on the Kinetic Energy Distribution Model of the Rocking Rigid Body within the System of Mass Points. *Buildings* **2024**, *14*, 2119. https://doi.org/10.3390/buildings14072119.
14. Ma, Y.; Yang, J.; Ma, X. Response Modification Factor of High-Strength Steel Frames with D-Eccentric Brace Using the IDA Method. *Buildings* **2024**, *14*, 1619. https://doi.org/10.3390/buildings14061619.

buildings

MDPI

Article

Rapid Damage Assessment and Bayesian-Based Debris Prediction for Building Clusters Against Earthquakes

Xiaowei Zheng [1,*], Yaozu Hou [1], Jie Cheng [1], Shuai Xu [1] and Wenming Wang [2]

1 School of Mechanics and Civil Engineering, China University of Mining and Technology, Xuzhou 221116, China; hyz89672025@163.com (Y.H.); chengjie7050@163.com (J.C.); ts24030237p31@cumt.edu.cn (S.X.)
2 Key Laboratory of Building Structural Retrofitting & Underground Space Engineering, Ministry of Education, Shandong Jianzhu University, Jinan 250101, China; wangwenmingmr@sdjzu.edu.cn
* Correspondence: xwzheng@cumt.edu.cn

Abstract: In the whole service life of building clusters, they will encounter multiple hazards, including the disaster chain of earthquakes and building debris. The falling debris may block the post-earthquake roads and even severely affect the evacuation, emergency, and recovery operations. It is of great significance to develop a surrogate model for predicting seismic responses of building clusters as well as a prediction model of post-earthquake debris. This paper presents a general methodology for developing a surrogate model for rapid seismic responses calculation of building clusters and probabilistic prediction model of debris width. Firstly, the building cluster is divided into several types of representative buildings according to the building function. Secondly, the finite element (FE) method and discrete element (DE) method are, respectively, used to generate the data pool of structural floor responses and debris width. Finally, with the structural response data of maximum floor displacement, a surrogate model for rapidly calculating seismic responses of structures is developed based on the XGBoost algorithm, achieving $R^2 > 0.99$ for floor displacements and $R^2 = 0.989$ for maximum inter-story drift ratio (MIDR) predictions. In addition, an unbiased probabilistic prediction model for debris width of blockage is established with Bayesian updating rule, reducing the standard deviation of model error by 60% (from $\sigma = 10.2$ to $\sigma = 4.1$). The presented models are applied to evaluate the seismic damage of the campus building complex in China University of Mining and Technology, and then to estimate the range of post-earthquake falling debris. The results indicate that the surrogate model reduces computational time by over 90% compared to traditional nonlinear time-history analysis. The application in this paper is helpful for the development of disaster prevention and mitigation policies as well as the post-earthquake rescue and evacuation strategies for urban building complexes.

Keywords: surrogate model; debris; building clusters; Bayes theorem; resilience

1. Introduction

There have been significant differences in the seismic resistance of structures in building clusters. It is a challenging task to rapidly compute the seismic response of regional building structures. In addition, the debris accumulation from the nonstructural components or adjacent collapsed buildings might block roads and even seriously affect post-earthquake emergency rescue operations [1,2]. As reported, there were about 13 million cubic yards of debris produced in the 2010 Haiti earthquake event, which vastly hindered emergency and recovery operations [3]. Thereby, it is necessary to develop a surrogate model for rapid calculation of seismic structural responses as well as a prediction model for post-earthquake debris width.

For building clusters against individual seismic events, a series of studies have been conducted on the seismic response calculation of buildings at the community or city level. The rapid estimation method for calculating the seismic responses of building clusters mainly includes the following: (1) FEMA-P58 [4] summarizes the seismic fragility database of common components such as beams, columns, and walls, which is beneficial for rapid assessment of the seismic fragility of structures. (2) The structure is simply simulated using a lumped mass model, and the mechanical behavior of a certain floor or several floors in buildings is simulated by a hysteresis model [5–7]. Then, the seismic response of structures can be quickly obtained by the nonlinear time-history analysis conducted on the simplified structural model. However, this method neglects the uncertainties related to the spatial characteristics and structural parameters of buildings.

Furthermore, artificial intelligence techniques are widely used in the rapid seismic response estimation of buildings under earthquakes [8–11]. Chen [10] employed the principal component analysis method to generate seismic factors, and 326 school buildings were obtained to investigate the relationship between seismic factors and the structural performance. Then, the artificial neural network (ANN) was used to develop a surrogate model for estimating the seismic responses of school buildings. Sergio Ruggieri et al. [11] presented an automated procedure for fragility calculations for masonry buildings at the urban scale, which is beneficial for rapidly obtaining the seismic responses of a large number of archetype models. Stojadinovic et al. [12] proposed a framework for rapid seismic loss assessment of building clusters based on a machine learning method. Similarly, Zhang et al. [13] and Zhou et al. [14] devoted attention to the surrogate model for rapid calculation of seismic responses of buildings.

Furthermore, the estimation of the debris width is an important reference for planning post-earthquake evacuation and rescue routes. To this purpose, both the American standard ASCE 7-22 [15] and the Chinese standard "Standard for urban planning on comprehensive disaster resistance and prevention (GB/T51327-2018)" [16] specify the threshold of the falling debris width as the maximum distance affected by falling debris between 1/2 and 2/3 of building height [16]. Similar research [17–19] has been devoted to developing the distribution law of debris after an earthquake. However, to the authors' best knowledge, the theoretical formulas for calculating the debris width from collapsed buildings in the current literature are mainly deterministic versions. The deterministic prediction models fail to account for the influences of various damage levels of buildings subjected to earthquakes, and they describe the random distribution law of post-earthquake debris. Therefore, it is necessary to develop a probabilistic prediction model for the post-earthquake debris width.

This paper presents a framework to develop the surrogate model for rapid calculation of seismic responses of structures as well as the prediction model of post-earthquake debris width. Section 2 discusses the general concept of this presented framework. In Section 3, a surrogate model of rapid seismic response calculation is developed for the multi-story reinforced concrete (RC) frame structures. Section 4 presents a Bayesian-based prediction model of debris width for the RC frame structures based on the virtual data obtained by the physics engine technique on the Blender platform. Finally, the main conclusions are drawn in the last section.

2. Framework for Developing the Surrogate Model for Rapid Seismic Responses Calculation of Building Clusters and Probabilistic Prediction Model of Debris Width

Urban building clusters comprise structures characterized by heterogeneous structural configurations, functional purposes, heights, and construction periods. Despite their inherent variability, these buildings possess shared attributes. Given that both FE analysis and DE analysis are computationally intensive, this study introduces a method for classifying the building cluster into representative structural typologies. Subsequently, the seismic-induced responses of building clusters are calculated by a surrogate model trained by the machine learning method according to the virtual data obtained from the finite element (FE) analysis of a representative building structures set.

The general concept of this presented methodology for developing a surrogate model for rapid seismic responses calculation of building clusters, as well as a Bayesian-based prediction model of post-earthquake debris width, are plotted in Figure 1.

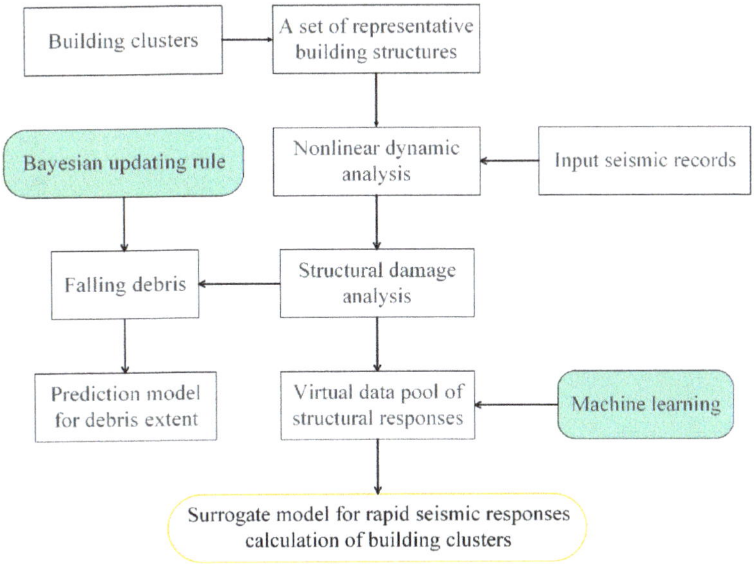

Figure 1. General concept of this presented framework.

As previously pointed out, the FE analysis is a time-consuming process. The trade-offs between computing speed, generalizability, and accuracy are balanced by the following:

(i) Using the nonlinear time-history analysis (NTHA) on the refined FE models of buildings on OpenSees platform to generate high-quality training data.

(ii) A surrogate model is trained with the XGBOOST algorithm based on the data obtained by the NTHA method.

(iii) The trained surrogate model is used to rapidly calculate the seismic-induced structural responses of building clusters replacing the NTHA.

According to the debris distribution data and Bayesian updating rules, the probabilistic prediction model for post-earthquake debris width is expressed as follows [20–23]:

$$D(\mathbf{x}, \Theta) = d + \gamma(\mathbf{x}, \boldsymbol{\theta}) + \sigma\varepsilon \tag{1}$$

where d is the deterministic predictive model for debris width that can be described by existing traditional models; $\gamma(\mathbf{x}, \boldsymbol{\theta})$ is the linear correction term; $\boldsymbol{\theta}$ is the coefficient vector in the correction term; $\sigma\varepsilon$ is the model error following a normal distribution with a mean value of 0 and a standard deviation of σ; $\boldsymbol{\Theta} = (\boldsymbol{\theta}, \sigma)$ represents the unknown model parameters. The error correction term can be rewritten as follows [24]:

$$\gamma(\mathbf{x}, \boldsymbol{\theta}) = \sum_{i=1}^{n} \theta_i h_i(\mathbf{x}) \tag{2}$$

in which h_i is the normalized variable or explanatory function that may affect the stochastic distribution of building debris. To capture potential biases in this probabilistic prediction model independent of the variable $\mathbf{x}$, we set $h_1(\mathbf{x}) = 1.0$ [24]. According to the Bayesian updating rule [23], the posterior probability density function (PDF) of the unknown model parameters $\boldsymbol{\Theta}$ is given by

$$f'(\boldsymbol{\Theta}) = kL(\boldsymbol{\Theta})f(\boldsymbol{\Theta}) \tag{3}$$

where the normalization coefficient $k = [\int L(\boldsymbol{\Theta})f(\boldsymbol{\Theta})\mathrm{d}\boldsymbol{\Theta}]^{-1}$ is used to ensure the integration of posterior probability density function $f'(\boldsymbol{\Theta})$ equaling 1.0; $L(\boldsymbol{\Theta})$ is the likelihood function. To reflect the lack of prior information, the noninformative prior probability density function of the unknown parameter $\boldsymbol{\Theta}$ can be expressed as follows [20]:

$$f(\boldsymbol{\Theta}) \propto \frac{1}{\sigma} \tag{4}$$

Upon determining $f'(\boldsymbol{\Theta})$, the Markov chain Monte Carlo (MCMC) simulation can be used to obtain the estimation of the unknown model parameters. The Metropolis–Hastings algorithm is commonly used to realize the MCMC sampling by the following steps [25]. Firstly, assume an arbitrary initial value $\boldsymbol{\Theta}_0$ for the unknown parameters satisfying $f'(\boldsymbol{\Theta}_0 \mid \text{Data})>0$. Secondly, the proposal distribution $q(\boldsymbol{\Theta}' \mid \boldsymbol{\Theta}_0)$ is used to generate a new candidate value $\boldsymbol{\Theta}'$ from the initial value $\boldsymbol{\Theta}_0$. If $\boldsymbol{\Theta}'$ follows the proposal distribution, we accept this point as the new state $\boldsymbol{\Theta}_1$ of the chain. Otherwise, $\boldsymbol{\Theta}'$ is rejected, then we set $\boldsymbol{\Theta}_1 = \boldsymbol{\Theta}_0$. Thirdly, we repeat the above process to obtain the chain for $\boldsymbol{\Theta}$.

Furthermore, the uncertainty associated with unknown model parameters $\boldsymbol{\Theta}$ can be represented by their probability density functions developed by the Bayesian updating rule. Then, uncertainties in unknown parameters can be quantified by integrating over the domain of unknown parameters based on full probability theory.

3. The Surrogate Model and Debris Width Prediction Model for Multi-Story RC Frame Structures

3.1. Input Seismic Ground Motion Records

To conduct the nonlinear dynamic analysis, 100 seismic records were downloaded from the PEER-NGA database (http://peer.berkeley.edu/nga/) [26]. Additionally, four seismic motion records from the Wenchuan earthquake catalogue were obtained from China Strong Motion Center. Totally, there are 104 seismic records used for finite element analysis.

The comparison between the elastic response spectrum of these seismic records and the code-prescribed spectra are plotted in Figure 2. It can be observed that the average of elastic response spectrum lies within 0.8 to 1.2 times the code-prescribed value, meaning that these selected seismic records satisfy the requirements of the Chinese seismic code [26].

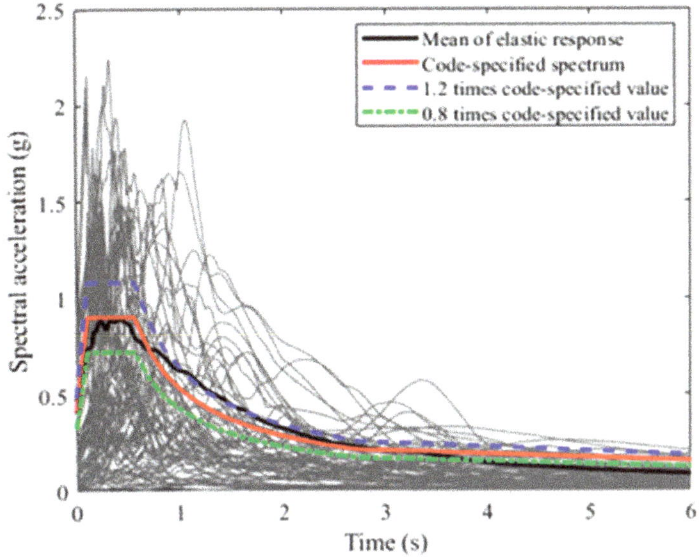

Figure 2. The elastic response and code-specified spectra of these ground motion records.

3.2. Machine-Learning-Based Surrogate Model

Referring to the "Code for Seismic Design of Buildings" (GB 50011-2010) [26], three multi-story RC frame structures were designed on the ETABS platform. The profile of these three representative buildings is shown in Table 1.

Table 1. Profile of representative building structures.

Categories	Function	Height (m)	Section (mm)
Five floors structures	Teaching building	Ground floor: 3.9 Other floors: 3.6 Total height: 18.3	Beams: 350 × 500 Reinforcement ratio: 2.285% Columns: 600 × 600 Reinforcement ratio: 1.111%
Five floors structures	Office building	Ground floor: 4.5 Other floors: 3.6 Total height: 18.9	Beams: 600 × 800 Reinforcement ratio: 1.308%; Columns: 800 × 800 Reinforcement ratio: 0.981%;
Six floors structures	Dormitory building	Ground floor: 3.6, Other floors: 3.3 Total height: 23.4	Beams: 400 × 600 Reinforcement ratio: 1.667%; Columns: 700 × 700 Reinforcement ratio: 0.816%

The OpenSees platform is used to construct the FE model of multi-story RC frame structures. The Concrete01 and Steel02 model are used to describe the mechanical behavior of concrete and steels, respectively. Both the beam and column members are simulated by the "dispBeamColumn" element. The FE models of these three representative building structures used, respectively, as teaching building, office building, and dormitory building, are illustrated in Figure 3.

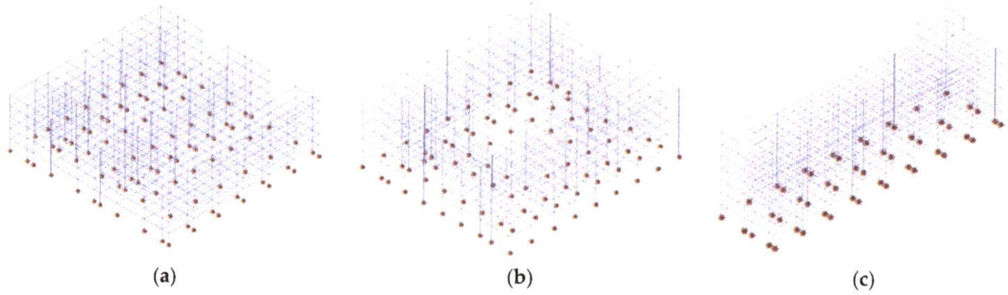

Figure 3. The FE models of typical building structures. (a) Teaching building. (b) Office building. (c) Dormitory building.

In order to verify the modeling accuracy on OpenSees, the teaching building is taken as an example to compare the natural vibration period obtained by ETABS and OpenSees platforms, as listed in Table 2. In this table, T_e represents the natural vibration period of the structure calculated by ETABS; conversely, T_o denotes the results determined by OpenSees.

Table 2. The preceding six-order natural periods of the school building obtained by ETABS and OpenSees.

Natural Period of Vibration	1	2	3	4	5	6
T_e	0.972	0.865	0.836	0.807	0.791	0.756
T_o	0.997	0.890	0.857	0.835	0.834	0.750
ϵ	2.54%	2.90%	2.48%	3.39%	5.49%	4.50%

It can be seen in Table 2 that the FE analysis of RC frame structure on OpenSees platform has high accuracy and can be used to establish the virtual data pool for training the surrogate model. In order to consider the uncertainty from various sources, the mechanical properties of concrete and steel materials, as well as the damping ratios, are regarded as random variables. The statistical characteristics and probability distributions of these random variables are summarized in Table 3.

Table 3. Probabilistic description of the uncertain parameters.

Parameter	Mean Value	COV/%	Probability Distribution	Upper Limit	Lower Limit	Source
f_y (MPa)	363.8	5	Lognormal	$1.1 f_{y,mean}$	$0.9 f_{y,mean}$	Melchers [27]
α_s (−)	0.03	20	Lognormal	0.04	0.02	Barbato et al. [28]
E (Mpa)	201,000	3.3	Lognormal	214,266	187,734	Barbato et al. [28]
ξ (−)	0.05	40	Normal	$1.4\ \xi_{mean}$	$0.6\ \xi_{mean}$	Porter et al. [29]

The dynamic nonlinear analysis of structures under earthquakes is a time-consuming process. To balance the computational efficiency and accuracy, 30 FE models of structures were generated by the LHS method. Considering both the teaching building and office building having 5 floors, only 10 FE models were generated for the teaching building. Finally, a total of 70 FE models of RC frame structures were used for dynamic analysis, that is, 7280 (= 70 × 104) nonlinear time-history analyses were performed on the OpenSees platform. When the uncertain variables are taken as an average value, the maximum inter-story drift angle and the maximum lateral displacement of each floor of these three representative building structures under earthquake are shown in Figure 4.

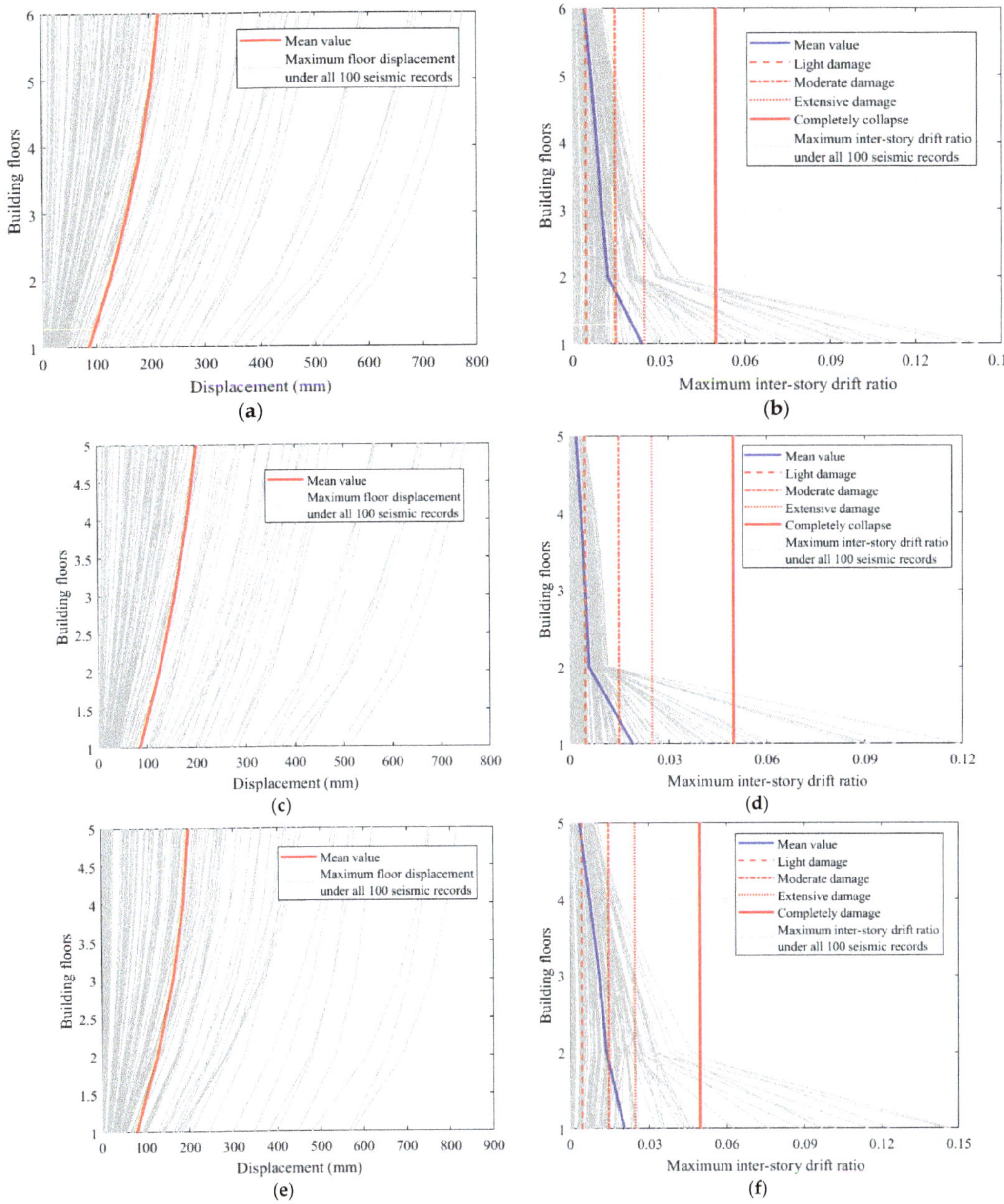

Figure 4. Structural response of representative buildings under all seismic records. (**a**) Dormitory building: maximum lateral floor displacement. (**b**) Dormitory building: maximum inter-story drift ratio. (**c**) Office building: maximum lateral floor displacement. (**d**) Office building: maximum inter-story drift ratio. (**e**) Teaching building: maximum lateral floor displacement. (**f**) Teaching building: maximum inter-story drift ratio.

Herein, the maximum inter-story drift ratio (MIDR) is applied to define the damage state of the RC frame structure, in which slight damage: MIDR > 0.5%; moderate damage: MIDR > 1.5%; serious damage: MIDR > 2.5%; and complete collapse: MIDR > 5.0% [30]. Figure 4 indicates that the representative building structures will experience serious damage or even collapse when subjected to earthquakes. In addition, the uncertainty of random variables has a significant impact on the seismic structural responses.

The input features used for training the surrogate model for rapidly calculating the seismic-induced structural responses mainly include building information, uncertainty variables, and intensity measurement of input ground motions, as shown in Table 4. The eigenvalues of the output layer are the maximum lateral floor displacements and the maximum inter-story drift ratio. In training the surrogate model, 70% of the total sample was selected as the training set, 20% of the sample is the testing set, and the remaining 10% of data is the verification set. We consider the advantages of the XGBoost algorithm in solving the following problems: (i) Handling high-dimensional data by gradient boosting and regularization, (ii) high computational efficiency for large datasets, and (iii) high accuracy in regression tasks. The XGBoost algorithm is used to train this surrogate model, in which the optimal network parameters are determined by the Newton–Raphson optimization algorithm and Trap Avoidance Operator (TAO) [31].

Table 4. Input features of network.

Eigenvalue Type	Parameter	Data Format
Earthquake	PGA, Δt, duration.	Numeric data
Uncertainty parameters	Damping ratio, modulus of elasticity, yield strength, post-yield stiffness ratio.	Numeric data
Architectural information	Use function.	Text data
	X and Y direction natural period of vibration, maximum column spacing, span number, beam-column reinforcement ratio, floor height, number of building floors.	Numeric data

In this paper, the accuracy of the surrogate model is evaluated by parameters of mean absolute error (MAE), mean square error (MSE), root mean square error (RMSE), mean absolute percentage error (MAPE), and coefficient of determination (R^2), as summarized in Table 5. It can be observed that the surrogate model for predicting the maximum floor drift and the maximum inter-story drift ratio responses under seismic loads has high accuracy.

Table 5. Goodness of fit for these surrogate models.

Output Characteristic	Set	MAE	MSE	RMSE	MAPE	R^2
Floor displacement	Training	3.186	22.890	4.784	0.050	0.9990
	Testing	5.048	65.998	8.124	0.072	0.9972
MIDR	Training	3.252×10^{-4}	3.727×10^{-7}	6.105×10^{-4}	0.054	0.9983
	Testing	6.042×10^{-4}	2.509×10^{-6}	1.584×10^{-3}	0.075	0.9890

Taking the MIDR of each floor in representative buildings, the regression results of the surrogate models are plotted in Figure 5. It can be found that there is high goodness of fit of this surrogate model with a correlation coefficient larger than 99%, which can be applied to replace the FE analysis. However, the surrogate model presented in this current study cannot generate the real-time structural responses, which is the main limitation of this paper.

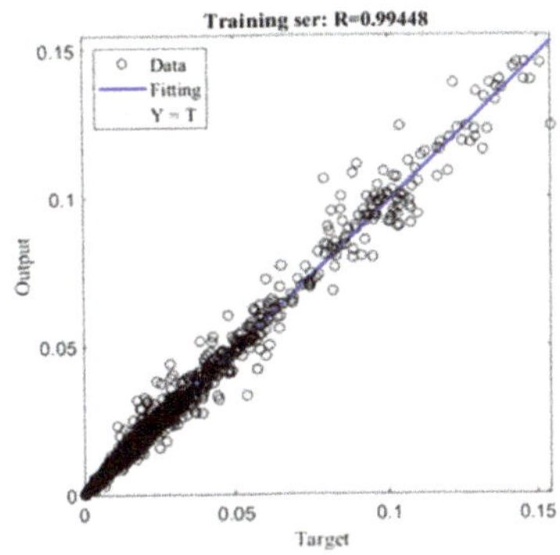

Figure 5. Regression results of surrogate models.

To further show the superiority of the XGBoost algorithm, the comparison between the predicted and target MIDR is illustrated in Figure 6.

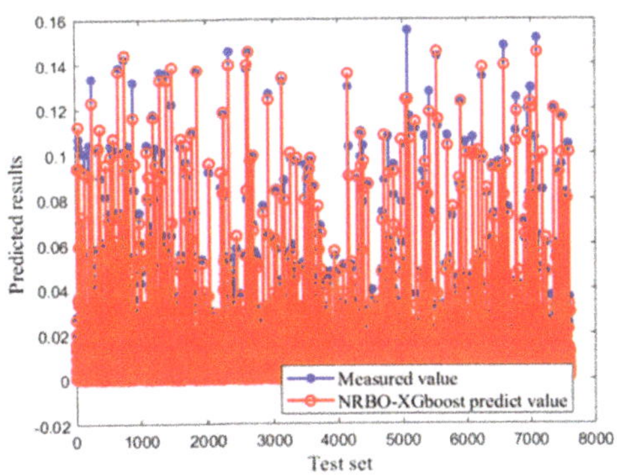

Figure 6. Comparison of predicted and target values obtained by surrogate models.

Figure 6 indicates that most of the predicted values are similar to the target values. Thus, this trained surrogate model will be used for the following seismic damage assessment of building clusters.

3.3. Bayesian-Based Prediction Model of Debris Width

In this subsection, a multi-story RC frame structure is used as a case to generate the virtual debris data by Bullet Constraints Builder (BCB) for Blender to develop the Bayesian-based prediction model for the debris width. This office building has a span of 6 m, with a first-floor height of 4.2 m and other-floor height of 3.5 m, as shown in Figure 7.

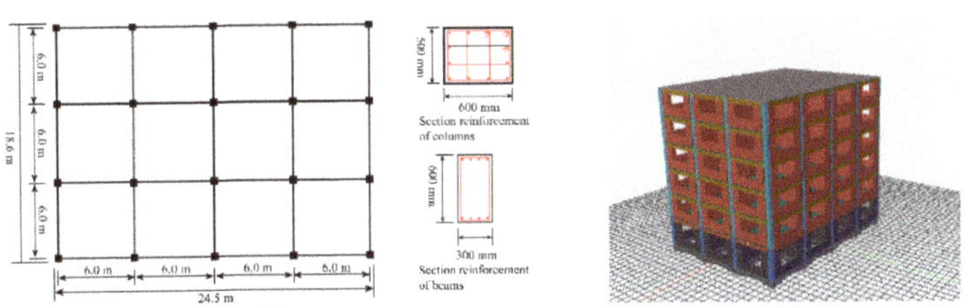

Figure 7. The design information and discrete element model of this 6-story RCF structure. (**a**) Dimension and reinforcement details. (**b**) Discrete element model.

The reinforcement condition of beams is characterized as the longitudinal reinforcements with a diameter of 22 mm and stirrup bars of 8 mm diameter spaced at 150 mm intervals. The columns incorporate longitudinal reinforcements of 25 mm diameter and stirrup bars of 8 mm diameter spaced at 100 mm intervals. In addition, the HRB400 steel is used for all rebars and the concrete cover is 30 mm of beams and columns. The reinforcement details of the beams and columns are illustrated in Figure 7a. The floor slabs are designed with bidirectional reinforcement, with 8 mm diameter steel bars spaced 180 mm apart, made of HRB335 steel. The concrete grade for slabs is C30, with concrete cover of 30 mm. The layout of openings in the first floor and the building's discrete element model are shown in Figure 7b. In modeling this building by the BCB, the representative value of gravity load on all floors roof is 750 kg/m^2.

The BCB physics engine is applied to simulate the dynamic response of this six-story RC frame structure under 104 seismic motions bidirectional loading at a ratio of 1:0.85, respectively, for the X-axis and the Y-axis. Then, 104 post-earthquake building debris distribution maps are obtained, as shown in Figure 8.

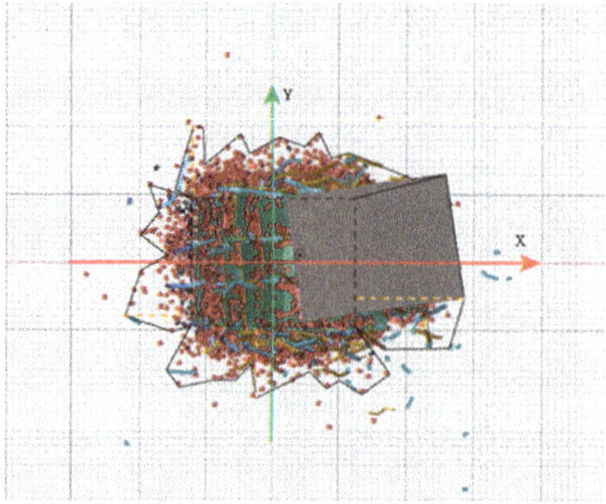

Figure 8. Distribution of post-earthquake building falling debris.

In Figure 8, the maximum vertical distance from the building footprint to the edge of debris width is defined as the maximum debris width; the area of the region enclosed by the debris envelope is denoted as $A_{\mathrm{b,direction}}$, in which the subscript direction refers to the orientation of the debris around the building, namely, up, down, left, and right. In addition, the side length of one small square in Figure 8 is 1 m. The average distance is computed by the ratio of A_{b} to the side length of the building footprint in the corresponding direction. The statistical characteristics of the debris width are shown in Figure 9.

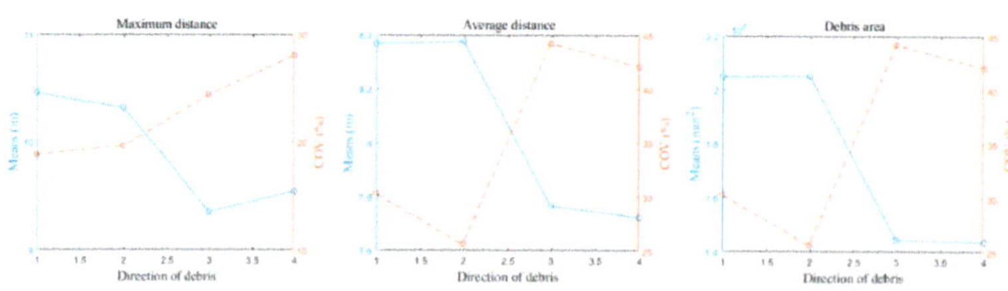

Figure 9. Statistical characteristics of post-earthquake debris width.

It can be observed that there is a significant variability in debris width for different directions. The coefficient of variation (COV) of debris width ranges from 10% to 45%. Generally speaking, the post-earthquake debris width shows prominent random characteristics. Considering the characteristics of the traditional model, a model proposed by He et al. [32]

is selected as the deterministic term in the prediction model. For the prediction model of maximum debris width, Equation (1) can be further rewritten as

$$D(\mathbf{x},\Theta) = 0.58H^{1.06}\mathrm{PGA}^{0.5} + \mathrm{PGA}(\theta_1 + \theta_2 A_b) + \sigma\varepsilon \quad (5)$$

The normalized area of debris width and PGA are the random variables involved in the error correction term. The area of debris width is normalized by the area of the building plane. Herein, only the maximum distance of debris width in the left direction is presented as an example, and the posterior statistics of unknown model parameters are summarized in Table 6.

Table 6. Posterior statistics of unknown parameters.

Parameters	Average Value	Standard Deviation	Correlation Coefficient Matrix		
			θ_1	θ_2	σ
θ_1	−8.226	4.198	1	−0.98	−0.01
θ_2	16.161	8.322	−0.98	1	0.01
σ	4.116	0.289	−0.01	0.01	1

The values presented in the third column of Table 6 indicate that the model parameters exhibit significant dispersion. In other words, if the model parameters are taken as deterministic values, it will inevitably introduce considerable uncertainty into subsequent analyses. Therefore, it is necessary to develop a probability distribution of the unknown model parameters.

To further illustrate the advantage of the Bayesian-based method in reducing the impact of uncertainty, the prior and posterior probability density functions (PDFs) of the model error are illustrated in Figure 10. It can be found that the posterior PDF is more concentrated compared to the prior one, meaning that the Bayesian-based prediction model holds smaller dispersion.

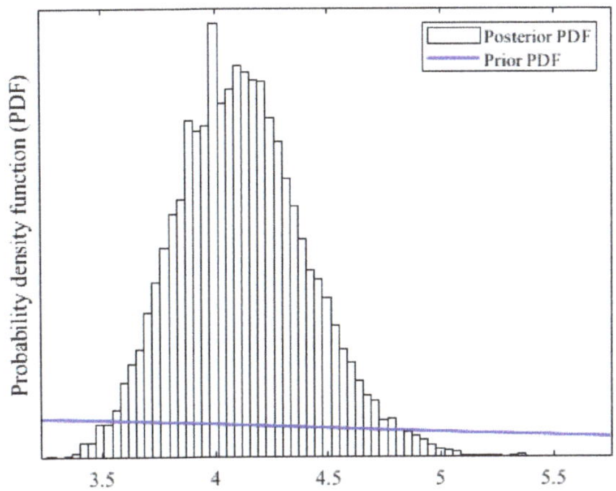

Figure 10. Comparison between prior and posterior probability density functions.

The Bayesian-based prediction model of the maximum debris width in the left direction is shown in Figure 11.

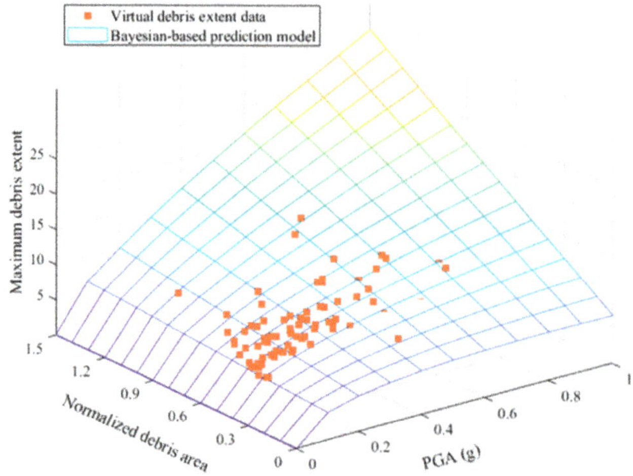

Figure 11. Bayesian-based prediction model for the maximum debris width.

4. Seismic Damage Assessment for Building Clusters

In this section, the 2008 Wenchuan earthquake with PGA = 203.450 cm/s^2 and Δt = 0.002 is assumed to occur in Xuzhou City, Jiangsu Province. The reinforced concrete building clusters on the campus of China University of Mining and Technology are used as a case study to demonstrate the superiority of the proposed framework in rapidly assessing structural damage and predicting post-earthquake debris distribution patterns.

The trained surrogate model is applied to assess the structural damage of the RC buildings in the campus of China University of Mining and Technology, as shown in Figure 12. In this figure, the colors of blue, green, orange, and red, respectively, denote the light damage, moderate damage, severe damage, and complete collapse.

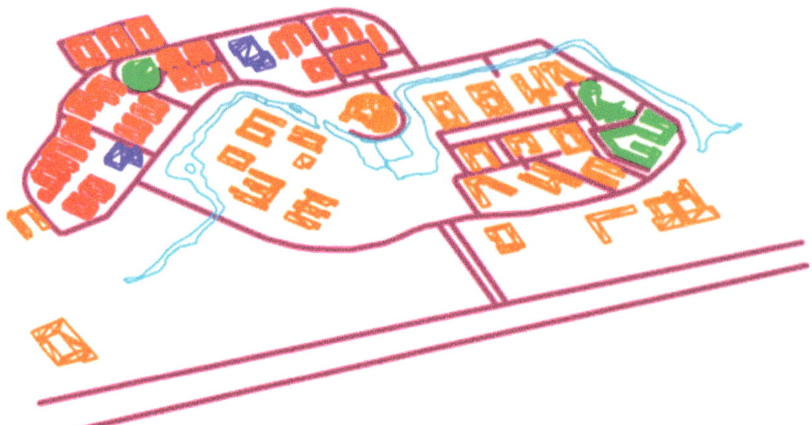

Figure 12. Damage distribution of campus buildings.

Then, the Bayesian-based prediction model is used to evaluate the post-earthquake debris width distributed in the university campus, as shown in Figure 13.

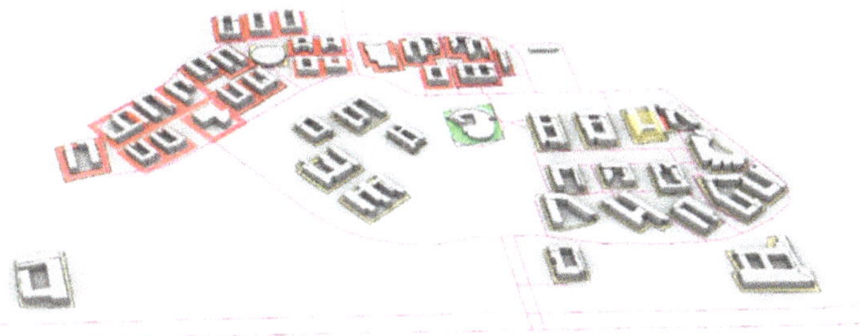

Figure 13. The debris extent distribution in the university campus.

Combining the results presented in Figures 12 and 13, it can be found that under the action of the Wenchuan earthquake, the dormitory building complex (the red area in the figures) will be severely damaged. Furthermore, the falling debris will block 45% of the adjacent roads (shown in Figure 13), meaning that it may be difficult to carry out post-earthquake rescue and evacuation operations. This application highlights the method's effectiveness in real-world scenarios by quantifying damage states and debris blockage risks.

5. Conclusions

This study presents a novel framework integrating a machine-learning-based surrogate model and a Bayesian probabilistic approach to rapidly assess seismic responses of building clusters and predict post-earthquake debris width. The surrogate model, trained by the XGBoost algorithm based on finite element (FE) simulation data, enables efficient seismic response calculations for multi-story reinforced concrete frame structures. The Bayesian-based prediction model, incorporating uncertainty quantification and virtual debris data generated via physics engines, provides probabilistic estimates of debris width. The main conclusions are as follows:

(1) The XGBoost-based surrogate model achieves exceptional accuracy in predicting structural responses (e.g., $R^2 > 0.99$ for floor displacement and $R^2 = 0.989$ for maximum inter-story drift ratio), reducing computational time by over 90% compared to traditional nonlinear time-history analyses. This surrogate model enables rapid damage assessment for large-scale building clusters, which is critical for real-time emergency decision making.

(2) The Bayesian-updated model significantly reduces uncertainty in debris width prediction, narrowing the standard deviation of model error by 60%, from prior $\sigma = 10.2$ to posterior $\sigma = 4.1$. This probabilistic approach outperforms deterministic methods, providing 95% confidence intervals for debris blockage distances, which is essential for evacuation route planning.

(3) Application to a campus building cluster under the Wenchuan earthquake event with PGA = 203.45 cm/s^2 reveals severe damage (MIDR > 2.5%) in 35% of dormitory buildings, with debris blocking 45% of adjacent roads. These results highlight the utility of this presented framework in identifying high-risk zones and optimizing post-disaster rescue strategies.

In future research, we will expand the virtual data for buildings with various characteristics to enlarge the virtual data pool, and the LSTM or Transformer algorithms will be applied to update the trained surrogate model that can obtain the real-time responses of structures. In addition, the experimental results and satellite image data of collapsed buildings will be used to enhance the realism and suitability of debris prediction models in various urban settings.

Author Contributions: Conceptualization, methodology, resources, analysis, and writing: X.Z.; data curation: Y.H., J.C., and S.X.; constructive comments: W.W. All authors have read and agreed to the published version of the manuscript.

Funding: The authors sincerely appreciate the financial support of the National Natural Science Foundation of China (no. 52208511).

Data Availability Statement: Some or all data, models, or codes that support the findings of this study are available from the corresponding author upon reasonable request.

Conflicts of Interest: The authors declare that there are no conflicts of interest.

References

1. Feng, K.; Li, Q.; Ellingwood, B.R. Post-earthquake modelling of transportation networks using an agent-based model. *Struct. Infrastruct. Eng.* **2020**, *16*, 1578–1592. [CrossRef]
2. Byun, J.-E.; D'Ayala, D. Urban seismic resilience mapping: A transportation network in Istanbul, Turkey. *Sci. Rep.* **2022**, *12*, 8188. [CrossRef] [PubMed]
3. Sediek Omar, A.; El-Tawil, S.; McCormick, J. Seismic Debris Field for Collapsed RC Moment Resisting Frame Buildings. *J. Struct. Eng.* **2021**, *147*, 04021045. [CrossRef]
4. *FEMA-P-58*; Seismic Performance Assessment of Buildings, Methodology and Implementation. Federal Emergency Management Agency (FEMA): Washington, DC, USA, 2018.
5. Zhang, B.; Xiong, F.; Lu, Y.; Ge, Q.; Liu, Y.; Mei, Z.; Ran, M. Regional seismic damage analysis considering soil–structure cluster interaction using lumped parameter models: A case study of Sichuan University Wangjiang Campus buildings. *Bull. Earthq. Eng.* **2021**, *19*, 4289–4310. [CrossRef]
6. Xu, Z.; Wu, Y.; Lu, X.Z.; Jin, X.L. Photo-realistic visualization of seismic dynamic responses of urban building clusters based on oblique aerial photography. *Adv. Eng. Inform.* **2020**, *43*, 101025. [CrossRef]
7. Hayashi, K.; Saito, T.; Horioka, T.; Sato, E. Implementation of real-time seismic diagnostic system on emergency management center buildings: System introduction and operational status on municipal government office buildings. *J. Civ. Struct. Health Monit.* **2019**, *9*, 529–541. [CrossRef]
8. Morfidis, K.; Stefanidou, S.; Markogiannaki, O. A Rapid Seismic Damage Assessment (RASDA) Tool for RC Buildings Based on an Artificial Intelligence Algorithm. *Appl. Sci.* **2023**, *13*, 5100. [CrossRef]
9. Kalakonas, P.; Silva, V. Seismic vulnerability modelling of building portfolios using artificial neural networks. *Earthq. Eng. Struct. Dyn.* **2022**, *51*, 310–327. [CrossRef]
10. Chen, C.S. Seismic performance assessments of school buildings in Taiwan using artificial intelligence theories. *Eng. Comput.* **2020**, *37*, 3321–3343. [CrossRef]
11. Ruggieri, S.; Liguori, F.S.; Leggieri, V.; Bilotta, A.; Madeo, A.; Casolo, S.; Uva, G. An archetype-based automated procedure to derive global-local seismic fragility of masonry building aggregates: META-FORMA-XL. *Int. J. Disaster Risk Reduct.* **2023**, *95*, 103903. [CrossRef]
12. Stojadinovic, Z.; Kovacevic, M.; Marinkovic, D.; Stojadinovic, B. Rapid earthquake loss assessment based on machine learning and representative sampling. *Earthq. Spectra* **2022**, *38*, 152–177. [CrossRef]
13. Zhang, C.; Wen, W.; Zhai, C.; Jia, J.; Zhou, B. Structural nonlinear seismic time-history response prediction of urban-scale reinforced concrete frames based on deep learning. *Eng. Struct.* **2024**, *317*, 118702. [CrossRef]
14. Zhou, Y.; Meng, S.; Lou, Y.; Kong, Q. Physics-Informed Deep Learning-Based Real-Time Structural Response Prediction Method. *Engineering* **2024**, *35*, 140–157. [CrossRef]
15. *ASCE7-22*; Minimum Design Loads and Associated Criteria for Buildings and Other Structures. American Society of Civil Engineers (ASCE): Reston, VA, USA, 2022.
16. *GB/T51327-2018*; Standard for Urban Planning on Comprehensive Disaster Resistance and Prevention. China Architecture & Building Press: Beijing, China, 2018.
17. Lu, X.; Yang, Z.; Chea, C.; Guan, H. Experimental study on earthquake-induced falling debris of exterior infill walls and its impact to pedestrian evacuation. *Int. J. Disaster Risk Reduct.* **2020**, *43*, 101372. [CrossRef]
18. Wang, W.; Zhang, N.; Wang, L.; Wang, Z.; Ma, D. A Study of Influence Distance and Road Safety Avoidance Distance from Postearthquake Building Debris Accumulation. *Adv. Civ. Eng.* **2020**, *2020*, 1–17. [CrossRef]
19. Zhou, Y.; Chen, H.; Lin, X.; Chen, Z. Research on Emergency Evacuation Route Planning Method based on Building Groups Seismic Damage Simulation. *J. Disaster Prev. Mitig. Eng.* **2022**, *42*, 1174–1182.
20. Zheng, X.-W.; Li, H.-N.; Shi, Z.-Q. Hybrid AI-Bayesian-based demand models and fragility estimates for tall buildings against multi-hazard of earthquakes and winds. *Thin-Walled Struct.* **2023**, *187*, 110749. [CrossRef]

21. Zheng, X.-W.; Li, H.-N.; Gardoni, P. Hybrid Bayesian-Copula-based risk assessment for tall buildings subject to wind loads considering various uncertainties. *Reliab. Eng. Syst. Saf.* **2023**, *233*, 109100. [CrossRef]
22. Zheng, X.-W.; Li, H.-N.; Lv, H.-L.; Huo, L.-S.; Zhang, Y.-Y. Bayesian-based seismic resilience assessment for high-rise buildings with the uncertainty in various variables. *J. Build. Eng.* **2022**, *51*, 104321. [CrossRef]
23. Zheng, X.-W.; Li, H.-N.; Gardoni, P. Probabilistic seismic demand models and life-cycle fragility estimates for high-rise buildings. *J. Struct. Eng.* **2021**, *147*, 04021216. [CrossRef]
24. Gardoni, P.; Der Kiureghian, A.; Mosalam, K.M. Probabilistic capacity models and fragility estimates for reinforced concrete columns based on experimental observations. *J. Eng. Mech.* **2002**, *128*, 1024–1038. [CrossRef]
25. Berg, B.A.; Billoire, A. *Markov Chain Monte Carlo Simulations*; John Wiley & Sons, Inc.: New York, NY, USA, 2008.
26. *GB50011-2010*; Code for Seismic Design of Buildings. China Architecture and Building Press: Beijing, China, 2016.
27. Melchers, R.E. *Structural Reliability Analysis and Prediction*; John Wiley: New York, NY, USA, 1999.
28. Barbato, M.; Gu, Q.; Conte, J.P. Probabilistic Pushover analysis of structural and soil-structure systems. *J. Struct. Eng.* **2010**, *136*, 1330–1341. [CrossRef]
29. Porter, K.A.; Beck, J.L.; Shaikhutdinov, R.V.; Earthquake, P. *Investigation of Sensitivity of Building Loss Estimates to Major Uncertain Variables for the Van Nuys Testbed*; Pacific Earthquake Engineering Research Center: Berkeley, CA, USA, 2002.
30. Wen, Y.K.; Kang, Y.J. Minimum building life-cycle cost design criteria. I: Methodology. *J. Struct. Eng.* **2001**, *127*, 330–337. [CrossRef]
31. Sowmya, R.; Premkumar, M.; Jangir, P. Newton-Raphson-based optimizer: A new population-based metaheuristic algorithm for continuous optimization problems. *Eng. Appl. Artif. Intell.* **2024**, *128*, 107532. [CrossRef]
32. He, Y.; Zhai, C.; Wen, W. Probabilistic prediction of post-earthquake rubble range for RC frame structures. *Soil Dyn. Earthq. Eng.* **2023**, *171*, 107927. [CrossRef]

buildings

MDPI

Article

Mechanical Properties of Precast Recycled Concrete Thermal Insulation Panels with GFRP Connectors

Xiuling Li [1,2,*], Haodong Sun [2], Tianxuan Zhang [2], Tongxing Bu [2], Haoming Yu [2], Jiaxin Sun [2] and Hu Feng [3]

1 School of Civil Engineering, Shandong Jiaotong University, Jinan 250357, China
2 Key Laboratory of Building Structural Retrofitting and Underground Space Engineering, Ministry of Education, School of Civil Engineering, Shandong Jianzhu University, Jinan 250101, China; horizonsun77@163.com (H.S.)
3 School of Civil Engineering, Zhengzhou University, Zhengzhou 450001, China; fenghu@zzu.edu.cn
* Correspondence: dlutiem@163.com

Abstract: To improve both the composite performance of precast thermal insulation wall panels and the environmental sustainability of the structure, this study employs recycled concrete, and introduces an innovative four-footstool Glass Fiber Reinforced Plastic (GFRP) connector to join the inner and outer panels of precast thermal insulation wall systems. The experimental program included pull-out, shear, and bending tests to compare the performance of wall panels equipped with traditional Thermomass MS connectors and the novel GFRP connectors, using both conventional and fully recycled concrete. The results indicate that, when paired with recycled concrete, the GFRP connectors exhibited a 14.8% higher pull-out bearing capacity than the traditional connectors. Additionally, shear tests demonstrated that the GFRP connectors offered a 20.6% improvement in shear resistance compared to the Thermomass MS connectors. The bending strength of panels with GFRP connectors also showed an enhancement, with a 16.5% increase in flexural strength relative to those using traditional connectors. Notably, the GFRP connectors contributed to a more uniform crack distribution under loading, thereby improving the overall structural integrity. A reduction factor γ for the GFRP four-footstool connector was proposed based on a fully composite model, and the analysis of the composite degree calculation showed that the recycled concrete sample using the new GFRP connector had the highest composite degree.

Keywords: GFRP connectors; precast structure; recycled concrete; mechanical performance; thermal insulation wall panels

1. Introduction

In the context of sustainable architecture and energy-efficient design, prefabricated insulated exterior wall panels have gained increasing attention due to their superior thermal insulation properties, ease of construction, and environmental benefits [1–3]. Compared to traditional cast-in-place walls, prefabricated concrete sandwich panels significantly reduce construction time, minimize on-site wet work, lower construction risks, and enhance overall efficiency. These panels have become a core element in modern energy-efficient building systems, effectively reducing building energy consumption and improving overall sustainability [4–6].

Recently, the popularity of prefabricated insulated wall panels has risen due to their energy efficiency and construction advantages. However, traditional systems using connectors, such as the Thermomass MS needle-type connector, may have limitations that hinder their overall performance. These limitations include poor load transfer efficiency, particularly at the interface between the concrete wythes and the insulation layer, as well as susceptibility to stress concentrations [7–9]. The relatively small contact area between the connector and the concrete can lead to uneven load distribution, resulting in localized failures and reduced durability over time [10,11]. Additionally, traditional connectors often fail to ensure optimal composite action between the concrete and insulation, compromising the overall structural integrity of the wall panel [12,13]. These shortcomings highlight the need for improved connectors to enhance the performance and longevity of prefabricated insulated wall panels [14,15]. In particular, at the interface between the two concrete layers, composite behavior can lead to uneven mechanical performance distribution, affecting both load-bearing capacity and durability. The degree of composite action in sandwich panels depends on the shear connector configuration and the connection method used [16,17]. Effective connectors are crucial to ensuring the stability, durability, and energy performance of the interface between the insulation material and the concrete structure [18–21]. Therefore, developing connectors that meet structural and energy efficiency requirements is vital for advancing this field.

For instance, Wang et al. [22] proposed an innovative GFRP shell panel and studied the effects of mesh thickness, mesh spacing, mesh height, and skin thickness on axial stiffness, displacement ductility, and energy dissipation. They developed corresponding models to predict the strength of this type of wall panel. Choi et al. [23] performed monotonic loading and wind-induced cyclic loading tests on full-scale wall panels, demonstrating that all specimens met the usage standards. Benayoune et al. [24] examined the structural performance of precast concrete sandwich panels (PCSP) under bending, using both experimental and theoretical methods, and proposed a two-dimensional finite element model to assess the composite action in the PCSP system. Sharaf et al. [25] investigated a novel sandwich panel with two GFRP shells laminated onto the exterior of a prefabricated polyurethane foam core. Their results show that panels with soft cores were vulnerable to local effects under concentrated loads, experiencing inward wrinkling of the compressed surface at lower ultimate loads. Zhi et al. [26] introduced a new type of shear connector made from steel and GFRP materials. They conducted flexural tests on 12 full-scale precast concrete sandwich panels to assess their composite behavior. Naito et al. [27] examined 14 different shear connector types, quantified their failure modes, and established simplified multilinear strength curves for each type. Kang et al. [28] proposed a design model for sandwich wall panels (SWPs) reinforced with GFRP grids to predict flexural failure, simplifying the calculation process for such structures.

In the design of PCSP, composite action refers to the interaction between the two concrete wythes and the intermediate insulation layer in the sandwich panel. Based on the degree of composite effect, PCSPs can be classified into three design types: non-composite (NC), partially composite (PC), and fully composite (FC) [29,30]. The degree of composite action between the two concrete interlayers is critical for determining the mechanical properties, bearing capacity, and stiffness of the panel. The composite effect depends on the connectors, which enable the concrete and insulation layers to collaboratively bear external loads. Stronger composite effects lead to better collaboration between the layers, thereby improving strength and stiffness [31–33]. Although many connectors only provide partial composite effects, advanced connectors, such as high-strength steel or fiberglass-reinforced plastic, can significantly improve performance, reduce material use, and lower long-term construction and maintenance costs [34,35].

To address the limitations of existing connectors, this study proposes a novel four-footstool GFRP connector. Made from GFRP material, this connector exhibits high strength, low thermal conductivity, and excellent corrosion resistance, contributing to improved thermal insulation performance. Compared to the conventional Thermomass MS needle-type connector, the four-footstool connector increases the contact area with the concrete wall panel, thereby enhancing mechanical performance. Manufactured from epoxy resin-based GFRP, this new connector offers greater durability and efficiency than traditional connectors.

To evaluate the performance of the newly designed connector, this study conducted pull-out, shear, and flexural tests to compare the mechanical properties of the four-footstool GFRP connector with those of the conventional Thermomass MS needle-type connector. The primary focus was on analyzing the load–displacement behavior, ultimate strength, and failure modes of the two connector types under different concrete materials (C35 and recycled concrete). The experimental results are expected to provide deeper insights into the advantages of the novel connector in terms of structural performance, offering theoretical support for optimizing prefabricated insulated wall panel systems. Finally, through theoretical analysis, this study calculates the degree of composite action in wall panels using the new connector, and proposes a modification coefficient for the flexural strength calculation of external wall panels with the GFRP four-footstool connector. The introduction of this modification coefficient, γ, was validated by experimental testing, where the adjusted flexural strength using γ closely matched the experimentally measured values. This adjustment factor simplifies the design process, as engineers can apply the coefficient to the traditional flexural strength formula to account for the improved composite action due to the GFRP four-footstool connectors.

2. Experimental Program

2.1. Design of Connectors

Currently, most connectors used in industrial production for wall panels are Thermomass MS needle-type connectors, as shown in Figure 1a. These connectors, made from glass fiber, offer advantages such as high strength, excellent corrosion resistance, and low thermal conductivity. However, they have certain limitations in composite performance. The main issue lies in the relatively small contact area between the connector and the concrete layers, which can lead to inefficient load transfer and stress concentration at the interface between the concrete wythes and the insulation layer. This can result in suboptimal composite action and a higher likelihood of localized failure.

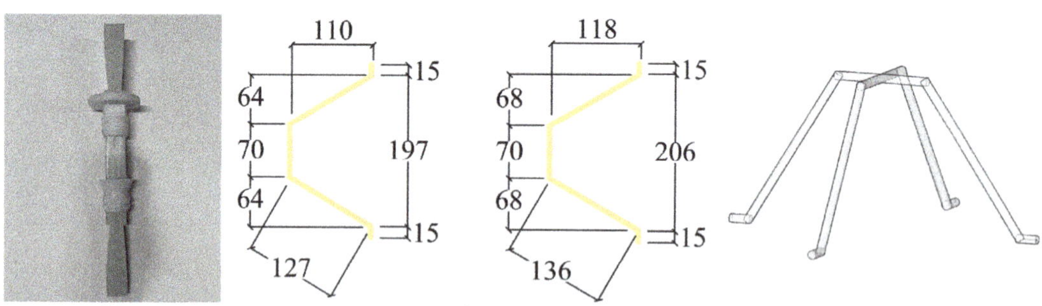

(a) (b) (c)

Figure 1. Design of connectors: (**a**) Thermomass MS needle type connector; (**b**) GFRP four-footstool connector design drawing; (**c**) GFRP four-footstool connector combination diagram.

To overcome these limitations, a novel GFRP four-footstool connector was designed, as shown in Figure 1b. The four-footstool connector is formed by the cross-combination of two similar open trapezoidal connectors, with the assembled configuration illustrated in Figure 1c. This cross-combination design significantly increases the contact area between the concrete layers and the insulation, improving load transfer and ensuring better stress distribution. By increasing the effective contact area, the GFRP four-footstool connector enhances the composite action between the two concrete layers and the insulation, which leads to improved load-bearing capacity and overall structural performance. The four-footstool connector can effectively distribute the load and avoid excessive material usage while ensuring the required strength and stability. Additionally, the unique four-footstool geometry provides better mechanical interlocking between the concrete and insulation layers. The combination of two trapezoidal shapes helps to optimize the connection, providing a more robust and stable composite structure. This design minimizes the risk of failure at the connector interface and addresses the composite performance shortcomings of the Thermomass MS needle-type connectors, making the GFRP four-footstool connector a more efficient and reliable choice for precast insulated wall panels.

The GFRP four-footstool connector is entirely made of epoxy resin-based GFRP material, with a cylindrical diameter of 8 mm. The initial 8 mm diameter of the GFRP four-footstool connector was chosen based on practical design considerations and typical connector sizes. However, further testing with larger diameters (e.g., 10 mm, 12 mm, 14 mm) will be conducted to assess their impact on the mechanical performance of the connector, the load-bearing capacity, and the overall efficiency. It is manufactured through thermoplastic processing using custom factory molds. The molds and finished products are shown in Figure 2.

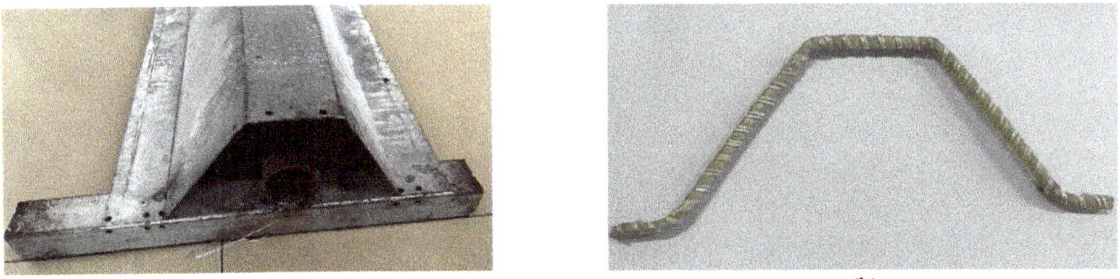

(a) (b)

Figure 2. Connector mold and finished product: (**a**) GFRP four-footstool connector mold; (**b**) GFRP four-footstool connector.

2.2. Design of Specimens

A total of four groups of comparative pull-out tests were designed for the connectors of prefabricated thermal insulation exterior wall panels. One specimen was fabricated and tested for each group, with variables including concrete materials (ordinary concrete and fully recycled concrete) and connector types (Thermomass MS needle type connectors and GFRP four-footstool connectors). The anchor lengths at both ends of the Thermomass MS needle type connectors were 35 mm, while the internal and external anchor lengths of the GFRP four-footstool connectors were 34 mm, which is essentially consistent with the Thermomass MS type. The thickness of the insulation layer used in the tests was 50 mm for all specimens. The specimen numbers and specific parameters are listed in Table 1. One specimen was designed for each group of tests. The dimensions and reinforcement of the specimens are shown in Figure 3.

A total of four groups of comparative shear tests were designed for the connectors of prefabricated thermal insulation exterior wall panels. The dimensions, materials, and anchorage lengths of the two types of connectors were consistent with those used in the pull-out tests. The specimen numbers and specific parameters are listed in Table 2. The specimen design is shown in Figure 4.

Table 1. Parameters of pull-out test.

Specimen	Connector Type	Exterior Panel Material	Interior Panel Material
POG-C [1]	GFRP Four-Footstool	C35	C35
POG-R [2]	GFRP Four-Footstool	RC	RC
POT-C [3]	Thermomass MS	C35	C35
POT-R [4]	Thermomass MS	RC	RC

[1] "POG-C" is pull-out GFRP four-footstool connector-C35. [2] "POG-R" is pull-out GFRP four-footstool connector—recycled concrete. [3] "POT-C" is pull-out Thermomass connector—C35. [4] "POT-R" is pull-out Thermomass connector—recycled concrete.

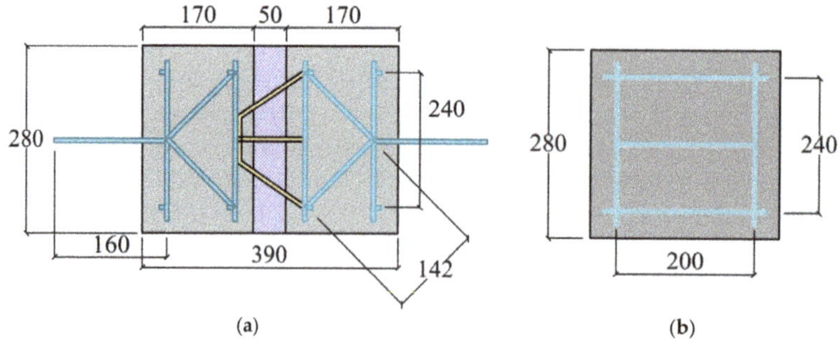

Figure 3. Construction design of pull-out specimen: (**a**) front view; (**b**) top view.

Table 2. Design parameters of shear test.

Specimen	Connector Type	Inner and Outer Panel Materials
CG-C [1]	GFRP four-footstool	C35
CG-R [2]	GFRP four-footstool	RC
CT-C [3]	Thermomass MS	C35
CT-R [4]	Thermomass MS	RC

[1] "CG-C" is cut-GFRP four-footstool connector—C35. [2] "CG-R" is cut-GFRP four-footstool connector—recycled concrete. [3] "CT-C" is cut-Thermomass connector—C35. [4] "CT-R" is cut-Thermomass connector—recycled concrete.

Flexural test specimens of prefabricated insulated exterior wall hanging panels were designed and fabricated. The comparison parameters included the following: W-1 (C35 specimen with Thermomass MS needle type connector) and W-2 (C35 specimen with GFRP four-footstool connector); W-2 (using ordinary concrete) and W-4 (using fully recycled concrete). Both specimens used GFRP four-footstool connectors. W-3 and W-4 had the same concrete materials and connectors. W-3 (the insulation board was in the middle of the wall panel, and the thickness of the inner and outer wall panels was the same), W-4 (the outer wall thickness was 50 mm, and the inner wall thickness was 80 mm). The specific design parameters are shown in Table 3.

A total of four flexural test specimens were fabricated. The size of the specimens was 3200 mm × 1200 mm × 180 mm. The thickness of the inner and outer wall panels and the locations of the insulation boards of W-1, W-2 and W-4 were the same, with the thickness from the inside to the outside being 80 mm, 50 mm, and 50 mm, respectively. The thickness of W-3 from the inside to the outside was 65 mm, 50 mm, and 65 mm. W-1 used ordinary concrete with a Thermomass MS needle-type connector, W-2 used C35 with a GFRP four-footstool connector, and W-3 and W-4 both used fully recycled concrete with a GFRP four-footstool connector.

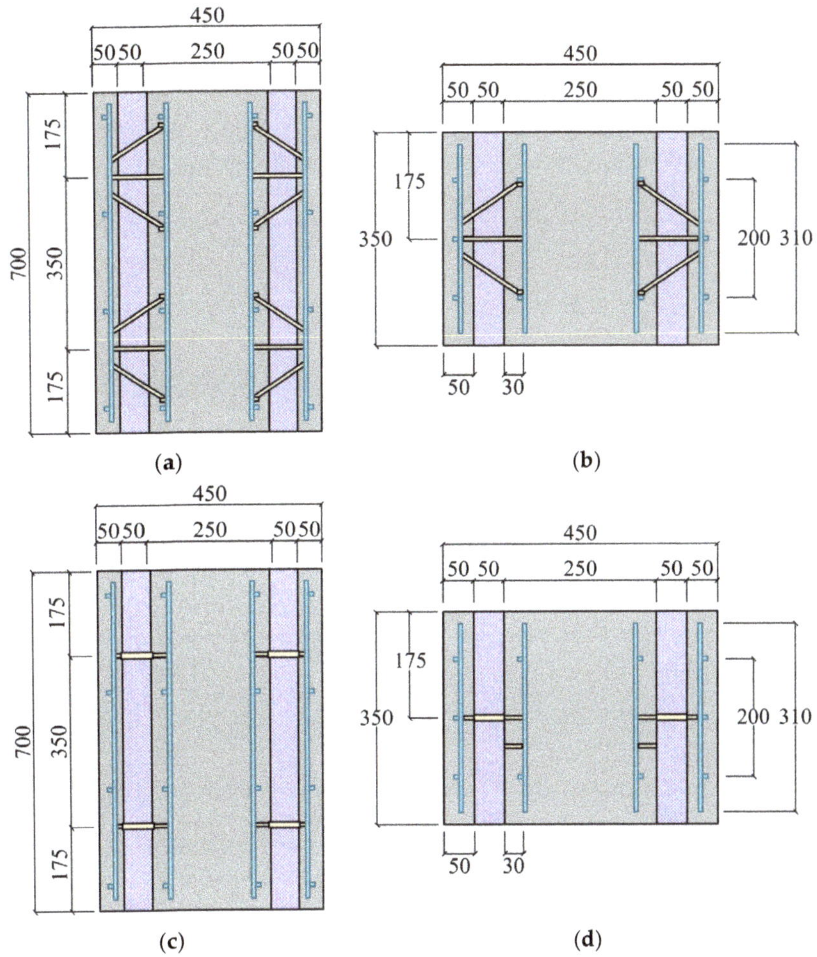

Figure 4. Construction design of shear specimen: (**a**) main view of GFRP specimen; (**b**) top view of GFRP specimen; (**c**) main view of Thermomass specimen; (**d**) top view of Thermomass specimen.

Table 3. Design parameter of flexural specimen of wall panels.

Specimen	Outer Panel/mm	Inner Panel/mm	Connector	Outer Material	Inner Material
W-1 [1]	50	80	Thermomass MS Type	C35	C35
W-2	50	80	GFRP Four-footstool	C35	C35
W-3	65	65	GFRP Four-footstool	RC	RC
W-4	50	80	GFRP Four-footstool	RC	RC

[1]. "W-1" is wall 1.

Both the inner and outer wall panels of the specimen were equipped with a single-layer steel mesh of Φ8@200. The connectors were evenly distributed at a spacing of 350 mm × 250 mm longitudinally and transversely. The specific dimensions and reinforcement diagram of the sample are shown in Figure 5.

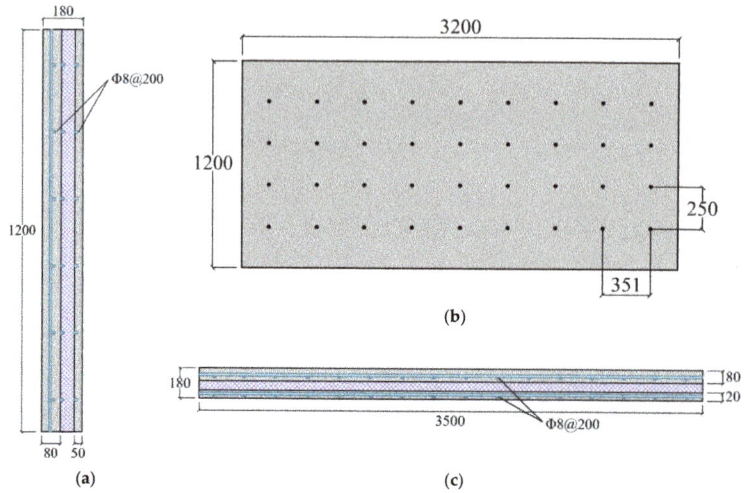

Figure 5. Construction design of flexural specimen: (**a**) side view; (**b**) top view; (**c**) front view.

2.3. Materials

To investigate the mechanical properties and thermal conductivity of fully recycled concrete, and to determine the appropriate mix proportion for the test, an orthogonal experiment was conducted, as shown in Table 4. The factors considered included the water–cement ratio, RCA replacement rate, RFA replacement rate, RCP replacement rate, and the amount of AOS air-entraining agent. A total of 16 test groups were established. The dosage of the water-reducing agent was determined based on the fluidity of the mix. The base mix used for the test was ordinary C35 concrete, as shown in Table 5. In the experiment, RCP was used to replace a portion of the cement in the base mix according to the design ratio, while the water-reducing agent and AOS air-entraining agent were added based on the mass of the cementitious materials (cement and RCP) in the mix, in accordance with the design specifications.

Table 4. Mix proportion design of orthogonal test.

Mixture	w/b [1]	RCA [2] Replacement Rate (%)	RFA [3] Replacement Rate (%)	RCP [4] Replacement Rate (%)	AOS Air-Entraining Agent [5] (%)
1	0.38	0	0	0	0
2	0.38	25	25	15	0.07
3	0.38	50	50	30	0.14
4	0.38	75	75	45	0.21
5	0.43	0	25	30	0.21
6	0.43	25	0	45	0.14
7	0.43	50	75	0	0.07
8	0.43	75	50	15	0
9	0.48	0	50	45	0.07
10	0.48	25	75	30	0
11	0.48	50	0	15	0.21
12	0.48	75	25	0	0.14
13	0.53	0	75	15	0.14
14	0.53	25	50	0	0.21
15	0.53	50	25	45	0
16	0.53	75	0	30	0.07

[1] "w/b" is water to binder ratio. [2] "RCA" is recycled concrete aggregate. [3] "RFA" is recycled fine aggregate. [4] "RCP" is recycled concrete powder. [5] "AOS air-entraining agent" is the amount of Alpha–Olefin sulfonate air-entraining agent.

Table 5. Mix proportion of C35 (kg/m^3).

Grade	Cement	NFA [1]	NCA [2]	Water	HRWR [3]
C35	440	700	1050	167.2	4.4

[1] "NFA" is natural fine aggregate. [2] "NCA" is natural coarse aggregate. [3] "HRWR" is a high-range water reducer used to improve the flow ability of concrete.

The compressive and flexural strength tests of fully recycled concrete were conducted as shown in Figure 6a,b. The specimens were placed on a universal testing machine and subjected to uniform loading during the test. At the conclusion of the test, the average strength of each group was recorded for comparison. Thermal conductivity tests were performed using a Hot Disk TPS2500S apparatus (Hot Disk Co., Ltd, Uppsala, Sweden). Prior to testing, the specimens were dried to constant weight in an oven at 80 °C. The probes were clamped using two test specimens to ensure the surfaces were flat, and the contact area was more than twice the size of the test probes. During the testing process, a constant current induced a temperature change, and the resistance change inside the probe caused a voltage drop. The data variation was recorded to accurately measure the thermal conductivity of the sample. The Hot Disk TPS2500S instrument is shown in Figure 6c.

(a) (b) (c)

Figure 6. Recycled concrete properties test: (**a**) Compressive strength test; (**b**) flexural strength tests; (**c**) Hot Disk TPS2500S thermal conductivity meter.

The compressive strength, flexural strength, thermal conductivity and specific heat capacity of 16 groups of test blocks were tested, respectively, and the test results are shown in Table 6. Figure 7 shows the effects of different factors on recycled concrete.

Table 6. Test results.

Mixture	Compressive Strength/MPa	Flexural Strength/MPa	Compression Ratio	Thermal Conductivity/W/m·K	Specific Heat Capacity/MJ/m^3·K
1	66.4	5.9	0.0889	1.2577	0.9814
2	29.6	4.1	0.1385	0.8152	0.4754
3	19.8	2.5	0.1263	0.7505	0.7922
4	16.9	3.2	0.1893	0.6402	0.7332
5	24.0	3.2	0.1333	0.9097	0.5927
6	19.5	3.0	0.1538	1.3423	1.2899
7	17.3	3.2	0.1850	0.6486	0.6028
8	45.8	4.6	0.1004	0.9844	0.8807
9	15.0	2.4	0.1600	0.6854	0.4811
10	33.5	4.4	0.1313	1.4563	1.3870
11	14.1	2.5	0.1773	0.6966	0.3978
12	17.9	3.2	0.1788	0.6860	0.5743
13	15.1	2.5	0.1656	0.5447	0.4738
14	18.4	2.7	0.1467	1.1039	0.9031
15	18.2	3.0	0.1648	0.7356	0.9177
16	17.2	2.4	0.1395	1.2613	1.0015

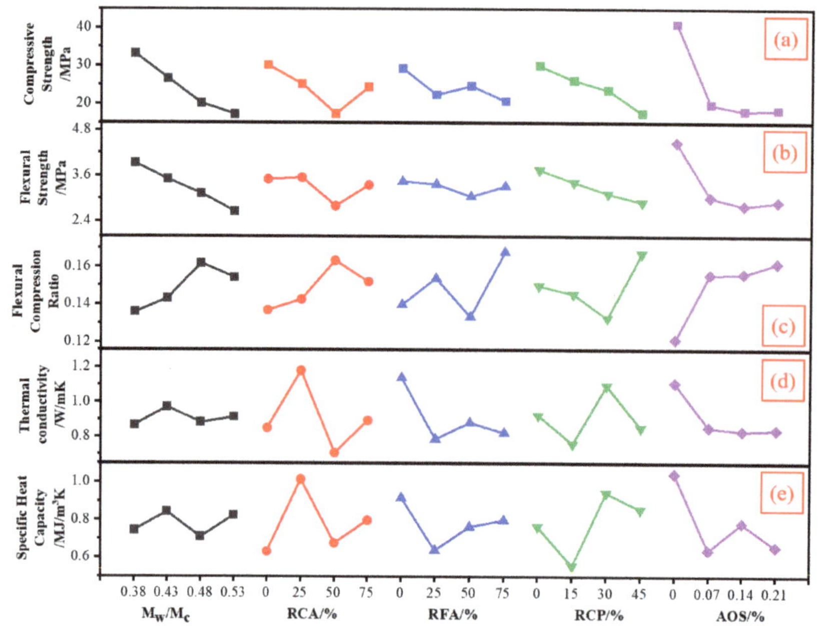

Figure 7. Effects of different factors on recycled concrete: (**a**) Compressive strength; (**b**) flexural strength; (**c**) flexural compression ratio; (**d**) thermal conductivity; (**e**) specific heat capacity.

We used Minitab (v19.1) software on the whole recycled concrete compressive strength, flexural strength and thermal conductivity coefficients of the three mechanical properties and thermal conductivity of the important indicators of the results of the comprehensive analysis. Table 7 shows the results of an analysis of the polarity of the factors, ranking the comprehensive performance of the whole recycled concrete in the order of precedence as regards the impacts of different factors as follows: AOS air-entraining agent > water to binder ratio > RCA replacement rate > RCP replacement rate > RFA replacement rate.

Table 7. Comprehensive performance range analysis.

	w/b [1]	RCA [2] Replacement Rate (%)	RFA [3] Replacement Rate (%)	RCP [4] Replacement Rate (%)	AOS Air-Entraining Agent [5] (%)
Average K1	12.655	11.491	11.296	11.558	15.52
Average K1	10.374	9.993	8.862	10.112	7.884
Average K1	8.044	6.953	9.56	9.281	7.235
Average K1	6.929	9.564	8.282	7.05	7.363
Range	5.727	4.539	3.014	4.508	8.284
Rank	2	3	5	4	1

[1] w/b: water to binder ratio. [2] RCA: recycled concrete aggregate. [3] RFA: recycled fine aggregate. [4] RCP: recycled concrete powder. [5] AOS air-entraining agent: the amount of Alpha–Olefin Sulfonate air-entraining agent.

Through polar analysis and a comprehensive evaluation of the actual conditions, the optimal mix design for fully recycled concrete was determined using an orthogonal test based on the analysis of eight ratio groups. The selected mix design includes a water–cement ratio of 0.43, a recycled concrete aggregate (RCA) replacement rate of 75%, an RCA replacement rate of 50%, a recycled concrete powder (RCP) replacement rate of 15%, and AOS dosage of 0%. Table 8 illustrates the mix proportions for recycled concrete. The compressive strength of the fully recycled concrete prepared using this mix design reaches 45.8 MPa, with a flexural strength of 4.6 MPa and a thermal conductivity of 0.9844, which is 21.7% lower than that of ordinary concrete. These mechanical properties meet the requirements of most projects, while maximizing the use of recycled materials, thereby contributing positively to the reduction in carbon emissions.

Table 8. Mix proportion of recycled concrete (kg/m^3).

Grade	Cement	RCP [1]	Sand	RFA [2]	Aggregate	RCA [3]	Water	HRWR [4]
RC	374	66	350	350	262.5	787.5	189.2	4.4

[1] "RCP" is recycled concrete powder. [2] "RFA" is recycled fine aggregate. [3] "RCA" is recycled concrete aggregate. [4] "HRWR" is a high-range water reducer used to improve the flow ability of concrete.

2.4. Preparation of Specimens

The production of shear test specimens first involves cutting the insulation board into a rectangular shape of 350 mm × 700 mm, dividing it into two squares and positioning the midpoint, and opening holes to install connectors (Figure 8a). Next, make the specimen mold (Figure 8b), pour vertically, arrange the steel mesh, and place the insulation board, connectors, and steel mesh in order (Figure 8c). Subsequently, C35 concrete and fully recycled concrete are mixed and poured (Figure 8d), compacted and smoothed, and finally covered with plastic wrap for 28 days of natural curing. At the same time, cubic test blocks are made for strength testing.

(a) (b) (c) (d)

Figure 8. Production process of shear specimen: (**a**) positioning and installation of connectors; (**b**) sample mold production; (**c**) place insulation board steel mesh; (**d**) concreting.

The size of the insulation board used in the flexural test is 1200 mm × 600 mm, and a single piece cannot cover the entire exterior wall hanging board, so it needs to be spliced. The splicing quality, including the alignment and bonding strength of the insulation boards, was critical for ensuring effective composite action. Any misalignment or weakness in the splicing process could lead to localized load transfer inefficiencies and affect the overall bending performance. Additionally, the GFRP four-footstool connectors were installed at specific locations in the panel. The installation of these connectors required precise alignment and secure placement to ensure optimal load transfer between the concrete and insulation layers. Improper installation could have led to poor mechanical interlocking, diminishing the overall composite behavior and impacting the bending strength of the panel. Both factors—the splicing of the insulation and the installation of the GFRP connectors—were carefully controlled to assess their impacts on the flexural performance of the wall panels.

Firstly, locate the position of the connector according to the design drawings, slot and insert the GFRP four-footstool type connector to ensure its height is consistent with the design, and fix it with epoxy resin adhesive. The test piece is poured horizontally in layers, and a wooden mold with dimensions of 3200 mm × 1200 mm × 180 mm is made. The inner wall panel concrete is poured and compacted first, and then the insulation board for installing the connectors is laid in sequence to ensure that the connectors are inserted into the concrete. Next, the steel mesh of the exterior wall panel is erected on the connecting pieces, and the concrete pouring of the exterior wall panel is completed before being compacted by vibration. Finally, the surface is smoothed and cured for 28 days. The entire construction process is shown in Figure 9.

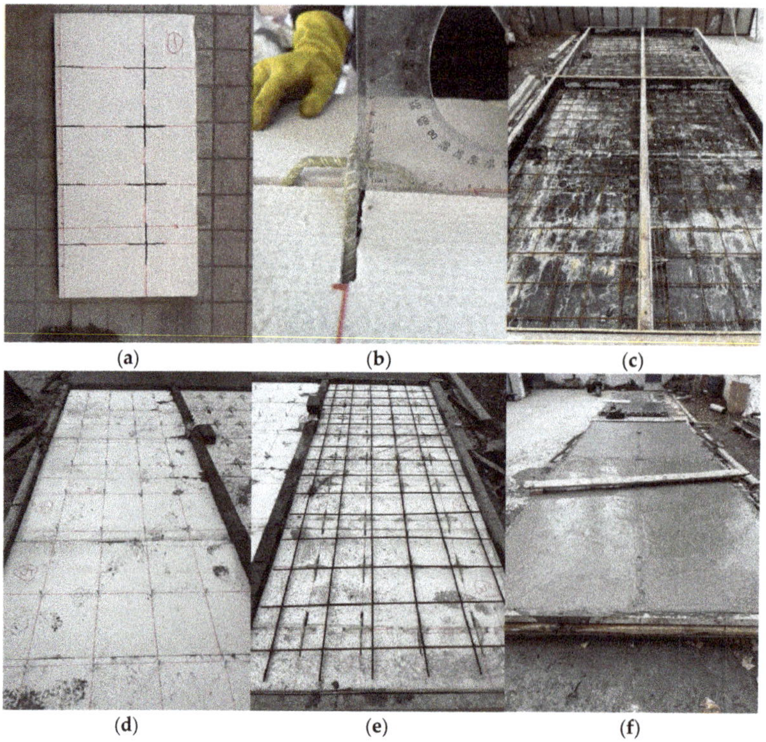

Figure 9. Production process of flexural specimen: (**a**) Insulation board number positioning slotting; (**b**) installation of connectors; (**c**) template production and steel mesh installation; (**d**) installation of connectors and insulation boards; (**e**) laying of outer steel mesh; (**f**) smearing and finishing.

2.5. Test Setup

The pull-out tests were carried out using a microcomputer controlled electro-hydraulic servo universal testing machine model SANS SHT4106 (MTS Industrial Systems (China) Co., Ltd., Shenzhen, China). The specific loading device is shown in Figure 10a,b. During the test, the protruding bars at both ends of the specimen were clamped by the upper and lower clamps of the testing machine. The upper crossbeam of the testing machine was kept fixed and the lower crossbeam was moved downward to generate the tension force. The tests were performed using the displacement-controlled unidirectional loading method. Based on previous studies and experimental considerations, a displacement-controlled loading rate of 0.2 mm/min was selected for the bending tests until the failure of the specimen to ensure the accurate measurement of specimen deformation and to avoid dynamic effects. This rate provided a slow enough load increase to ensure that critical damage characteristics were captured while maintaining a controlled test duration.

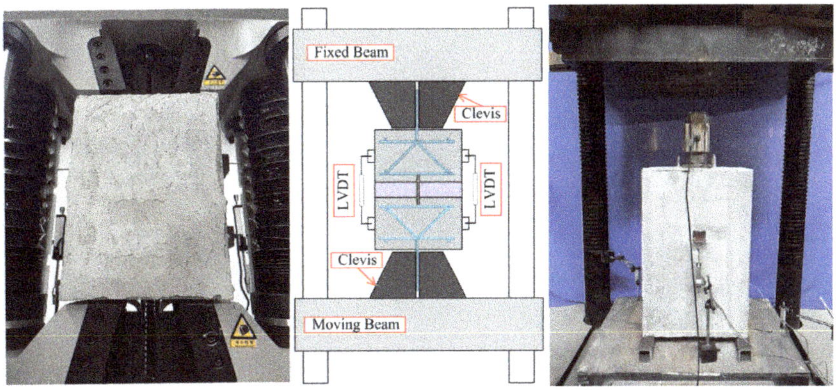

(a) (b) (c)

Figure 10. Test setup: (**a**) pull-out test; (**b**) measurement diagram; (**c**) shear test.

The shear tests were conducted using a 500 T pressure testing machine (MTS Industrial Systems (China) Co., Ltd., Shenzhen, China). The specific loading setup is shown in Figure 10c. Two square steel tubes (60 mm × 60 mm × 10 mm) were placed under the bottom of the exterior wall panels as supports, lifting the exterior wall panels on both sides. During the test, the central interior wall panel was subjected to force, and the lifted portion provided sufficient downward displacement space. The loading trolley applied a constant loading speed of 2 mm/min until the load dropped to 85% of the peak value, at which point the specimen was declared failed, and the test ended.

The flexural test employs a four-point support system with symmetrical two-point loading. This method was selected because it ensures a uniform load distribution, creating a well-defined region of pure bending within the panel. To conduct the test, two I-shaped steel beams, with dimensions matching those of the sample, are placed 600 mm apart, positioned symmetrically on both sides at the midpoint of the specimen. A layer of fine sand is applied to the contact surface between the steel beams and the specimen to ensure uniform contact. A distribution beam is placed horizontally at the center between the two steel beams. At the center of the distribution beam, a pressure sensor and a manual hydraulic jack are positioned for loading, as shown in Figure 11.

The specimen is subjected to monotonic loading controlled by the applied load, with the grading monitored via the pressure sensor display. This monotonic loading method was chosen to progressively increase the load until failure, providing clear data on the ultimate flexural strength of panels, failure modes, and deformation characteristics. Before formal testing, a preload of 5 kN is applied to ensure uniform contact between the specimen and the test apparatus, and to verify that the instrument readings are accurate. After this, the load is reduced to zero, and formal testing begins. During formal testing, the load is incremented in 2 kN steps, with each step held for 5 min. This increment size was selected to provide a clear measurement of the response of specimen to loading without unnecessarily extending the testing duration. Although smaller increments could provide more detailed data, the 2 kN step size offered sufficient resolution to capture key behavioral changes in the specimen. Testing continues until the load of specimen drops below 85% of the peak value without recovery, or until the deflection reaches 1/50 of the plate span length. At this point, the specimen is considered damaged, and the test is terminated.

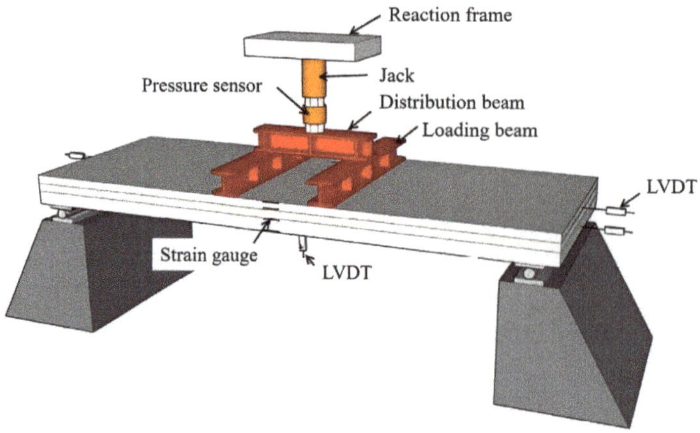

(a)

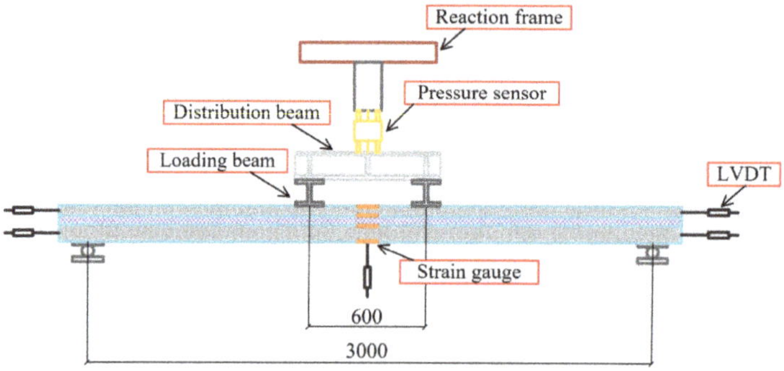

(b)

Figure 11. Test setup: (a) loading device 3D model diagram; (b) load the device main view.

3. Experimental Results and Discussion

3.1. Drawing Property

3.1.1. Drawing Failure Modes

In the initial loading stage, the POT-C bears a relatively small load and does not show significant changes. When approaching the peak load, as the relative displacement increased, the bonding surface between the insulation board and the lower concrete began to separate, and cracks rapidly developed. After continued loading, the cracks expanded and eventually, under peak load, the lower anchorage end of the connector suffered pull-out failure, as shown in Figure 12a; the main body of the connector broke, and the bearing capacity of the specimen suddenly dropped, showing brittle failure. The test ended. The failure mode of POT-R is similar to that of POT-C, with the only difference being the concrete material, which ultimately exhibits brittle failure, as shown in Figure 12b. The POG-C was made of ordinary C35 concrete, and the loading process was similar to that of POT-C. However, after the load reached its peak, the GFRP connector underwent fiber fracture, and the concrete at the anchorage end of the connector peeled off extensively. The bearing capacity of the specimen decreased to the failure load, showing ductile failure, as shown in Figure 12c. The POG-R was similar to POG-C, using fully recycled concrete. During the loading process, it experienced punching and splitting failure, and ultimately exhibited ductile failure, as shown in Figure 12d.

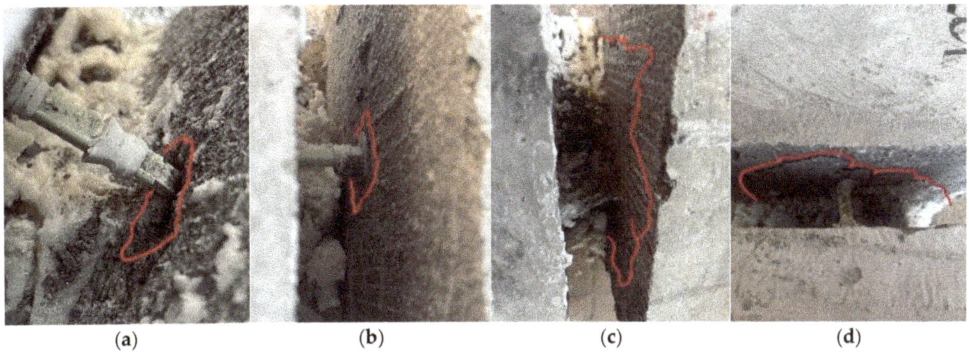

Figure 12. Failure state of specimen pull-out test: (**a**) POT-C; (**b**) POT-R; (**c**) POG-C; (**d**) POG-R.

3.1.2. Displacement Under Load

To compare the tensile strengths of different connectors and concrete materials, load displacement curves of four specimens were plotted, as shown in Figure 13. From the curve, it can be seen that POT-C and POT-R using Thermomass MS-type needle connectors exhibited a cliff-like steep drop after reaching peak load, showing brittle failure. The specimens using GFRP four-footstool-type connectors experienced load fluctuations after peak load, ultimately resulting in ductile failure. The gradual deformation of GFRP under load allowed energy absorption and displacement beyond the peak, causing load fluctuations as the connector underwent progressive deformation. The cross-combination design of the connector led to localized deformations, causing stress redistribution and further load fluctuations, characteristic of ductile materials. All the specimens showed a linear increase in the stage of load rise, but during this process, the foam insulation board showed a slight steep drop when separated from the concrete.

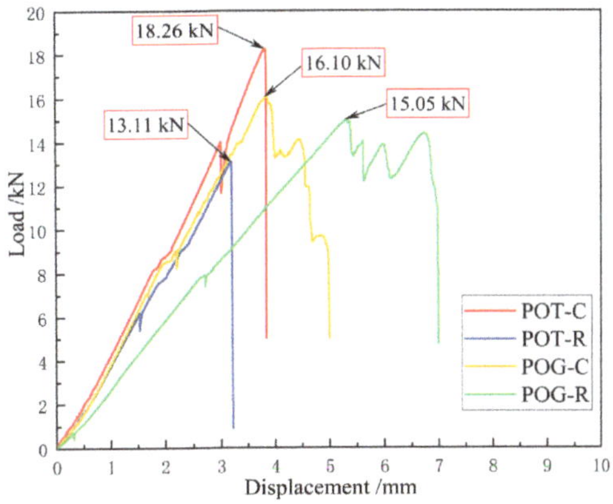

Figure 13. Load–displacement curves of specimen pull-out test.

Through comparative analysis of connectors and concrete materials, it was found that the displacement of POG-C and POG-R during tensile failure was significantly greater than that of POT-C and POT-R. In terms of ultimate tensile bearing capacity, the tensile bearing capacity of POT-C was higher than that of POG-C, while the bearing capacity of POG-R was higher than that of POT-R. Overall, C35 specimens exhibited stronger tensile strength due to their higher strength, which resulted in a tighter interaction between the connectors and the concrete.

3.2. Shear Behaviour

3.2.1. Shear Failure Modes

During the loading process of CT-C, the initial load was relatively small, and the connectors did not show significant deformation. The concrete in the insulation board area slightly cracked. As the load gradually increased, the shear bearing capacity of the connector approached its limit. With a "bang" sound, indicating fracture, obvious cracks appeared on the surface of the connector, and bending failure occurred, as shown in Figure 14a, resulting in a rapid decrease in the bearing capacity of the specimen, and ultimately brittle failure. The loading process of CT-R was similar to that of CT-C, but due to the use of fully recycled concrete, the shear force borne by the connectors caused the rapid expansion of concrete cracks in the insulation board area, and the failure modes of the connectors also manifested as shear and bending failure, as shown in Figure 14b.

(a) (b) (c) (d)

Figure 14. Failure state of specimen shear test: (**a**) CT-C; (**b**) CT-R; (**c**) CG-C; (**d**) CG-R.

During the loading process of CG-C, the behaviors of the connecting components became more complex. Firstly, the right connecting component bore a load approaching its limit, followed by the sound of the GFRP reinforcement fiber fracturing, and the connecting component showed obvious bending and fracture. As the load continued to increase, the left connector also eventually failed, with the failure modes of shear and tensile fracture, as shown in Figure 14c, indicating the brittle failure of the connector under bidirectional stress. The failure process of CG-R was similar to that of CG-C, but due to the lower compressive strength of fully recycled concrete, larger concrete crushing areas were generated in the connectors under stress, and the fibers of the connectors showed more obvious fractures and bifurcations, resulting in further increases in the failure area, as shown in Figure 14d.

3.2.2. Load-Displacement Curve

By analyzing the load displacement curve (Figure 15), it can be found that the trend of the shear failure load displacement curve was similar for specimens using the same connectors. The CT-C and CT-R of Thermomass MS-type needle connectors exhibited a unimodal state, with both connectors jointly bearing the load. After failure occurred, the specimens lost their load-bearing capacity. The CG-C and CG-R of GFRP four-footstool-type connectors exhibited a bimodal state, showing two failures. Firstly, the two connectors on the same side failed, resulting in a sudden drop in bearing capacity. Then, the connector on the other side failed, forming a second peak, and the load decreased and remained at a relatively small level. The first peak represents the elastic phase, where the connector transfers were loaded without significant deformation. After the first peak, localized plastic deformations occurred, causing a temporary reduction in load-bearing capacity. The second peak indicates force reorganization and stress redistribution, briefly restoring load capacity before final failure. This pattern highlights the ability of the cross-combination design of the GFRP material and the connector to absorb energy and maintain load transfer despite localized failures.

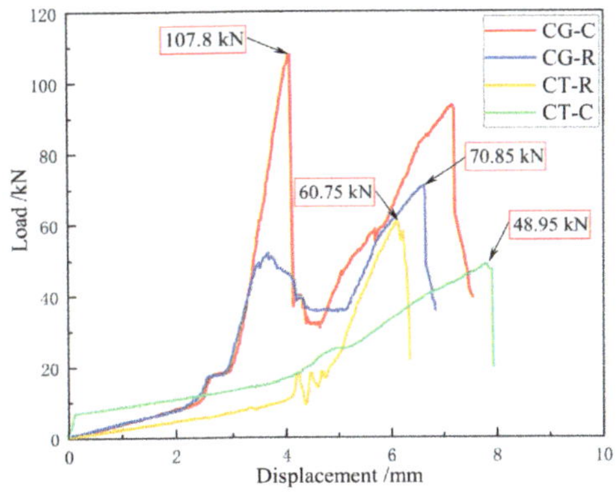

Figure 15. Load–displacement curves of each specimen shear test.

In terms of the types of connectors, specimens using GFRP four-footstool-type connectors had significantly higher shear bearing capacity than specimens using Thermomass MS-type connectors, whether in ordinary concrete or recycled concrete. The ultimate shear-bearing capacity of CG-C was 107.80 kN, which is about 1.2 times higher than that of CT-C. In the recycled concrete specimens, the ultimate shear bearing capacity of CG-R was 70.85 kN, which is 16.62% higher than that of CT-R. In addition, the displacement of GFRP four-footstool-type connectors during failure was smaller, and their comprehensive performances were better than those of Thermomass MS-type connectors. The influence of concrete type was manifested in that, when using the same connectors, the shear bearing capacity of C35 specimens was significantly higher than that of recycled concrete specimens. The main reason is that the compressive strength of C35 was higher, resulting in the higher shear strength of specimens when subjected to shear force.

3.3. Flexural Behaviour

3.3.1. Flexural Failure Modes

The flexural behavior results are shown in Table 9. As shown in Figure 16a, W-1 did not undergo significant damage during the initial loading stage, but when the load reached 10 kN, the specimen cracked, and a 50 mm-long crack appeared at the mid span of the specimen. As the load increased, the cracks gradually expanded. When the load reached 22 kN, new cracks appeared at the bottom of the specimen, and the mid span deflection increased sharply to 35.89 mm. When further loaded to 26 kN, the specimen entered the yielding stage, with a sharp decrease in load to 21.69 kN and a rapid increase in mid-span deflection to 62.03 mm, ultimately indicating failure. As shown in Figure 16b, W-2 showed a similar trend, but cracking occurred at a later stage, and the crack distribution was more widespread. When the load reached 34 kN, the crack width of the specimen significantly increased, and the crack extended to various areas of the specimen. When the final load reached 41.75 kN, the specimen failed. As shown in Figure 16c, the loading processes of W-3 and W-4 were relatively stable, and the crack propagation was relatively uniform. When W-3 reached the failure standard specified in the specifications, the load was 47.89 kN, and W-4 had a load of 44.73 kN at a deflection of 58 mm. W-4 showed no significant slip phenomenon throughout the entire testing process, demonstrating a good combination of inner and outer wall panel structures, as shown in Figure 16d.

Table 9. Results of flexural test.

Specimen	Cracking Load/kN	Ultimate Load/kN
W-1	10.00	26.00
W-2	14.00	41.75
W-3	12.00	47.89
W-4	12.00	44.73

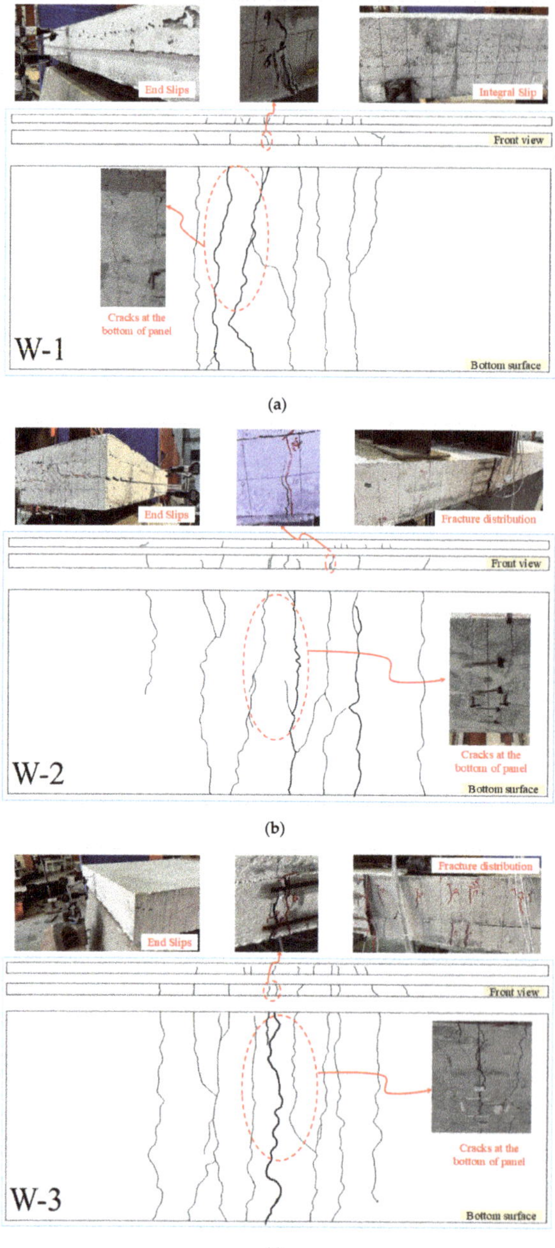

Figure 16. *Cont.*

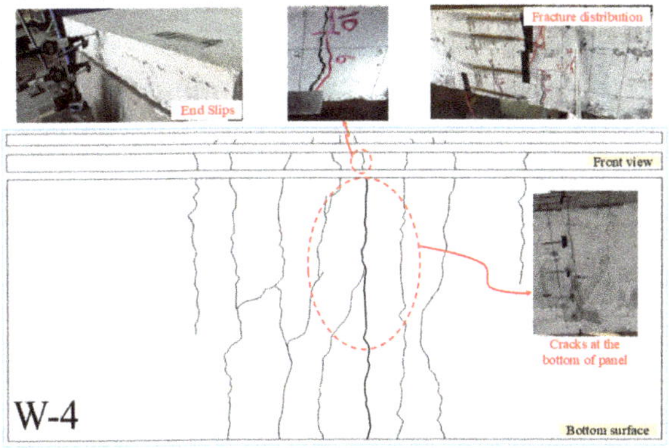

(d)

Figure 16. Crack diagram of specimen bending test: (**a**) W-1; (**b**) W-2; (**c**) W-3; (**d**) W-4.

3.3.2. Cracking Behavior

As shown in Figure 16a, W-1 used a Thermomass MS-type connector, and the cracks were mainly concentrated in the middle of the bottom plate, with an average crack spacing of 15.71 cm. The main crack was located about 5 cm to the left of the mid-span, indicating that the stress on this connector was relatively concentrated. As shown in Figure 16b, the W-2 used GFRP four-footstool-type connectors, and the crack distribution was more uniform than in W-1, with an average crack spacing of 23.57 cm, an increase of 50% compared to W-1. The main crack was located about 10 cm to the right of the bottom span, indicating that GFRP four-footstool-type connectors can better disperse the stress on the bottom concrete, improving the distribution of cracks. As shown in Figure 16c, the crack spacing of W-3 was the smallest, with an average of 14.44 cm. The main crack was located in the center of the bottom plate, and the dense cracks were due to the small thickness of the bottom plate and poor crack resistance of the specimen. As shown in Figure 16d, the crack interval of W-4 was 20.63 cm, similar to W-2. The main crack was located about 15 cm to the right of the mid-span of the bottom plate, showing a relatively uniform crack distribution, indicating a more balanced stress situation.

3.3.3. Load–Mid-Span Deflection

As shown in Figure 17, the primary difference between samples W-1 and W-2 lies in the type of connector used. Sample W-1 employed ordinary concrete and Thermomass MS connectors, while W-2 utilized GFRP four-footstool connectors. The load–deflection curve for W-2 is notably higher than that for W-1, with the maximum bending capacity of W-2 being 60.58% greater than that of W-1. This indicates that GFRP four-footstool connectors provided superior shear-bearing capacity, and effectively resisted misalignment and sliding between the inner and outer wall panels, thereby enhancing the bending capacity. W-1 entered the yield stage after reaching a cracking load of 10 kN, and its curve slope decreased rapidly, while the slope of the W-2 curve remained relatively stable, delaying the yield stage. This demonstrates the advantage of GFRP four-footstool connectors in improving bending capacity.

For samples W-2 and W-4, both using GFRP four-footstool connectors, the key difference is that W-4 incorporated recycled concrete, while W-2 used ordinary concrete. The load–deflection curve for W-4 is similar to that of W-2, but the curve of W-4 shows a faster decrease in slope, suggesting that the strength of recycled concrete is lower, resulting in weaker resistance to bending loads. However, after the yielding stage, the flexural capacity of W-4 exceeded that of W-2 by 7.14%, indicating that recycled concrete demonstrated good flexural strength under certain conditions. Overall, the mechanical properties of W-4 (with recycled concrete) were comparable to those of W-2, suggesting that the recycled content had minimal effect on the performance under the test conditions. This can be attributed to factors such as strength requirements met through processing, mix design adjustments, and similar properties between W-2 and W-4. Therefore, with appropriate optimization, recycled concrete can perform similarly to ordinary concrete.

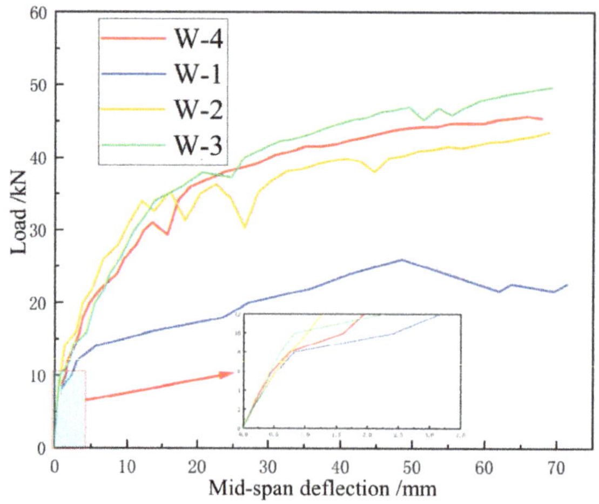

Figure 17. Load–mid-span deflection curve of specimen bending test.

In the comparison between W-3 and W-4, both of which used GFRP four-footstool connectors and recycled concrete, W-3 featured equal thicknesses for the inner and outer wall panels, whereas W-4 included a 50 mm-thick insulation board between the inner and outer panels. Although the load–displacement curves for the two samples followed a similar trend, the bending capacity of W-3 was 3.25 kN, 7.26% lower than that of W-4. This indicates that the position of the insulation board and the thickness of the wall panels significantly affected the flexural bearing capacity, with the structural configuration of W-4 demonstrating superior performance.

3.3.4. Load-Wythe Slip Responses

From the comprehensive comparison of load slip curves, as shown in Figure 18, it can be seen that samples with the same connector type exhibited similar curve shapes and changing trends. When the mid-span of the specimen reached the failure deflection, the relative slip between the inner and outer wall panels of W-1 using the Thermomass MS connector was the largest, which was 129.47% higher than that of W-2 using a GFRP four-footstool connector. This phenomenon indicates that GFRP four-footstool type connectors significantly improve the combination performance of exterior wall panels and reduce the slip between interior and exterior wall panels. In contrast, the relative slip ratio of sample W-4 increased by 81.75% compared to W-2, further demonstrating the significant influence of concrete strength on the performance of the wall panel combination. Finally, compared with sample W-4, the relative slip rate of sample W-3 only increased by 5.41%, indicating that the position of the insulation board had little effect on the performance of the wall panel combination.

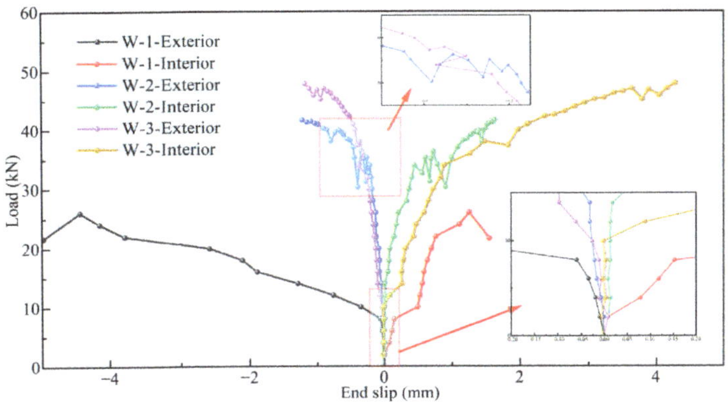

Figure 18. Load–slip curve of specimen.

3.3.5. Calculation of Combination Degree and Flexural Bearing Capacity

This section sets up equivalent models for fully composite sandwich wall panels and non-composite sandwich wall panels, and calculates the bending capacity of both models separately. By using strength calculation theory, we here determine the degree of combination of each specimen. In the model of non-composite wall panels, during the loading process of the experiment, the inner and outer walls are treated as two independent concrete slabs without interaction. The calculation model can be regarded as the superposition of the strength of two reinforced concrete slabs with different thicknesses, as shown in Figure 19.

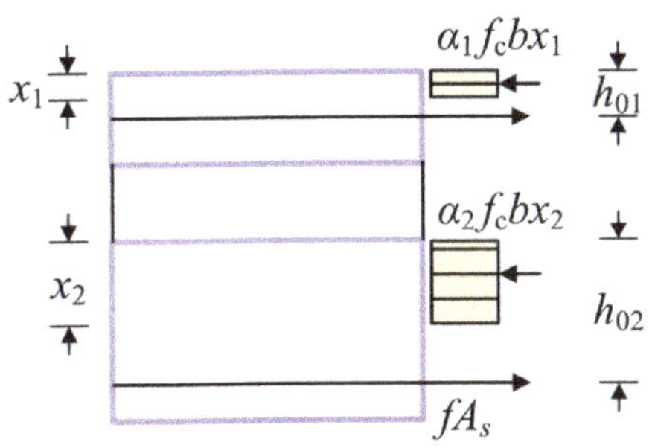

Figure 19. Non-composite wall panel calculation diagram.

According to the method specified in the "Code for Design of Concrete Structures" GB50010-2010 [36], the bending moment and load of the concrete panel of the sandwich wall panel in this test were calculated. The bending moment calculation formulas are as follows:

$$M_{nc} = M_{nc1} + M_{nc2} \tag{1}$$

$$M_{nc1} = fA_s(h_{01} - 0.5x_1) \tag{2}$$

$$M_{nc2} = fA_s(h_{02} - 0.5x_2) \tag{3}$$

where A_s is the sum of the cross-sectional areas of the tensile reinforcement, with the reinforcement distribution being consistent between the inner and outer wall panels, where $A_s = 301.44\ \text{mm}^2$. h_0 denotes the effective height of the section, while f is the strength of the reinforcement. The yield strength standard value f_{yk} and the ultimate strength standard value f_{stk} of HRB400-grade steel bars are substituted into f, from which the yield moment and ultimate moment of the individual concrete panel can be determined. x represents the height of the compression zone, and the specimen width b is 1200 mm.

In this experiment, the dimensions of the inner and outer wall panels of specimens W-1, W-2, and W-4 are taken as consistent. The inner wall panel has a thickness of 80 mm, and the outer wall panel has a thickness of 50 mm for calculations. The wall panel of specimen W-3 has an equal thickness for both the inner and outer panels, which is 65 mm, with the same reinforcement layout. Only one side is calculated, and the moment for W-3 is obtained by multiplying this by 2.

$$F_{nc} = \frac{8M_{nc}}{l} \tag{4}$$

The load on the fully composite wall panel is entirely transmitted through the connectors in the middle insulation layer. The two separate wall panels can be treated as a single entity under load, with no relative displacement between the panels. The calculation process is identical to that of solid slabs, as shown in Figure 20.

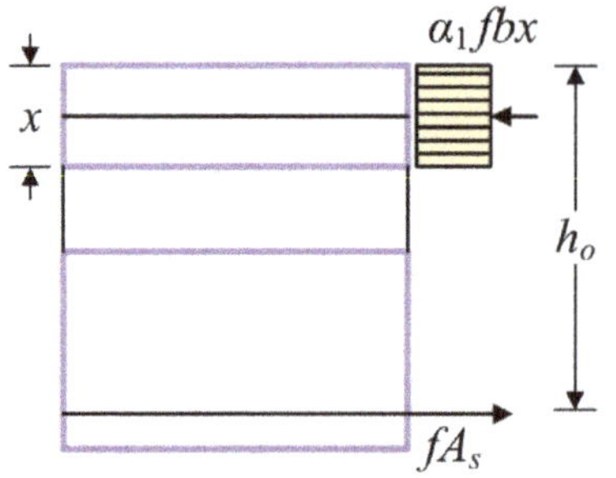

Figure 20. Completely combined wall panel calculation diagram.

Similarly, according to the calculation formula for double-reinforced cross-sections in the "Code for Design of Concrete Structures" (GB50010-2010) [36], the inner and outer wall panels are considered as a single entity and calculated using the following formula:

$$M_{fc} = fA_s(h_0 - 0.5x) \tag{5}$$

where A_s is the cross-sectional area of the tensile reinforcement; h_0 is the effective height of the section, where h_0 = 155 mm; f is the reinforcement strength. By substituting the yield strength standard value f_{yk} and the ultimate strength standard value f_{stk} of HRB400-grade steel into f, the yield moment and ultimate moment of the individual concrete panel can be determined. x is the height of the compression zone, given by $x = \frac{fA_s}{\alpha_1 f_c b}$, and b is the specimen width, which is 1200 mm.

In this experiment, the dimensions of the inner and outer wall panels of specimens W-1, W-2, and W-4 are consistent, with the inner wall panel having a thickness of 80 mm and the outer wall panel a thickness of 50 mm for calculations. For W-3, the inner and outer wall panels have equal thicknesses of 65 mm, with the same reinforcement layout. Only one side is calculated, and the moment for W-3 is obtained by multiplying this by 2. The bending moments for each wall panel are calculated separately according to their respective dimensions.

The bending capacity is then determined based on the simply supported beam four-point bending Formula (6), as follows:

$$F_{fc} = \frac{8M_{fc}}{l} \tag{6}$$

The composite degree of the wall panel is represented by the composite index κ, which is calculated according to the Formula (7). Table 10 provides the statistical results for the composite degree calculations at the yield load and ultimate load states for each experimental wall panel.

$$\kappa = \frac{F_{exp} - F_{nc}}{F_{fc} - F_{nc}} \times 100 \quad (7)$$

where F_{exp} is the experimentally measured load.

Table 10. Combination degree of wall panels.

Specimen	Yield Load/kN	Ultimate Load/kN	κ (Yield)/%	κ (Ultimate)/%
W-1	22.00	26.00	−7.14	−15.15
W-2	34.00	41.75	40.71	30.74
W-3	38.30	47.89	58.77	49.93
W-4	35.78	44.73	48.79	40.80

From the data in Table 10, it can be seen that the bonding degree between sample W-1 and the Thermomass MS connector was the lowest. In contrast, the adhesion between sample W-2 and the GFRP four-footstool connector reached 40.71% under yield load and 30.74% under ultimate load, which values are 47.85% and 45.89% higher than those of W-1, respectively. This difference is mainly due to the fact that during the bending process of the specimen, the part between the loading points was a pure bending section, and there was significant friction between the outer side of the loading point and the supporting ends of the inner and outer wall panels, resulting in higher shear forces on the connecting components. Due to the stronger shear strength of the GFRP four-footstool connector compared to the Thermomass MS connector, the bonding degree of W-2 was higher. Sample W-3 had the highest degree of combination, with the same thicknesses of inner and outer wall panels. During the bending process, the concrete structure with thicker upper outer wall panels could withstand greater pressure, thus achieving the strongest bonding effect.

3.3.6. Modification of Flexural Strength Calculation

Due to the differences in the types and structural forms of connectors used for exterior wall cladding, as well as their varying degrees of combination, the calculation of bending capacity becomes relatively complex. To simplify the calculation, this section proposes a correction factor for the flexural strength of GFRP four-footstool-type connectors for exterior wall cladding, in order to facilitate the calculation of flexural strength. Based on the comparison between the bending strength calculated by the fully combined model in the previous section and the measured strength, the correction factor has here been determined. The specific data are shown in Table 11.

Table 11. Correction value for flexural strength calculation of wall panels.

Specimen	Experimental Flexural Strength F_G/kN	Calculation of Flexural Strength F_{fc}/kN	Correction Factor/γ	Correction Value/Test Value
W-2	41.75	65.52	0.68	1.07
W-3	47.89	65.22	0.68	0.93
W-4	44.73	65.22	0.68	0.99

From the results in Table 11, it can be seen that when calculating the flexural strength of GFRP four-footstool connector wall panels, the flexural bearing capacity can first be calculated based on the calculation Formulas (5) and (6) of the fully composite wall panel model. Next, one must multiply the calculation result by the combination degree reduction factor $\gamma = 0.68$ to obtain the corrected flexural strength. The revised calculation formula is shown in Formula (8), and the deviation between the bending capacity calculated using this method and the measured value is small, making the calculation process more convenient.

$$F_G = \gamma \cdot \frac{8M_{fc}}{l} \tag{8}$$

4. Conclusions

This article proposes a new type of GFRP four-footstool connector and combines it with a common Thermomass MS needle connector to conduct experimental research on the pull-out, shear, and full-size wall panel bending performance of prefabricated insulation exterior wall panels made of fully recycled concrete and ordinary C35 concrete. Through experimental analysis, the following conclusions can be drawn.

The pull-out test revealed significant differences in the failure modes of two types of connectors. The specimen using the Thermomass MS connector exhibited brittle failure under the ultimate load, while the specimen with the GFRP four-footstool connector demonstrated enhanced ductility. These specimens showed greater relative displacement and a larger safety margin before failure, ultimately undergoing a ductile failure mode. Notably, the tensile strength of samples using recycled concrete increased by 14.8% compared to ordinary concrete, highlighting the positive impact of recycled materials on performance.

In the shear test, the Thermomass MS connector failed due to the connector itself, while samples made of both ordinary concrete and fully recycled concrete failed due to bending when the connector reached its maximum capacity. In contrast, the specimens with GFRP four-footstool connectors exhibited more complex failure behavior, including damage to the connector material and crushing of the concrete in the compression zone. Among all the samples, the one with the GFRP four-footstool connector demonstrated the highest ultimate shear bearing capacity.

In the bending test, crack formation in the specimen with the Thermomass MS connector was concentrated in the central area, while the sample with the GFRP four-footstool connector showed a 50% increase in crack spacing, indicating an improvement in crack distribution. The sample with the GFRP four-footstool connector exhibited a more uniform crack pattern and a wider stress range, with its maximum bending capacity being 60.58% higher than that of the Thermomass MS connector. The difference in failure mode and bearing capacity between the ordinary concrete and recycled concrete specimens was minimal, indicating that the GFRP four-footstool connector performed similarly to the joint action of both concrete types.

A comparative analysis of specimens with varying thicknesses of inner and outer wall panels revealed the influence of insulation board placement on bending performance. Although the early load–deflection curves were similar, the flexural capacity of specimens with equal concrete thickness on both sides of the insulation board was 7.26% higher than that of specimens with unequal thicknesses. This suggests that the positioning of the insulation layer affects the structural behavior of the wall panel, with the insulation layer ideally placed at the center during design.

Compared to the Thermomass MS connector sample, the relative slip observed in the specimen using the GFRP four-footstool connector increased significantly by 129.47%, highlighting the role of the GFRP connector in enhancing the overall connection between components. This finding emphasizes the importance of selecting appropriate connector types to improve the performance of prefabricated wall panels. To simplify design calculations, a reduction factor (γ) for the combination degree of GFRP four-footstool connectors is proposed, offering a practical approach for future design considerations.

Author Contributions: Methodology, X.L.; Software, X.L.; Validation, X.L., H.S. and T.Z.; Formal analysis, X.L.; Investigation, X.L., H.S., T.Z., T.B., H.Y., J.S. and H.F.; Resources, X.L.; Data curation, X.L.; Writing—original draft, H.S. and T.Z.; Writing—review & editing, X.L. and H.F.; Visualization, X.L.; Supervision, X.L.; Project administration, X.L.; Funding acquisition, X.L. All authors have read and agreed to the published version of the manuscript.

Funding: This research was funded by the National Natural Science Foundation of China [grant No. 52278507]; the Natural Science Foundation of Shandong Province [grant No. ZR2020ME245]; and the Major Scientific & Technological Innovation Project of Shandong Province [grant No. 2021CXGC011204].

Data Availability Statement: The original contributions presented in this study are included in the article. Further inquiries can be directed to the corresponding author.

Conflicts of Interest: The authors declare that we have no known competing financial interests or personal relationships that could have appeared to influence the work reported in this paper.

References

1. Ding, X.; Yu, J.; Lin, J.; Chen, Z.; Li, J. Experimental investigations of prefabricated lightweight self-insulating foamed concrete wall panels. *Structures* **2024**, *61*, 106001. [CrossRef]
2. Palani, H.; Karatas, A. Investigating the disparities between experimental and computational analyses of thermal performance in prefabricated wall panels. *Appl. Therm. Eng.* **2024**, *236*, 121568. [CrossRef]
3. Yu, S.; Liu, Y.; Wang, D.; Bahaj, A.; Wu, Y.; Liu, J. Review of thermal and environmental performance of prefabricated buildings: Implications to emission reductions in China. *Renew. Sustain. Energy Rev.* **2021**, *137*, 110472. [CrossRef]
4. Tawil, H.; Tan, C.G.; Sulong, N.H.R.; Nazri, F.M.; Sherif, M.M.; El-Shafie, A. Mechanical and Thermal Properties of Composite Precast Concrete Sandwich Panels: A Review. *Buildings* **2022**, *12*, 1429. [CrossRef]
5. Oliveira, T.; Carvalho, J.; Mendes, J.; Souza, G.; Carvalho, V.; Peixoto, R. Precast concrete sandwich panels (PCSP): An analytical review and evaluation of CO_2 equivalent. *Constr. Build. Mater.* **2022**, *358*, 129424. [CrossRef]
6. Tushar, Q.; Zhang, G.; Bhuiyan, M.A.; Navaratnam, S.; Giustozzi, F.; Hou, L. Retrofit of Building Façade Using Precast Sandwich Panel: An Integrated Thermal and Environmental Assessment on BIM-Based LCA. *Buildings* **2022**, *12*, 2098. [CrossRef]
7. Ghazy, M.F.; Abd Elaty, M.A.A.; Sakr, M.A.E.; Arafa, E.M. Thermo-Mechanical Performance of Sustainable Lightweight Sandwich Panels Utilizing Ultra-High-Performance Fiber-Reinforced Concrete. *Buildings* **2025**, *15*, 593. [CrossRef]
8. Zou, J.; Cao, B.; Zeng, X.; Zhang, Y. Lateral Performance of Composite Wall with Cold-Formed Thin-Walled Steel–Concrete Sandwich Panel. *Buildings* **2024**, *14*, 2928. [CrossRef]
9. Alves, L.F.; Pessôa, P.d.M.; Jardim, P.I.L.G.; Faustino, E.; dos Santos, H.F.; Lahr, F.A.R.; de Almeida, D.H.; Christoforo, A.L. Numerical Evaluation of the Influence of Using Carbon-Fiber-Reinforced Polymer Rebars as Shear Connectors for Cross-Laminated Timber–Concrete Panels. *Buildings* **2024**, *14*, 2178. [CrossRef]
10. Jiang, H.; Xie, L.; Chang, F.; Cao, Y.; Guo, Z. Study on the Composite Performance of Sandwich Wall Panels with SGCCs. *Buildings* **2025**, *15*, 65. [CrossRef]
11. Wang, X.; Sun, H.; Hou, Y.; Li, Y. Study on Seismic Performance of Steel Frame Installed New-Type Lightweight Concrete Composite Exterior Wallboard. *Buildings* **2024**, *14*, 2224. [CrossRef]
12. Silva, J.K.F.; Lameiras, R.d.M. Experimental Analysis of the Mechanical Behavior of Shear Connectors for Precast Sandwich Wall Panels When Subjected to the Push-Out Tests. *Buildings* **2024**, *14*, 3246. [CrossRef]
13. Barbosa, K.; Silva, W.T.M.; Silva, R.; Vital, W.; Bezerra, L.M. Experimental Investigation of Axially Loaded Precast Sandwich Panels. *Buildings* **2023**, *13*, 1993. [CrossRef]
14. Sevil Yaman, T.; Lucier, G. Shear Transfer Mechanism between CFRP Grid and EPS Rigid Foam Insulation of Precast Concrete Sandwich Panels. *Buildings* **2023**, *13*, 928. [CrossRef]
15. Liu, X.; Wang, X.; Yang, T.; Wu, Z. The Shear Behavior of Insulated Precast Concrete Sandwich Panels Reinforced with BFRP. *Buildings* **2022**, *12*, 1326. [CrossRef]
16. Lou, X.; Xue, W.; Bai, H.; Li, Y.; Huang, Q. Shear behavior of stainless-steel plate connectors for insulated precast concrete sandwich panels. *Structures* **2022**, *44*, 1046–1056. [CrossRef]

17. Zhao, M.; Chong, X.; Huang, J.-Q.; Jiang, Q.; Feng, Y.-L.; Chang, Y. Experimental and numerical investigations on the axial compression performance of precast geopolymer concrete sandwich wall panel. *J. Build. Eng.* **2024**, *95*, 110303. [CrossRef]
18. O'Hegarty, R.; Kinnane, O. Review of precast concrete sandwich panels and their innovations. *Constr. Build. Mater.* **2020**, *233*, 117145. [CrossRef]
19. Bian, Y.; Xiong, F.; Liu, Y.; Ge, Q. Flexural performance of novel ECC-RC composite sandwich panels. *Eng. Struct.* **2023**, *292*, 116547. [CrossRef]
20. Huang, J.-Q.; Dai, J.-G. Flexural performance of precast geopolymer concrete sandwich panel enabled by FRP connector. *Compos. Struct.* **2020**, *248*, 112563. [CrossRef]
21. Kim, J.; You, Y.-C. Composite Behavior of a Novel Insulated Concrete Sandwich Wall Panel Reinforced with GFRP Shear Grids: Effects of Insulation Types. *Materials* **2015**, *8*, 899–913. [CrossRef] [PubMed]
22. Wang, L.; Wu, Z.; Liu, W.; Wan, L. Structural behavior of load-bearing sandwich wall panels with GFRP skin and a foam-web core. *Sci. Eng. Compos. Mater.* **2018**, *25*, 173–188. [CrossRef]
23. Choi, I.; Kim, J.; You, Y. Effect of cyclic loading on composite behavior of insulated concrete sandwich wall panels with GFRP shear connectors. *Compos. Part B Eng.* **2016**, *96*, 7–19. [CrossRef]
24. Benayoune, A.; Abdul, S.; Trikha, D.; Ali, A.A.; Ellinna, S.H.M. Flexural behaviour of pre-cast concrete sandwich composite panel-Experimental and theoretical investigations. *Constr. Build. Mater.* **2008**, *22*, 580–592. [CrossRef]
25. Sharaf, T.; Shawkat, W.; Fam, A. Structural Performance of Sandwich Wall Panels with Different Foam Core Densities in One-way Bending. *J. Compos. Mater.* **2010**, *44*, 2249–2263. [CrossRef]
26. Zhi, Q.; Guo, Z. Experimental evaluation of precast concrete sandwich wall panels with steel–glass fiber–reinforced polymer shear connectors. *Adv. Struct. Eng.* **2017**, *20*, 1476–1492. [CrossRef]
27. Naito, C.; Hoemann, J.; Beacraft, M.; Bewick, B. Performance and Characterization of Shear Ties for Use in Insulated Precast Concrete Sandwich Wall Panels. *J. Struct. Eng.* **2012**, *138*, 52–61. [CrossRef]
28. Kang, W.; Kim, J. Reliability-based flexural design models for concrete sandwich wall panels with continuous GFRP shear connectors. *Compos. Part B Eng.* **2016**, *89*, 340–351. [CrossRef]
29. Xiong, F.; Bian, Y.; Liu, Y.; Ge, Q.; Deng, C. Data-driven prediction method for flexural performance of ECC composite sandwich panels. *Structures* **2024**, *70*, 107524. [CrossRef]
30. He, J.; Xu, Z.; Zhang, L.; Lin, Z.; Hu, Z.; Li, Q.; Dong, Y. Shaking table tests and seismic assessment of a full-scale precast concrete sandwich wall panel structure with bolt connections. *Eng. Struct.* **2023**, *278*, 115543. [CrossRef]
31. Choi, W.; Jang, S.; Yun, H. Design properties of insulated precast concrete sandwich panels with composite shear connectors. *Compos. Part B Eng.* **2019**, *157*, 36–42. [CrossRef]
32. Zhang, M.; Feng, W.; Chen, K.; Li, B. Flexural Behavior of a New Precast Insulation Mortar Sandwich Panel. *Appl. Sci.* **2024**, *14*, 2071. [CrossRef]
33. Zhao, K.; Fan, Z.; Zhang, Y.; Xu, Y.; Liu, S. Experimental Study on Seismic Performance of Prefabricated Monolithic Concrete–Polystyrene Panel Composite Wall Panels. *Buildings* **2024**, *14*, 442. [CrossRef]
34. Chen, D.; Li, K.; Yuan, Z.; Cheng, B.; Kang, X. Shear Behavior of FRP Connectors in Precast Sandwich Insulation Wall Panels. *Buildings* **2022**, *12*, 1095. [CrossRef]
35. Xie, J.; Chen, F.; Zhao, J.; Lu, P.; Liu, F.; Li, L. Flexural behaviour of full-scale precast recycled concrete sandwich panels with BFRP connectors. *J. Build. Eng.* **2022**, *56*, 104816. [CrossRef]
36. *GB 50010-2010*; Chinese Standard Code for Design of Concrete Structures. China Architecture & Building Press: Beijing, China, 2015.

buildings

MDPI

Article

Simplified FE-Based Post-Earthquake Vulnerability Assessment of a Partially Collapsed Historic Mosque

Rüya Kılıç Demircan

Department of Construction Technology, Boyabat Vocational School, Sinop University, 57000 Sinop, Türkiye; ruyakilic86@gmail.com or rkdemircan@sinop.edu.tr

Abstract: On 6 February 2023, two major earthquakes struck southeastern Türkiye along the East Anatolian Fault, causing widespread structural damage, including the partial collapse of the historic Habibi Neccar Mosque in Antakya. This study presents a simulation-based approach to rapidly assess the seismic vulnerability of this partially damaged historic masonry structure. Due to the complexity and urgent condition of such heritage buildings, a simplified finite element (FE) modeling methodology is employed to evaluate structural behavior and support immediate stabilization decisions. Response spectrum analysis is applied to simulate and interpret stress distribution and deformation patterns in both undamaged and damaged states. The simulation results highlight significant tensile stress concentrations exceeding 0.2 MPa at dome–arch joints and vaults—primary indicators of localized failures. Additionally, the analysis reveals increased out-of-plane deformations and the influence of soil amplification in the remaining walls, both of which further compromise the structural integrity of the building. The findings demonstrate that simplified FE simulations can serve as practical and efficient tools for early seismic assessment of historic structures, contributing to rapid decision making, risk mitigation, and cultural heritage preservation in earthquake-prone areas.

Keywords: historic masonry structure; finite element modeling; post-earthquake assessment; structural simulation

1. Introduction

On 6 February 2023, at 04:17 (01:17 GMT) and 13:24 (10:24 GMT), two large earthquakes shocked southern and southeastern Türkiye. The moment magnitudes (Mw) of these earthquakes were 7.7 Mw and 7.6 Mw, and occurred on the Eastern Anatolian Fault [1]. The epicenter of the first earthquake was the Pazarcık district of Kahramanmaraş province, and the depth was 8.6 km. A second earthquake occurred 9 h later in the Elbistan district of Kahramanmaraş province with a depth of 7 km. In addition to hundreds of aftershocks following the first two earthquakes, another large earthquake with a magnitude of 6.4 Mw and a depth of 22 km occurred in the Defne district of Hatay province on 20 February 2023 at 20:04 (17:04 GMT) [2,3]. Figure 1 marks the locations of these earthquakes.

These earthquakes were felt and effective in an expansive geography, including Lebanon, Cyprus, Iraq, Israel, Jordan, and Egypt, in addition to Türkiye and Syria, covering an area of approximately 350,000 km^2. In this wide region where more than 13 million people live in Türkiye, more than 50 thousand people lost their lives, and more than 100 thousand people were injured. Well over 40 thousand buildings were destroyed during the earthquakes, and 300 thousand buildings were found to be heavily damaged and in urgent need of demolition [3]. Many historical buildings, including Gaziantep Citadel,

Habib-i Neccar Mosque, Kahramanmaraş Great Mosque, Adıyaman Great Mosque, and the Latin Catholic Church in İskenderun, were severely damaged or destroyed. There are more than 8000 registered historical buildings and monuments in the area affected by the earthquake, some of which are on the UNESCO World Heritage List [4].

Figure 1. Epicenters of 6 February 2023, Kahramanmaraş-Pazarcık (Mw = 7.7), Kahramanmaraş-Elbistan (Mw = 7.6) and 10 February 2023 Defne-Hatay (Mw = 6.4) Earthquakes.

The Kahramanmaraş earthquakes have been extensively studied in the scientific community. Baltzopoulos et al. [5] analyzed ground motion around Gaziantep during the Mw 7.7 earthquake on 6 February 2023. Karataş et al. [6] proposed a systematic method for assessing post-earthquake damage at Antep Castle. Papazafeiropoulos and Plevris examined the effects of strong ground motions on building performance during the same event, while Binici et al. evaluated the performance of reinforced concrete (RC) buildings and provided recommendations for performance-based design [5–8]. In addition, repair and strengthening recommendations are given to eliminate the deformations that will occur in these structures [9,10]. Yurdakul et al. used 3D finite element modeling to assess the seismic performance of a 12th-century masonry minaret in Bayburt, identifying potential damage from tensile stresses caused by the Erzincan, Kocaeli-Düzce, and Van-Erciş earthquakes [11]. Kocaman analyzed the Molla Siyah Mosque using nonlinear dynamic analyses with ground motion records from the 1992 Erzincan, Cape Mendocino, and 1995 Kobe earthquakes, focusing on stress, displacement, and collapse mechanisms [12]. Schiavoni et al. studied the Church of Santa Maria Annunziata with advanced modeling techniques, including data from the February 2023 earthquakes in Türkiye, highlighting the role of material properties in seismic performance [9].

The seismic vulnerability of historical masonry structures in Türkiye and the Middle East has become a critical area of study. Kaya and Özkan [13] explored factors influencing the resilience of such structures in Türkiye, offering preservation and strengthening strategies. Aydın and Yılmaz [14] reviewed advances in failure analysis under seismic loads, evaluating modeling methods, while Toker and Alkan [15] analyzed post-earthquake damage in the Middle East, focusing on ground conditions, materials, and strengthening approaches. Recent studies emphasize developing effective seismic assessment methods that balance simplicity and accuracy to identify vulnerabilities and improve the resilience of historic structures, such as churches and mosques. These efforts aim to protect cultural heritage by tailoring preservation strategies to the unique characteristics of masonry structures [13–17]. Also, Valente's studies [18–20] indicate that slender bell towers in historical masonry churches suffer significant damage due to interactions with adjacent structures. While isolated towers develop initial cracks at the base, integrated towers experience cracking at connection points. These findings highlight the importance of considering structural interactions for accurate damage and collapse predictions [18–20].

Although the exterior masonry walls of many historic buildings appear as regular cut stone, their internal structure is often irregular and complex. Studies focusing on the structural behavior of such buildings under risk, modeling through simulations, and risk mitigation measures provide valuable guidance [21–23]. Gönen et al. highlighted how uncertainty in estimating the modulus of elasticity for masonry impacts numerical analysis results [24], as it is nearly impossible to define this property precisely in models [25–27]. Moreover, unlike reinforced concrete or steel, masonry materials lack true continuity due to inherent cracks, complicating accurate material definitions in structural modeling. Additionally, the heterogeneous nature of historical masonry materials, resulting from variations in stone types, mortar quality, and construction techniques, further complicates numerical modeling and seismic assessments. Research on heterogeneous material characterization and advanced modeling approaches plays a crucial role in improving the accuracy of structural analyses and developing more reliable assessment methods for historical buildings [28–35].

Traditional engineering assessments for historical masonry structures, especially those that are partially damaged, often require highly detailed finite element models with complex material definitions and computationally intensive nonlinear analyses. However, such sophisticated methods are not always practical for urgent post-earthquake evaluations, where rapid and effective stabilization decisions are needed. In this context, the originality of this study lies not in proposing a new analysis method, but in adapting an established approach within a simplified FE framework tailored for the emergency evaluation of partially collapsed historical structures. This methodology bridges the gap between detailed nonlinear models and rapid post-disaster needs, providing a practical and interdisciplinary decision-making tool. In this study, a simplified FE modeling approach is adopted to provide an efficient and accessible structural assessment. This method focuses on identifying critical stress concentrations and deformation patterns in both the undamaged and damaged states of the structure, ensuring timely intervention and stabilization efforts. This study examines the primary advantages and limitations of the simplified modeling methodology, contrasting it with conventional finite element methods to determine its suitable application range for post-earthquake emergency assessment.

Moreover, post-earthquake assessments of historical structures involve not only engineers but also experts from various disciplines, including architects, art historians, and restorers. Since these professionals may not always interpret structural results purely from an engineering perspective, a simplified and rapid evaluation method facilitates interdisciplinary decision making and improves coordination in emergency response efforts. By balancing computational efficiency with practical applicability, this approach offers a valuable tool for the preliminary evaluation of seismic damage in cultural heritage structures.

This research addresses the specific problems posed by partially collapsed historic structures, unlike previous studies that mostly focus on intact historical buildings or general seismic assessments. The proposed analytical approach links extensive, resource-intensive evaluations with the urgent need for stabilization measures in post-disaster contexts. This method enables efficient and swift evaluations of seismic hazards in cultural heritage sites by prioritizing computational efficiency and practical applicability.

This study uses the Antakya Habibi Neccar Mosque as a case study to demonstrate the need for prompt evaluation and fortification of partially collapsed historical edifices after seismic occurrences. The main objective is to create a computationally efficient structural analysis method to assist in emergency stabilization and conservation planning for masonry structures damaged by earthquakes. This study introduces a simpler FE modeling strategy that supports rapid and dependable structural evaluations, in contrast to conventional methods that require comprehensive nonlinear analysis and in situ testing. Response spectrum analysis is employed to run numerical simulations for both the intact and compromised states of the mosque. The results indicate that tensile stress concentrations above 0.2 MPa at dome–arch joints, vaults, and piers led to localized collapses, which were amplified by soil-strengthening effects. The results confirm that an efficient analytical method can provide significant preliminary insight into the seismic vulnerability of heritage structures. This study's innovation resides in its amalgamation of swift numerical modeling and post-earthquake heritage conservation, creating a framework that connects computational efficiency with practical applications. This methodology facilitates the establishment of standardized techniques for the protection of vulnerable cultural heritage edifices in seismically active areas, informing subsequent retrofitting and conservation initiatives.

2. Case Study and Methodology: A Simplified Structural Analysis Approach for Urgent Assessments

2.1. Case Study: The Mosque and Its History

The Habibi Neccar Mosque, located in Antakya, was built in 636 and is the first mosque built within the borders of present-day Türkiye. Madrasah rooms surround the mosque, with the Habibi Neccar Tomb situated 4 m below the northeast corner. The fountain in the mosque garden is a work of the 19th century [36]. The mosque was partially destroyed and heavily damaged during the 7.7 Mw earthquake centered in Kahramanmaraş.

The Habibi Neccar Mosque was located on an ancient street that has existed since the foundation of Antakya, connecting the two ends of the city on a southwest–northeast axis. The structure has a rectangular plan with approximate dimensions of 30 m in width and 20 m in depth, including the last congregation area. The overall height of the building is approximately 13.5 m, and it reaches up to 25 m at the apex of the central dome. Based on visual inspection and reference to similar Ottoman-era structures, the wall thickness ranges between 0.9 and 1.2 m, and the vault thickness was assumed to be 0.35 m during modeling. As seen in Figure 2, the mosque, which is easily recognizable due to its larger mass compared to the other buildings around it and its location on the street, has the appearance of a complex consisting of madrasah rooms and a fountain. These parts are located around a trapezoidal courtyard and are bounded by the road to the east and south while connecting with neighboring parcels and roads in other directions [37]. The entrance to the mosque is through the crown door with a round arch with an inscription as well as a relief and pointed arch. As seen in Figure 3, the mosque has large, pointed arches on both sides and an interlaced vault. The circular main dome of the Habibi Neccar Mosque is supported by four large, pointed arches and pendants. There are 12 windows in the part forming the pulley of the main dome [38,39]. Figure 4 shows the dome, vaults, pendentives, arches, and walls in the schematic plan describing the general structural form of the mosque.

Figure 2. General view of Habibi Neccar Mosque before it was partially collapsed. This figure has been adapted from "Türkiye Kültür Portalı-Hatay Habibi Neccar Camisi" [40].

Figure 3. Interior view of the mosque before it was partially collapsed. This figure has been adapted from "Türkiye Kültür Portalı-Hatay Habibi Neccar Camisi" [40].

Figure 4. Schematic plan and general views of the mosque.

Damage Observations

Masonry walls are brittle and anisotropic, with high compressive strength but significantly lower tensile strength, making them particularly vulnerable to horizontal seismic forces [41]. Built using masonry stone elements, the Habibi Neccar Mosque is part of a larger complex that includes two mausoleums, madrasah sections, and a fountain. As seen in Figures 5 and 6, the complex does not follow a strict linear alignment, which may have influenced the structural response during the earthquake.

Figure 5. Habibi Neccar Mosque street lines. The figure has been adapted from "Hatay Governorship-Anadolu'nun ilk camisi" website [42].

Figure 6. Top view of Habibi Neccar Mosque. The figure has been adapted from "Hatay Governorship-Anadolu'nun ilk camisi" website [42].

The mosque sustained severe damage during two large earthquakes (Mw 7.7 and Mw 7.6) on 6 February 2023, causing partial collapse of the structure and complete collapse of the minaret (Figures 7 and 8). The minaret, located southwest of the mosque, collapsed onto the adjacent buildings, affecting the walls of the complex. However, the surrounding complex walls and the fountain remained undamaged, indicating that the damage intensity was more severe for the primary load-bearing elements of the mosque.

Figure 7. Complete collapse of the minaret and dome of the Habibi Neccar Mosque after the earthquake of Mw 7.7. The debris also affected the adjacent buildings on the southwest side. The figure has been adapted from the "DHA" website [43].

Shear cracks in the south and east walls caused by window openings can be seen in Figures 9 and 10, indicating the effects of lateral seismic forces. Figure 9 illustrates the formation of shear cracks along the southern and eastern walls, especially near the apertures. The cracks indicate the lateral seismic forces exerted on the unreinforced masonry walls during the earthquake. Figure 10 illustrates diagonal shear cracking on the southern wall of the mosque, especially in the vicinity of the window area. This damage indicates a common failure mode resulting from horizontal seismic forces impacting unreinforced masonry. The west wall of the mosque (Figure 11) collapsed completely, while the north and east arches supporting the dome survived. The damage pattern suggests a significant compromise in the mosque's structural integrity, with localized damage accumulating in the most vulnerable areas.

Figure 8. Collapse of the west wall and partial roof failure of the Habibi Neccar Mosque, with debris indicating the fall of the minaret onto adjacent structural elements. The figure has been adapted from the "DHA" website [43].

Figure 9. Aerial view of the Habibi Neccar Mosque showing visible shear cracks along the south walls. This figure has been adapted from the "DHA and istockphoto" website [43,44].

Figure 10. Diagonal shear cracks observed on the south wall of the mosque, extending through the masonry units and aligned with the window openings. This figure has been adapted from the "DHA and istockphoto" website [43,44].

Figure 11. Interior view showing the collapse of the mosque's west wall, while the north and east arches that support the dome remain standing. This figure has been adapted from the "DHA" website [43].

A detailed assessment of these damage patterns emphasizes the role of out-of-plane deformations, weak connections between structural elements, and the effects of ground shaking amplification. The progressive collapse seen in Figures 7–11 was likely caused by a mix of horizontal seismic forces, uneven mass distribution, and wall bonds that were already weak. These findings are consistent with damage patterns seen in other historic masonry structures subjected to strong seismic events [25,30]. In addition to visual damage indicators, quantitative results such as tensile stress concentrations (0.15–0.20 MPa), complete collapse of the west wall and eastern vault, and shear cracks aligned with lateral force concentrations near window openings were identified and discussed in the Results section.

2.2. Methodology: Structural Assessment of the Damaged Building and the Formation of the Numerical Model

The simplified numerical modeling approach used in this study provides a practical and efficient method for assessing the structural integrity of partially collapsed historical masonry buildings, such as the Habibi Neccar Mosque, following the 6 February 2023, earthquakes. This method addresses the unique challenges of historical masonry structures, including complex geometry and material properties, while producing reliable results to guide emergency interventions. Compared to more detailed modeling techniques incorporating nonlinear analyses and advanced material characterization, this approach offers an effective alternative for rapid structural assessments. The sequence of steps followed in the structural evaluation process is summarized in Figure 12 to enhance clarity and reproducibility. The validity of this simplified approach was supported by comparing the FE model results of the undamaged structure with the actual damage patterns observed in the field after the earthquake.

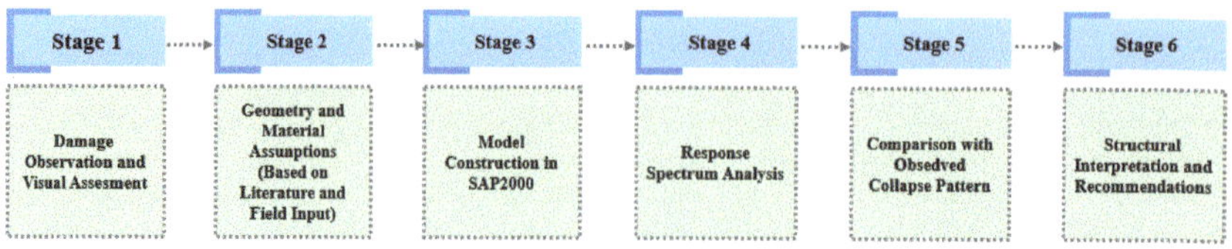

Figure 12. Flowchart of the simplified post-earthquake structural assessment approach.

Interventions to improve the earthquake performance of the building or to reduce the risk of damage require both a managerial and a technical approach. Technical strategies include interventions such as restoration of possible damages and defects in the structural system, increasing the strength and stiffness of the structure, examining the lateral displacement capacity, increasing the energy consumption capacity, or reducing the seismic susceptibility. The main criterion for intervention decisions is to limit the damages to the structural elements in terms of structural safety [45,46]. On the other hand, whether the intervention can be carried out while the structure is in use, whether it is divided into phases, and whether the use of the structure is limited or changed according to the risk level are the main elements of the managerial strategy.

Research-based diagnosis enables the organization of measures and indicators necessary for restoration and seismic retrofitting. Further research is needed to carefully analyze the unusual damage mechanisms observed in many historic buildings after the disaster and especially to correctly interpret the damage caused by previous structural retrofitting interventions. Furthermore, damage assessment and seismic vulnerability research allows a preparatory analysis for the design of temporary interventions and planning of the implementation method [47]. Systematic observation of emergency temporary interventions after previous disasters provides information on whether the methods chosen and the solutions considered appropriate sometimes contradict the safety measures taken on site. Due to their unusual characteristics, it is sometimes not possible to use conventional engineering solutions for emergency fortification of historic buildings. Therefore, specific and systematic research activities are required to optimize safety precaution techniques for damaged historic buildings by means of fast and simplified numerical calculation methods [47–49].

Following the 6 February 2023 earthquakes, an emergency structural analysis was carried out to determine the structural performance of the damaged Habibi Neccar Mosque during possible aftershocks. Since there were no drawings documenting the structural details of the building, and fieldwork opportunities were limited due to emergency conditions, new and accurate measurements could not be made. At this point, the importance and necessity of a facilitated numerical calculation method emerged once again.

Two numerical models were developed to represent the mosque before and after the earthquakes, ensuring accurate definition of stiffness and material properties. Special attention was given to the structural identification process, as experience plays a key role in modeling historical buildings. For instance, while the actual thickness of the main dome was accurately defined, the filling material on vaults and pendentives was considered only as additional weight. Such details are crucial, as digital documentation methods may sometimes overlook them, leading to discrepancies in stiffness and material representation. The pre-damage analysis results were validated by comparing the stress distribution with the observed damage patterns, a widely accepted methodology in structural assessment studies. To confirm the results, the calculated stress and deformation patterns were compared with actual damage assessments post-earthquake, including the collapse of the eastern vault and the failure zones at the dome–arch connections. In the damaged-state model, visual observations were used to define the collapse pattern, ensuring a realistic representation of the structural failure mechanisms.

Seismic Data for Station Selection, Numerical Model and Analysis of Historical Mosque

Figure 13 below shows the location of Station 3131 in relation to the Habibi Neccar Mosque. Due to its proximity to the mosque and similar ground conditions and regional seismic activity, Station 3131 is a good reference point for seismic data. This area is designated as a high seismic activity zone in the Turkish Seismic Hazard Map (AFAD). Alluvial soil layers, which are known to amplify seismic waves, are present in both locations [13]. The effects of ground amplification are compounded by relatively high groundwater levels. Station 3131 was selected for the seismic investigation of this study due to its similar soil characteristics and proximity to the mosque [50].

Figure 13. Location of the Habibi Neccar Mosque and the AFAD strong motion station 3131.

This study performs 3D linear analyses using the FE method to observe the dynamic properties of the Habibi Neccar Mosque. The behavior of masonry is characterized by using shell elements that can have their stiffness modified by constitutive assumptions about the elastic behavior of the material. Shell elements were chosen over brick elements in the FE model of the Habibi Neccar Mosque due to their efficiency in modeling thin-walled masonry structures, such as vaults, domes, and arches. This approach ensures accurate in-plane and out-of-plane deformation analysis while keeping computational costs manageable. Shell elements are widely used in historical masonry studies, providing reliable seismic response assessments without excessive computational complexity. A typical FE modeling scheme is used based on the concepts of homogenized material.

Creating a numerical model to study the seismic behavior of historic masonry structures requires a difficult process due to the complexity of the geometrical configuration of the structure and material properties that are nonlinearly inelastic and cannot be fully defined by laboratory tests [51]. The prerequisite for acquiring reliable analysis results is to create a refined mesh. However, due to the large number of parameters and degrees of freedom involved in the definition of the numerical model, the refined mesh cannot be properly applied to complex structural systems. However, the linear elastic static and dynamic analysis of such structures with simplified numerical models provides considerable information about the overall behavior and the interaction between the basic elements of the structure [52].

To develop the structural analysis models, SAP 2000 v14 software has been used by combining several element types depending on the structural members of the mosque. Four-nodded quadrilateral or three-nodded triangular shell elements were selected for meshing domes, vaults, pendentives, arches, and walls. The FE mesh for the first model, which describes the undamaged state of the mosque, consists of 7632 shell elements and 7512 nodes, while the second model, which describes the partially collapsed state, consists of 5786 shell elements and 5885 nodes (Figures 14 and 15).

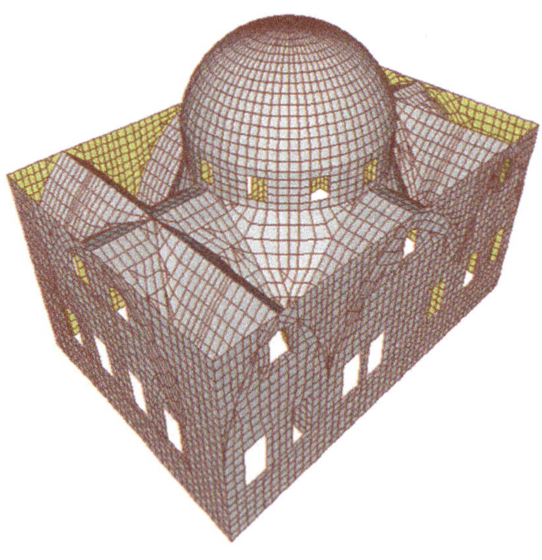

Figure 14. Numerical model for non-collapsed case.

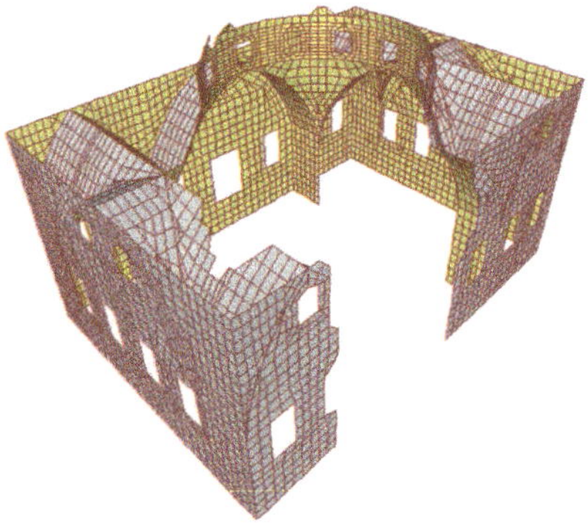

Figure 15. Numerical model for collapsed case.

In linear elastic numerical models, brick and stone masonry have been considered as linear isotropic materials with their properties shown in Table 1 [51,52]. The information given in Table 1 was obtained from the literature. These data were obtained by destructive test methods. The stone masonry pillars supporting the main arches, with cross-sectional dimensions of about 4 m × 2 m, were selected as massive in the models. However, these pillars surrounded by masonry stone were considered with a reduced modulus of elasticity in both analysis cases, due to infilled rubble material. Material properties were defined based on literature sources and destructive test data from comparable historical masonry buildings. The modulus of elasticity, unit weight, and unit mass values for the brick domes, stone walls, and piers are detailed in Table 1. These values reflect an isotropic material assumption to maintain computational efficiency while retaining consistency with historical material behavior.

Table 1. Material properties of the finite element model.

Element Type	Modulus of Elasticity E (kN/m^2)	Unit Weight (kN/m^3)
Brick dome and the pendantives (with mortar)	1,200,000 (1200 MPa)	24
Stone walls (with mortar)	450,000 (450 MPa)	24
Pillars (with inner filled materials)	200,000 (200 MPa)	34

The properties of the building materials were selected from data obtained from previous studies in the international literature [51,52]. The masonry was considered an isotropic material to simplify the analysis while maintaining computational efficiency, a common approach in large-scale historical structure modeling. The modulus of elasticity and unit weight values were determined by considering the masonry units and mortar as a single material.

Three different loads, namely gravity and earthquake loads, were applied to numerical models. In both analyses, calculations were made according to the damaged state of the building before and after the earthquake. The seismic effects provided by the Turkish Earthquake Code [53] were applied as two different load groups in the direction of the two main axes of the mosque perpendicular to each other (EQx: longitudinal, east–west direction) and (EQy: transversal, north–south direction).

Two load combinations, G + EQx (earthquake load in the X direction and vertical load) and G + EQy (earthquake load in the Y direction and vertical load), have been defined to assist in the evaluation of the results. This study conducted bi-directional seismic analyses to consider the combined impacts of ground motion in both the longitudinal (X) and transverse (Y) directions. The earthquake impacts were implemented separately in each direction as G + EQx (east–west) and G + EQy (north–south) load combinations, in line with seismic code guidelines. This method provides a more accurate evaluation of the mosque's dynamic reaction, considering the simultaneous impact of seismic pressures in orthogonal directions. In the response spectrum analysis, spectral calculation was performed for the first 30 modes. Some of the material properties considered in the finite element analysis of the Habibi Neccar Mosque are summarized in Table 1.

The spectrum curve used in the analysis of the Habibi Neccar Mosque is given in Figure 16. 3131 E-W is the spectral acceleration in the east–west direction, 3131 N-S is the spectral acceleration in the north–south direction, and 3131 U is the acceleration in the vertical direction. The earthquake acceleration causes the highest displacement in the horizontal direction, as shown by the dotted lines representing the east–west and north–south directions. In comparison, 3131 U, which corresponds to the vertical acceleration, has a lower magnitude in the figure. The spectrum curve was determined according to the values recommended for Antakya in the Turkish Building Earthquake Code. Figure 16 shows the differences between the design response spectra (DD1/DD2) and the measured response spectra (3131 E-W/3131 N-S). The DD1/DD2 spectra, derived from the Turkish Seismic Code [53], provide a simplified illustration of seismic demand, whereas the recorded spectra reflect the actual ground motion characteristics at Station 3131. The differences among these spectra highlight the impact of local soil conditions, site effects, and topography elements, which may result in the amplification or attenuation of seismic waves. The differences highlight the necessity of integrating site-specific seismic data in structural evaluations of historical masonry buildings. For example, while the design spectrum (DD-2) used for Antakya assumes a peak ground acceleration (PGA) of 1.11 g, the actual recorded PGA at Station 3131 was significantly lower, around 0.37 g. Additionally, the recorded spectral accelerations in both horizontal directions show sharper peaks at different periods compared to the smoother code-based curve. The evidence suggests that local soil conditions and topography caused site-specific amplification, which should be considered when evaluating seismic demand on the structure. In the dynamic analysis performed according to the condition of the building before the damage, the elastic spectrum was determined by considering the soft soil type (type ZE) containing loose sand and clay layer properties and the DD-2 earthquake ground motion level 2 (DD-2) category in the latest Turkish Earthquake Code. DD-2 earthquake ground motion is characterized by infrequent earthquake ground motion with a 10% probability of exceeding the spectral magnitudes in 50 years and a corresponding recurrence period of 475 years. This earthquake ground motion is also called standard design earthquake ground motion. Accordingly, for Antakya, the spectral acceleration coefficient (PGA) is taken as 1.11, which is equivalent to 111% of the gravitational acceleration.

In the 6 February 2023 earthquakes, the ground accelerations measured at five different stations around Antakya, whose locations are marked in Figure 17, vary between 123% g and 37% g, as shown in Table 2. In this case, it is not possible to include the PGA value at the location of the mosque in the calculations in a precise way. In fact, this is a very complex criterion, and it varies according to the ground conditions and the actual geometry and stiffness of the structure. Therefore, it is considered that the calculations made according to the DD-2 earthquake ground motion, which is recommended as the "standard design ground motion" in the Turkish Earthquake Specification for the purpose of emergency stabilization of the mosque after the disaster, are extremely consistent.

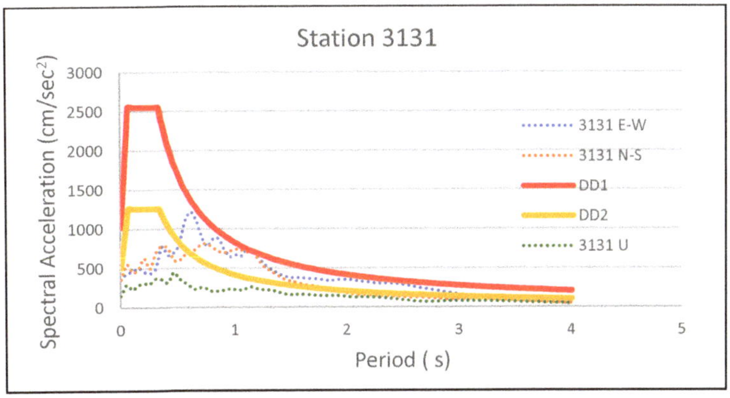

Figure 16. Response spectrum data of Station 3131.

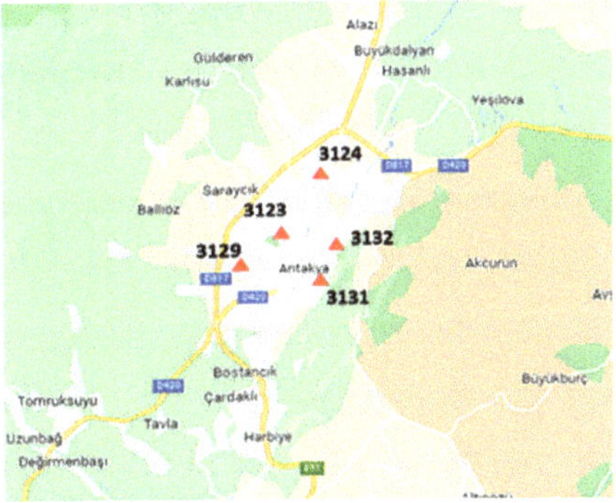

Figure 17. Location of the stations.

Table 2. The (PGA) ground accelerations measured at five different stations around Antakya.

Station No.	Location	PGA %g
3123	36.214° N 36.160° E	66.73
3124	36.239° N 36.172° E	65.04
3129	36.191° N 36.134° E	137.86
3131	36.191° N 36.163° E	37.19
3132	36.207° N 36.172° E	52.51

3. Results and Discussions: Assessment Method Based on Simplified Linear Dynamic Analysis

Numerical analyses were performed for the pre- and post-earthquake damage states of the Habibi Neccar Mosque to comprehensively assess its structural behavior and response to seismic forces. The pre-earthquake investigation established a benchmark for understanding the initial structural capacity of the mosque and identifying inherent risks. The post-earthquake investigation facilitated an assessment of how the partial collapse and damage affected the seismic performance of the structure. This dual approach is essential to guide both emergency stabilization efforts and long-term retrofit initiatives to ensure the effective preservation of the mosque as a cultural heritage site.

The periods and mode shapes obtained from the analyses of the mosque before and after the damage are presented in Figures 18 and 19, respectively. Tables 3 and 4 present the periods and mass participation ratios of the principal vibration modes in the transverse, longitudinal, and vertical directions for the non-collapsed and collapsed cases, respectively. As previously stated, the spectral calculation was conducted for the initial 30 modes included in the response spectrum. As can be seen in Table 2, 85% mass participation is achieved after Mode 22, which indicates that the number of modes selected for these analyses is sufficient. The modal analysis results show that only the first five modes contribute significantly to the dynamic response of the structure and capture the primary deformation patterns. The higher modes, including those up to and beyond mode 22, exhibited negligible involvement in the overall structural behavior. These mode shapes indicate that the structure exhibited significant torsional flexibility after the collapse, particularly due to the loss of stiffness in the northern and eastern walls. The shift from 0.56 s to 0.67 s in the first mode period confirms the increased deformability and reduced global stiffness of the structure. Additionally, the high out-of-plane deformations in the south and east walls observed in the first and second modes suggest a loss of diaphragm action and structural continuity in those regions. This is one of the objectives of the simplified calculation method. Analyses conducted with extensive finite element models encompassing a vast number of points necessitate high-performance computing hardware, resulting in prolonged analysis times. Due to the intricate geometry of historical masonry structures, inadequate integration of the finite element mesh may introduce unanticipated errors in matrix formulations and solutions.

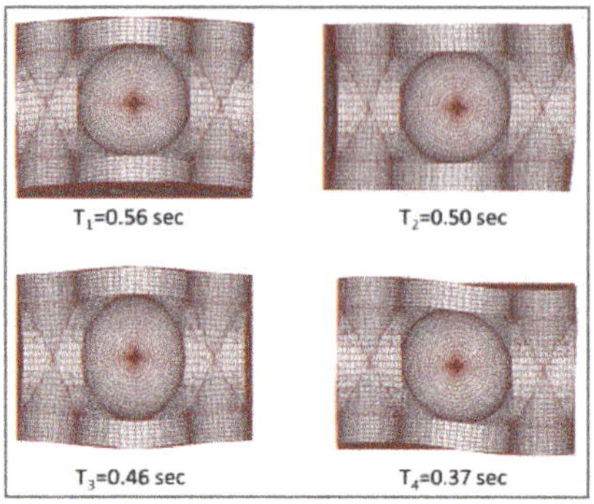

Figure 18. Periods and mode shapes of the 3D model for non-collapsed case.

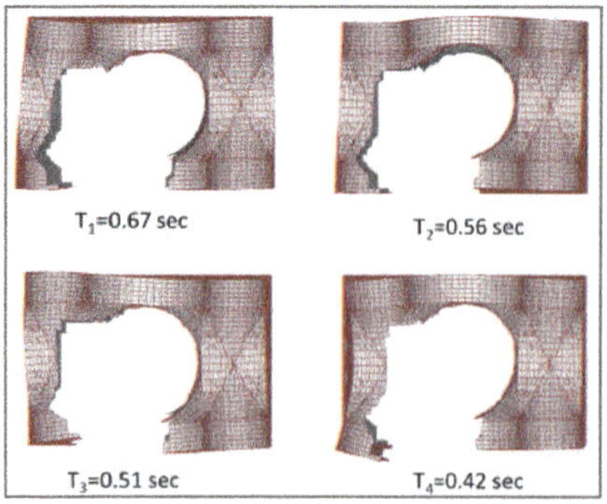

Figure 19. Periods and mode shapes of the 3D model for collapsed case.

Table 3. Periods and modal participating mass ratios for non-collapsed case (ΣMeff: cumulative effective modal mass participation ratio).

Mode Number	Period (s)	ΣMeff % in X-Longitudinal Direction	ΣMeff % in Y-Transversal Direction
1	0.56	0	75.05
2	0.50	78.05	75.05
3	0.46	78.05	75.05
4	0.36	78.07	75.06
22	0.16	84.12	83.13
23	0.15	84.14	83.13
24	0.15	84.38	83.13
25	0.14	84.38	83.20

Table 4. Periods and modal participating mass ratios for collapsed case (ΣMeff: cumulative effective modal mass participation ratio).

Mode Number	Period (s)	ΣMeff % in X-Longitudinal Direction	ΣMeff % in Y-Transversal Direction
1	0.67	6.25	15.89
2	0.56	44.92	32.05
3	0.51	68.97	37.59
4	0.42	69.36	69.79
27	0.13	83.16	83.83
28	0.13	83.25	84.01
29	0.13	83.31	84.04
30	0.12	83.38	84.18

As illustrated in Figure 18, the first mode of the undamaged building exhibits translational motion in the transverse direction, wherein the primary structural weakness is observed, accompanied by a notable out-of-plane deformation of the wall elements. The second mode shape illustrates a translation in the longitudinal direction, where the main body is observed to be robust. The third mode shape illustrates torsional deformations, thereby confirming the balanced distribution between the transverse and longitudinal structural elements of the mosque. The subsequent two mode shapes (not depicted herein) correspond to the second torsional mode (fourth mode) and the third torsional mode (fifth mode), respectively. Subsequently, the subsequent mode shapes are constituted by a combination of the transverse vibration mode and the torsional mode. The distribution of the mode shapes demonstrates that the perimeter walls, which exhibit low transversal and torsional stiffness, undergo significant out-of-plane deformations despite the presence of robust structural elements surrounding the building.

As stated in the previous paragraphs, the northern walls, main dome, eastern and northern main arches, and the largest part of the eastern vault were collapsed in the 6 February Türkiye earthquakes, thus giving the structure the appearance of a crescent with its open part facing north. As shown in Figure 19, in the first mode, the south walls experienced a considerable amount, or significant, out-of-plane deformation, while the east walls and more than half of the collapsed east vaults were translated outwards. In the second mode, the deformations at these locations have progressed similarly, but in the opposite direction. Considering that the modal shapes in modal analysis give important indications about the structural behavior of the building, this simplified analysis approach also critically guides urgent structural measures. The third and fourth modal shapes show the torsional deformations depending on the configuration of the transverse and longitudinal structural elements of the significantly collapsed mosque.

The periods listed comparatively in Tables 2 and 3 show important results for the simplified analyses performed for both cases (before and after demolition). The reason why the periods calculated in the analyses performed for the demolished state of the mosque are 10% higher is the regularization and weakening of the structural system. For instance, the first mode period increased from 0.56 s to 0.67 s after the collapse, indicating a reduction in global stiffness. Additionally, modal mass participation in the primary modes shifted significantly. In the undamaged model, over 75% of mass participated in the first two modes, whereas in the damaged case, this distribution was more scattered and delayed. These shifts point to a loss of regularity and a more flexible response due to partial structural degradation. Structural analyses performed for similar historical buildings in the scientific literature indicate that the mode shapes and periods depend on the exact material definition and the correct discretization of the geometry of the building. In this framework, in the context of emergency disaster management, it is essential to use experience-based data, visual observations, and statistics from official regulations in simplified structural analyses.

In seismic vulnerability assessments of historical masonry structures, comparing calculated stresses with accepted limit values is a key parameter. Due to uncertainties in material modeling and stress limits, direct validation remains challenging. However, by matching the stress patterns from the undamaged model with the actual way the structure collapsed, we can obtain a better estimate of the material's maximum stress limits, especially in emergency assessments after an earthquake.

Figures 20 and 21 illustrate the stress distribution under the applied design earthquake spectrum and the G + EQX and G + EQY load combinations. The results indicate that tensile stresses between 0.15 MPa and 0.20 MPa are concentrated at the dome–arch junctions and vaults, which are known as critical weak points in masonry structures. In the undamaged state, tensile stresses near the dome–arch joints reached around 0.20 MPa. In contrast, compressive stresses in the piers were over 2.5 MPa but still within safe limits. For the damaged condition, out-of-plane displacements at the upper part of the south wall reached nearly 8 mm under G + EQY loading. This suggests that the remaining wall segments are highly vulnerable, especially during possible aftershocks. These results are consistent with the observed damage and help explain the collapse pattern.

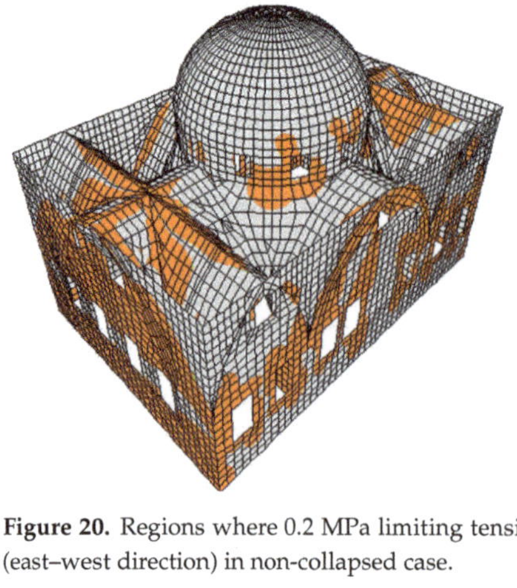

Figure 20. Regions where 0.2 MPa limiting tensile stresses are exceeded due to G + EQX loading (east–west direction) in non-collapsed case.

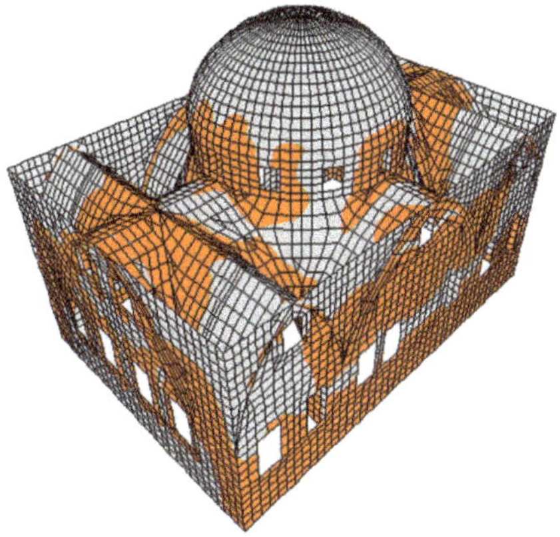

Figure 21. Regions where 0.2 MPa limiting tensile stresses are exceeded due to G + EQY loading (north–south) in non-collapsed cases.

The collapse of the mosque was not caused by isolated tensile failures alone, but rather by the progressive formation of a collapse mechanism, as cracks propagated and connected over time. While the eastern vault, which experienced tensile stresses of approximately 0.2 MPa, collapsed during the February 6 earthquakes, the western vault, subjected to similar stresses, remained intact, raising questions about additional influencing factors.

In addition to tensile stresses, compressive stress distributions were examined, particularly in piers and lower wall sections, to assess potential crushing failures. However, since masonry is significantly weaker in tension than in compression, the collapse was primarily driven by tensile failure. The observed damage pattern suggests that a limit state was exceeded when multiple tensile cracks merged, leading to a loss of structural stability.

Despite uncertainties in numerical modeling assumptions and seismic load estimations, the calculated tensile stress values (0.15–0.20 MPa) are consistent with widely accepted limits for historical masonry structures and generally align with the observed collapse pattern of the mosque. Previous research has identified comparable tensile strength values for unreinforced masonry. Asteris et al. [25], for example, reported that tensile capacities typically range from 0.1 to 0.3 MPa, influenced by the material properties and workmanship quality. Lourenço and Mele et al. have similarly noted that due to the inherently brittle and irregular nature of historical masonry, tensile failure generally initiates at stress levels below 0.3 MPa [25,30,51]. The tensile stresses obtained in the present study fall within this established range, reinforcing the plausibility of the numerical findings and corresponding damage patterns observed in the mosque.

In contrast to laboratory studies, the tensile stress capacity of masonry used in the construction of historic buildings is a hitherto unverified unknown for existing structures. Many ruined heritage structures have survived for hundreds of years, although unquestionably subjected to pure tensile stress. Therefore, it is not a correct approach to define the collapse sequence based only on tensile stresses. Initial cracks due to tensile stresses during an earthquake change the structural integrity of the building, and enormous compressive stresses may occur at unexpected locations. The usual limit for compressive stress in stone and brick masonry is between 2 MPa and 7 MPa, so the stress patterns shown in Figures 22 and 23 help us understand how the building might collapse. Considering that the undamaged building has had a certain regime of strength capacity for centuries based on its original architectural and structural configuration, major assumptions have been made when interpreting the results of this simplified analysis due to the complex properties of the masonry material. In this context, since the proposed simplified analysis method is for the emergency consolidation of historical buildings damaged by disasters, the stress distribution obtained in the analyses for the partially collapsed state of the mosque in Figures 24 and 25 will guide consolidation planning.

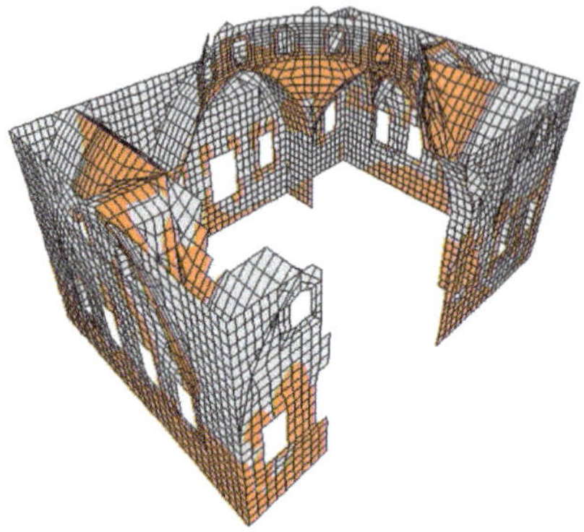

Figure 22. Regions where 0.2 MPa limiting tensile stresses are exceeded due to G + EQX loading (east–west direction) in collapsed case.

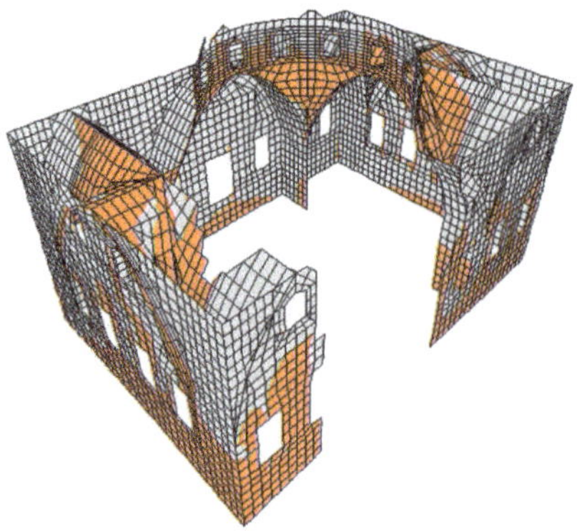

Figure 23. Regions where 0.2 MPa limiting tensile stresses are exceeded due to G + EQY loading (north–south) in collapsed case.

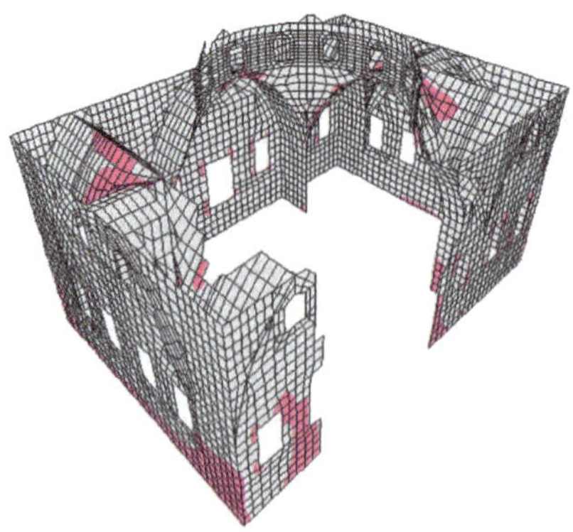

Figure 24. Regions of potential compressive failure identified in the collapsed state of the mosque under the G + EQX load combination (east–west direction). High compressive stress concentrations are observed along the lower wall segments and arch supports.

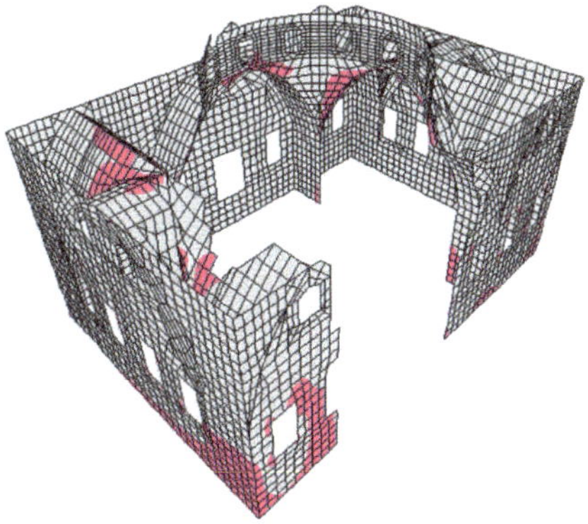

Figure 25. Regions of potential compressive failure identified in the collapsed state of the mosque under the G + EQY load combination (north–south direction). Significant compressive zones appear near the dome base and the central structural elements (Results are compared with field observations showing collapse at the same locations, validating the FE simulation.).

The partial collapse of the mosque has severely altered its structural continuity. In its undamaged state, the structural form of the dome, vaults, and pendentives has been disrupted. With the collapse of the dome and vaults, the peripheral walls (peripheral walls–exterior walls) were left unsupported, making them much more vulnerable to a possible aftershock. Based on both field observations and simulation results, the most severely damaged parts of the structure include the eastern vault, the dome–arch connection zones, and the unsupported western wall. These areas showed high tensile stresses and displacements, and their structural role is critical in maintaining overall stability. Their failure directly contributed to the collapse pattern seen after the earthquake. These results are consistent with the observed damage and help explain the collapse pattern. These results confirm that the undamaged FE model accurately predicted the collapse zones, validating the model's use for vulnerability assessment.

Ensuring the safety of the Habibi Neccar Mosque and its surroundings demands adherence to disaster and risk management principles, alongside a structural assessment of the damaged building in line with conservation and restoration practices for historic structures.

Figures 18–25 show the structural behavior of the mosque pre- and post-earthquake, emphasizing the impact of seismic forces on its integrity. The stress distribution prior to the earthquake was largely uniform, exhibiting regulated tensile and compressive stresses in key regions. The post-earthquake analysis reveals notable stress concentrations in specific regions where tensile stress surpasses the material's capacity, leading to localized failures and partial collapses, especially in the dome and upper wall sections.

Figures 20 and 21 indicate that the highest tensile stress concentrations were observed at the dome–arch intersections, the east and west vaults, and the upper sections of the piers. These regions exhibit increased vulnerability because of geometric discontinuities, material problems, and the impact of seismic forces. Field observations indicate that the eastern vault collapsed because of tensile stresses surpassing 0.2 MPa, whereas the western vault remained intact despite experiencing comparable stress levels. This discrepancy suggests potential variations in material properties or localized damage mechanisms. The stress concentrations aligned with the observed cracks and partial displacements at the openings of the north and east walls.

The results suggest that the collapse was not due to isolated tensile failures but rather a progressive failure mechanism involving the propagation and interconnection of multiple cracks, which ultimately surpassed a limit state and compromised the mosque's stability.

The findings offer a comprehensive insight into the impact of seismic forces on the structure, surpassing the information available from damage photographs alone.

The findings highlight the susceptibility of historical masonry structures to seismic forces and stress the necessity for focused restoration and retrofitting efforts. The identified critical zones in the post-earthquake context provide a basis for planning structural interventions aimed at restoring stability and improving future seismic performance. This analysis offers insights into damage mechanisms and establishes a framework for prioritizing retrofitting strategies in comparable historical structures.

4. Conclusions

Rapid post-earthquake assessment of historic masonry structures requires effective and reliable numerical modeling methods to aid in their prompt stabilization and long-term preservation and restoration planning. In this study, a simplified FE modeling approach is proposed to assess the structural integrity of the Habibi Neccar Mosque case study, which was severely damaged during the 6 February 2023 Kahramanmaraş earthquakes.

The results highlight several important aspects of the mosque's structural fragility:

- These findings not only support the accuracy of the simplified analysis but also offer theoretical insights into post-collapse behavior in historic masonry structures.
- Stress concentrations at the dome–arch joints and vaults exceeded material limits, contributing to local failures and partial collapse.
- Significant out-of-plane deformations were observed in the remaining walls, making them highly vulnerable to aftershocks and progressive damage.
- Ground amplification effects played a major role in increasing the seismic demand, further exacerbating the structural response and damage severity.

These results confirm that a simplified numerical modeling approach can provide rapid and valuable insights into the post-earthquake condition of historic structures. Such methods are essential for emergency decision making, ensuring timely stabilization measures and reducing the risk of further collapse.

In contrast to conventional finite element modeling techniques that require comprehensive geometric and material information, the simplified method introduced in this work provides expedited implementation, reduced processing demands, and adequate insights for emergency reactions. Nonetheless, it entails constraints, including linear material assumptions and a restricted ability to intricately simulate progressive failure causes. This trade-off renders it especially appropriate for preliminary evaluations, particularly when urgent structural assessments are required; however, time, data, or access to the structure is limited. For similar historic buildings with limited documentation and urgent stabilization needs, simplified FE-based modeling supported by measured ground motion data can offer valuable early-stage assessments. Such approaches are particularly useful in prioritizing structural interventions, identifying collapse-prone areas, and informing conservation strategies under seismic risk.

Future research should focus on the following aspects to improve the reliability and applicability of structural assessments for historic buildings:

- Creating and using nonlinear material modeling in a numerical setting: using detailed nonlinear analyses to show how damage builds up over time and how masonry units and mortar interact with each other.
- Improving damage assessment techniques: utilizing remote sensing, photogrammetry, and AI-based analysis to enhance the rapid evaluation of earthquake-damaged cultural heritage structures.
- Development of standardized emergency assessment protocols: establishing guidelines for the rapid assessment and stabilization of heritage sites in seismic zones.

By combining efficient computational methods with practical conservation strategies, these approaches can significantly enhance the earthquake resilience of historic buildings and ensure their preservation as vital cultural heritage assets.

Funding: This research received no external funding.

Data Availability Statement: The original contributions presented in this study are included in the article. Further inquiries can be directed to the author.

Acknowledgments: The author would like to sincerely thank Gülcan Acar for her generous permission to use photographic materials, and Demirören News Agency (DHA) for providing visual content support. Their contributions significantly enhanced the documentation of structural damage presented in this study.

Conflicts of Interest: The author declares no conflicts of interest.

References

1. Türkiye Cumhuriyeti Cumhurbaşkanlığı Strateji ve Bütçe Başkanlığı. *2023 Kahramanmaraş ve Hatay Depremleri Raporu*; Türkiye Cumhuriyeti Cumhurbaşkanlığı Strateji ve Bütçe Başkanlığı: Ankara, Türkiye, 2023.
2. Kocaman, İ. The effect of the Kahramanmaraş earthquakes (Mw 7.7 and Mw 7.6) on historical masonry mosques and minarets. *Eng. Fail. Anal.* **2023**, *149*, 107225. [CrossRef]
3. Guo, Y.; Li, H.; Liang, P.; Xiong, R.; Hu, C.; Xu, Y. Preliminary report of coseismic surface rupture (part) of Türkiye's MW7.8 earthquake by remote sensing interpretation. *Earthq. Res. Adv.* **2024**, *4*, 100219. [CrossRef]
4. Available online: https://whc.unesco.org/en/statesparties/tr (accessed on 12 May 2025).
5. Baltzopoulos, G.; Baraschino, R.; Chioccarelli, E.; Cito, P.; Vitale, A.; Iervolino, I. Near-source ground motion in the Mw7.8 Gaziantep (Türkiye) earthquake. *Earthq. Eng. Struct. Dyn.* **2023**, *52*, 3903–3912. [CrossRef]
6. Karataş, L.; Ateş, T.; Alptekin, A.; Dal, M.; Yakar, M. A systematic method for post-earthquake damage assessment: Case study of the Antep Castle, Türkiye. *Adv. Eng. Sci.* **2023**, *3*, 62–71.
7. Papazafeiropoulos, G.; Plevris, V. Kahramanmaraş—Gaziantep, Türkiye Mw 7.8 Earthquake on 6 February 2023: Strong ground motion and building response estimations. *Buildings* **2023**, *13*, 1194. [CrossRef]
8. Binici, B.; Yakut, A.; Kadas, K.; Demirel, O.; Akpinar, U.; Canbolat, A.; Yurtseven, F.; Oztaskin, O.; Aktas, S.; Canbay, E. Performance of RC buildings after Kahramanmaraş earthquakes: Lessons toward performance-based design. *Earthq. Eng. Eng. Vib.* **2023**, *22*, 883–894. [CrossRef]
9. Schiavoni, M.; Giordano, E.; Roscini, F.; Clementi, F. Numerical modeling of a majestic masonry structure: A comparison of advanced techniques. *Eng. Fail. Anal.* **2023**, *149*, 107293. [CrossRef]
10. Cosgun, T.; Uzdil, O.; Sayin, B.; Zengin, K.K. Seismic vulnerability assessment of a masonry structure and an FRP-strengthening proposal. *Case Stud. Constr. Mater.* **2022**, *17*, e01680. [CrossRef]
11. Yurdakul, M.; Yılmaz, F.; Artar, M.; Can, Ö.; Öner, E.; Daloğlu, A.T. Investigation of time-history response of a historical masonry minaret under seismic loads. *Structures* **2021**, *30*, 265–276. [CrossRef]
12. Kocaman, İ. Effect of restoration interventions on the seismic behavior of historical masonry buildings: The case of Molla Siyah mosque. *Eng. Fail. Anal.* **2023**, *148*, 107206. [CrossRef]
13. Kaya, M.; Özkan, T. Seismic vulnerability of historical masonry structures: The case of Turkey. *Quat. Int.* **2021**, *606*, 120–131. [CrossRef]
14. Aydın, K.; Yılmaz, H. Failure analysis of masonry structures under seismic loads: A review of recent developments. *Eng. Fail. Anal.* **2022**, *131*, 106706. [CrossRef]
15. Toker, B.; Alkan, R. Earthquake response of historical buildings: Case studies from the Middle East. *J. Earthq. Eng.* **2019**, *23*, 456–478. [CrossRef]
16. Kılıç Demircan, R. Determination of earthquake resistance of historical masonry inn with finite element analysis. *Bull. Earthq. Eng.* **2025**, *23*, 32693293. [CrossRef]
17. D'amato, M.; Michelangelo, L.; Daniela, D.F. Simplified Seismic Analyses of Ancient Churches in Matera's Landscape. *Int. J. Archit. Herit.* **2018**, *12*, 369–382. [CrossRef]
18. Valente, M. Earthquake response and damage patterns assessment of two historical masonry churches with bell tower. *Eng. Fail. Anal.* **2023**, *151*, 107418. [CrossRef]
19. Valente, M. Seismic behavior and damage assessment of two historical fortified masonry palaces with corner towers. *Eng. Fail. Anal.* **2022**, *134*, 106003. [CrossRef]

20. Valente, M. Seismic vulnerability assessment and earthquake response of slender historical masonry bell towers in South-East Lombardia. *Eng. Fail. Anal.* **2021**, *129*, 105656. [CrossRef]
21. Sood, R.; Salgado, R. Investigation of geosynthetic-reinforced soil performance under different conditions. *Open Constr. Build. Technol. J.* **2021**, *15*, 135–148. [CrossRef]
22. Kumar, R.; Ahmad, S. Seismic risk assessment of multi-story buildings using nonlinear static procedures. *J. Build. Perform.* **2018**, *13*, 189–204.
23. Lee, H.; Choi, J.; Kim, H. Performance assessment of seismic retrofitting techniques for structures subjected to aftershocks. *Arch. Civ. Mech. Eng.* **2021**, *21*, 312–328. [CrossRef]
24. Gonen, S.; Soyoz, S. Investigations on the elasticity modulus of stone masonry. *Structures* **2021**, *30*, 378–389. [CrossRef]
25. Asteris, P.G.; Chronopoulos, M.; Chrysostomou, C.; Varum, H.; Plevris, V.; Kyriakides, N.; Silva, V. Seismic vulnerability assessment of historical masonry structural systems. *Eng. Struct.* **2014**, *62–63*, 118–134. [CrossRef]
26. Mainstone, R.J. Structural Analysis, Structural Insights, and Historical Interpretation. *J. Soc. Archit. Hist.* **1997**, *56*, 316–340. [CrossRef]
27. Saleme Gidrão, G.M.; Carrazedo, R.; Bosse, R.M.; Silvestro, L.; Ribeiro, R.; de Souza, C.F.P. Numerical Modeling of the Dynamic Elastic Modulus of Concrete. *Materials* **2023**, *16*, 3955. [CrossRef]
28. Mallardo, V.; Malvezzi, R.; Milani, E.; Milani, G. Seismic vulnerability of historical masonry buildings: A case study in Ferrara. *Eng. Struct.* **2008**, *30*, 2223–2241. [CrossRef]
29. Romera, L.E.; Hernández, S.; Reinosa, J.M. Numerical characterization of the structural behaviour of the Basilica of Pilar in Zaragoza (Spain). Part 1: Global and local models. *Adv. Eng. Softw.* **2008**, *39*, 301–314. [CrossRef]
30. Mele, E.; De Luca, A.; Giordano, A. Modelling and analysis of a basilica under earthquake loading. *J. Cult. Herit.* **2003**, *4*, 355–367. [CrossRef]
31. Peña, F.; Lourenço, P.B.; Mendes, N.; Oliveira, D.V. Numerical models for the seismic assessment of an old masonry tower. *Eng. Struct.* **2010**, *32*, 1466–1478. [CrossRef]
32. Betti, M.; Vignoli, A. Modelling and analysis of a Romanesque church under earthquake loading: Assessment of seismic resistance. *Eng. Struct.* **2008**, *30*, 352–367. [CrossRef]
33. Betti, M.; Vignoli, A. Assessment of seismic resistance of a basilica-type church under earthquake loading: Modelling and analysis. *Adv. Eng. Softw.* **2008**, *39*, 258–283. [CrossRef]
34. Betti, M.; Galano, L.; Vignoli, A. Finite element modelling for seismic assessment of historic buildings. In *Earthquakes and Their Impact on Society*; Springer: Cham, Switzerland, 2016; pp. 1–14. [CrossRef]
35. Adami, C.E.; Vintzileou, E. Interventions to historic masonries: Investigation of the bond mechanism between stones or bricks and grouts. *Mater. Struct. Constr.* **2008**, *41*, 255–267. [CrossRef]
36. İstek, E. Seyyahlarin gözüyle antakya şehri (10–19. yüzyillar arasi). *Pamukkale Univ. J. Soc. Sci. Inst.* **2020**, 227–246. [CrossRef]
37. Diker, M.; Çeker, N. Religious Building for the City Identity: The Case of Antioch. *J. Plan.* **2017**, *27*, 180–192. [CrossRef]
38. Pınar, M. Antakya'da Restore Edilen ve Edilecek Yapılar Üzerinde Çevrenin Etkisi. *Turk. Stud.* **2015**, *11*, 579–604.
39. Eykay, İ.; Dalgın, T.; Çeken, H. The Evaluation of Religious Tourism Potential in Antakya. *J. Life Econ.* **2015**, *2*, 59–74. [CrossRef]
40. Habibi Neccar Camisi. Available online: https://www.kulturportali.gov.tr/turkiye/hatay/gezilecekyer/habib-i-neccar-camii (accessed on 2 January 2025).
41. Sarhosis, V.; Giarlelis, C.; Karakostas, C.; Smyrou, E.; Bal, I.E.; Valkaniotis, S.; Ganas, A. Observations from the March 2021 Thessaly Earthquakes: An earthquake engineering perspective for masonry structures. *Bull. Earthq. Eng.* **2022**, *20*, 5483–5515. [CrossRef]
42. Anadolu'nun İlk Camisi. Available online: http://www.hatay.gov.tr/anadolunun-ilk-camisi (accessed on 2 January 2025).
43. Habib-i Neccar. Available online: https://www.dha.com.tr/video/antakyadaki-14-asirlik-habib-i-neccar-camisi-depremde-yikildi-video-2204262 (accessed on 2 January 2025).
44. Antakya Depremi. Available online: https://www.istockphoto.com/tr/foto%C4%9Fraf/kahramanmaras-earthquake-gm1474099767-504056395 (accessed on 22 May 2025).
45. Cabalar, A.; Canbolat, A.; Akbulut, N.; Tercan, S.; Isik, H. Soil liquefaction potential in Kahramanmaras, Turkey. *Geomat. Nat. Hazards Risk* **2019**, *10*, 1822–1838. [CrossRef]
46. Lungu, D.; Arion, C. Seismic Risk and/or Real Estate Risk for the Bucharest Heritage Buildings. In Proceedings of the 15th World Conference on Earthquake Engineering (WCEE), Lisbon, Portugal, 24–28 September 2012; pp. 1–10.
47. Modena, C.; Valluzzi, M.R.; Da Porto, F.; Casarin, F. Structural aspects of the conservation of historic masonry constructions in seismic areas: Remedial measures and emergency actions. *Int. J. Archit. Herit.* **2011**, *5*, 539–558. [CrossRef]
48. Penna, A.; Morandi, P.; Rota, M.; Manzini, C.F.; da Porto, F.; Magenes, G. Performance of masonry buildings during the Emilia 2012 earthquake. *Bull. Earthq. Eng.* **2014**, *12*, 2255–2273. [CrossRef]

49. Bozyiğit, B.; Özdemir, A.; Donmez, K.; Dalgic, K.D.; Durgut, E.; Yesilyurt, C.; Dizgin, Y.; Yıldeniz, C.; Ispir, M.; Bedirhanoglu, I.; et al. Damage to monumental masonry buildings in Hatay and Osmaniye following the 2023 Turkey earthquake sequence: The role of wall geometry, construction quality, and material properties. *Earthq. Spectra* **2024**, *40*, 1870–1904. [CrossRef]
50. Baker, J.W. Measuring bias in structural response caused by ground motion scaling. In Proceedings of the 8th Pacific Conference on Earthquake Engineering, Singapore, 5–7 December 2007; p. 56.
51. Lourenço, P.B. Computations on historic masonry structures. *Prog. Struct. Eng. Mater.* **2002**, *4*, 301–319. [CrossRef]
52. Lourenço, P.B.; Mendes, N.; Ramos, L.F.; Oliveira, D.V. Analysis of Masonry Structures Without Box Behavior. *Int. J. Archit. Herit.* **2011**, *5*, 369–382. [CrossRef]
53. Turkey Building Earthquake Code (TBEC-2018). Available online: https://www.resmigazete.gov.tr/eskiler/2018/03/20180318M1-2.htm (accessed on 22 May 2025).

buildings

MDPI

Article

The Case Study of the Characteristic Analysis and Reinforcement Measures of RC Diaojiaolou Structures Under Different Seismic Intensities

Wenwu Zhong [1], Zhile Shu [2,*], Wenkai Feng [1], Xin Zhang [3], Xueye Ma [4] and Zheng Fei [3]

1 College of Environment and Civil Engineering, Chengdu University of Technology, Chengdu 610059, China; 212020085900035@stu.xhu.edu.cn (W.Z.); fengwenkai@cdut.cn (W.F.)
2 School of Emergency Management, Xihua University, Chengdu 610039, China
3 School of Architecture and Civil Engineering, Xihua University, Chengdu 610039, China; zhangxin028@mail.xhu.edu.cn (X.Z.); 212021085900028@stu.xhu.edu.cn (Z.F.)
4 College of Intelligent Construction and Environmental Engineering, Chengdu Textile College, Chengdu 611731, China; 18284537052@163.com
* Correspondence: zlshu2011@163.com; Tel.: +86-15998964464

Abstract: China is strengthening the construction of the disaster resistance capacity of its mountain buildings, which increases the demand for RC Diaojiaolou reinforcement technology. In this paper, the performance of RC Diaojiaolou structures (unreinforced and carbon-fiber cloth-reinforced) in an earthquake is studied by a physical model test. The results show that carbon-fiber cloth can effectively improve the seismic capacity. The natural vibration period and acceleration- and displacement-increment coefficients of DF and CDF conformed to the exponential law. The damage process can be divided into three stages: DS, YS, and PS. After reinforcement, the development law of the average value of the acceleration-increment coefficient changed from the N type to the V type, and the development law of the average value of the displacement-increment coefficient changed from the concave type to the V type. The Diaojiaolou was the least affected by the acceleration at I. The displacement deformation of DF was the least affected by the seismic waves at DZ1. The displacement deformation of CDF was the least affected by the seismic waves at I. These findings provide a theoretical basis for the seismic design of mountain Diaojiaolous.

Keywords: mountain building; RC Diaojiaolou; earthquake; carbon-fiber cloth; physical model

1. Introduction

China possesses diverse terrain, which can be categorized into five main types: plateaus, mountain ridges, plains, hills, and basins [1]. Additionally, China is highly susceptible to earthquakes. According to official data from the China Earthquake Networks Center, there were approximately 1280 earthquakes with magnitudes greater than 3 in China in 2024, including 7 earthquakes with magnitudes greater than 6 [2]. The north–south seismic belt (Sichuan–Yunnan–Gansu–Shaxi) accounts for about 48%, Xinjiang accounts for about 28%, and other regions account for about 24%. It can be seen that China's seismic activity is mainly distributed in mountainous areas, and the number of seismic activities above magnitude 3 in recent years has shown a gradual increasing trend (see Figure 1). In earthquake-prone areas, building collapse is the most important factor causing casualties. In the 2008 Wenchuan earthquake, the number of victims reached 69,226, the number of

 https://doi.org/10.3390/buildings15111795

injured reached 374,643, and the number of missing reached 17,923. Among the former, the number of deaths directly caused by the building collapse was about 53,000~55,000, accounting for more than 75% of the total number of deaths [3]. This has directly promoted the reform of the seismic design of Chinese buildings. Therefore, many scholars have carried out research on the impact of earthquakes on buildings.

Figure 1. The proportions of the terrain, the number of seismic activities, and RC Diaojiaolous.

When an earthquake occurs, energy is transmitted to the surface through seismic waves so that the building structure is subjected to complex dynamic loads, resulting in different degrees of damage or even the collapse of the building. Therefore, the characteristics of ground motion directly affect the structural response. For example, Heydarpour and Tehrani [4] studied the seismic responses of 21 horizontal-bending and vertical RC four-span bridges with different abutment types and found that compared with straight bridges, abutment characteristics and boundary conditions have a significant impact on the seismic response evaluation of curved bridges. Massumi and Gholami [5] estimated structural responses by using different combinations of principal component analysis and parameters. The results showed that frequency-related parameters can better predict damage criteria than time-related parameters. Pachla et al. [6] designed a five-story, irregularly shaped reinforced concrete (RC) building model to withstand earthquakes of different magnitudes. It was found that the earthquake that caused the building damage was characterized by high strength, a long duration, and high velocity. In the Lisbon earthquake in 1755, the correlation between building collapse and geological structure was first recorded [7]. Subsequently, many scholars have carried out a lot of research through the elastic rebound theory [8–10], the random vibration theory [11–14], and so on. The seismic response mechanisms of a wide range of building structures, such as reinforced concrete (RC) structures [15–17], steel structures [18–20], masonry structures [21–23], and timber structures [24–26], have been also studied in depth in recent years. Liu [27] carried out an experimental study on the use of steel braces on stilt buildings to reduce the irregularity of stiffness. It was found that the setting of steel braces can change the failure modes of stilt buildings and improve the ductility and deformation capacities of stilt buildings.

Li [28] proposed the lateral stiffness ratio of stilt buildings as the floor stiffness ratio (SSR) and found that the SSR is an important factor affecting the seismic failure mode of stilt buildings. With the development of manufacturing technology, the introduction of the electro–hydraulic-servo shaking-table has greatly facilitated the development of physical model tests. For example, Kumar [29] studied the influence of tunnel excavation on the responses of buildings with different types of frames at the foundation through shaking-table tests. Zhou [30] studied the seismic performance of a four-tower (frame shear wall) building through shaking-table tests and numerical simulation. In recent years, with the rapid development of artificial intelligence technology, machine learning [31–33] and digital twin technology [34,35] have gradually begun to realize the intelligent prediction and monitoring of earthquakes. At present, many scholars are constantly optimizing artificial intelligence and machine learning algorithms [36–39], which can be applied to predict the responses of building structures and improve disaster resistance in architectural design. At the same time, data-driven insights and advanced modeling are used to improve the safety of earthquake-prone areas.

In summary, many scholars have studied the influence of ground-motion characteristics on various building structures. At the same time, the seismic response mechanisms and seismic design-method improvement of various building structures are also being studied. However, there are few studies on the influence of ground-motion characteristics on RC Diaojiaolou (traditional Chinese stilted dwellings) structures. RC Diaojiaolous are mainly distributed in ethnic-minority gathering places and tourist areas such as southeastern Guizhou, northern Guangxi, and northeastern Yunnan and are concentrated in the north–south seismic belt. Also, China's Diaojiaolou, which has been passed down for thousands of years, is one of the oldest building types in China. The stilt design of the Diaojiaolou is not only the crystallization of the architectural wisdom of the mountain residents but also the inheritance and development of Chinese traditional culture. In the Luding earthquake, many Diaojiaolous and local residents were seriously injured. However, there are few special studies on China's Diaojiaolous for earthquakes. Therefore, in order to continue to inherit the architectural culture of the Diaojiaolou and reduce the risk of earthquakes in the above areas, this paper explores the influence of ground-motion characteristics on RC Diaojiaolou structures (without carbon-fiber cloth reinforcement and with carbon-fiber cloth reinforcement) through physical model tests. The research results can provide some reference values for the seismic design of ethnic-minority buildings in the mountainous areas of China. This not only enhances the emergency response capacity of China's ethnic-minority areas but also is an important manifestation of the firm implementation of the basic national policy of national unity.

2. Test Method

The physical model test research in this paper produced two RC-suspension frame models with a similarity ratio of 1:8 and the same parameters. Carbon-fiber cloth reinforcement measures were added to one of the models. The connection material between the carbon-fiber cloth and the Diaojiaolou was an epoxy resin adhesive. The basic structure of the epoxy resin was composed of an epoxy group and a resin group. During the curing process, the epoxy group reacted with the curing agent or hardener to form a three-dimensional network structure, which gave the epoxy resin excellent mechanical properties and chemical resistance. The RC Diaojiaolou model without the carbon-fiber cloth reinforcement measures was named DF. The RC Diaojiaolou model with the carbon-fiber cloth reinforcement measures was named CDF. Shaking-table tests were carried out on the two models to explore the influence of ground-motion characteristics on the two RC Diaojiaolous.

The shaking-table equipment was an electro–hydraulic-servo one-way seismic simulation shaking table produced by the ANCO Company in the United States. The size of the table was 3.0 (m) × 3.0 (m), the rated load was 10 (t), the maximum load was 20 (t), the peak acceleration was 1.5 (g), the test frequency was 0~50 (Hz) (Hertz), the maximum overturning moment was 300 (kN · m), and the maximum displacement was ±10 (cm). The material performances of C30 concrete and micro-concrete were tested by a 3000 (kN) microcomputer-controlled electro–hydraulic-servo pressure-testing machine. The performance test of a galvanized iron wire material was carried out by a 600 (kN) microcomputer-controlled electro–hydraulic-servo universal testing machine. The test equipment is shown in Figure 2.

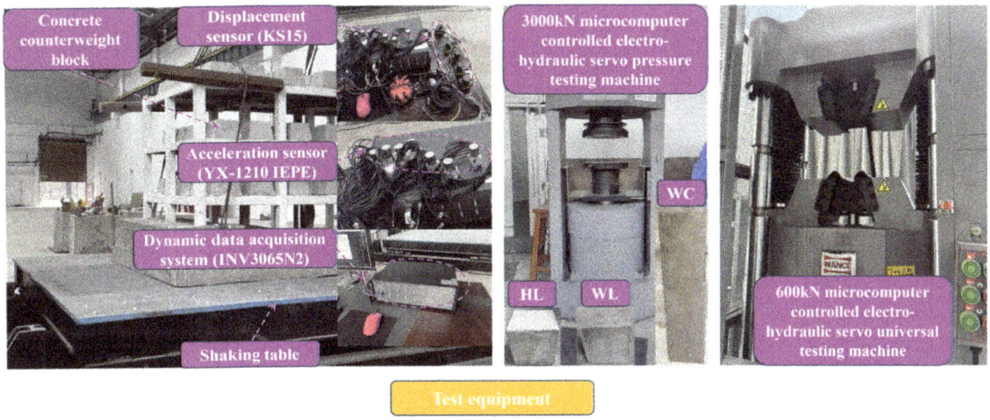

Figure 2. Test equipment.

2.1. Similarity Relation Calculation

We considered the size of the shaking table, the bearing range of the shaking table, and the similarity of the physical characteristics between the physical model and the prototype structure. According to the similarity theory [40], the geometric similarity constant of the physical model was determined to be 1/8, the stress-similarity constant of the micro-concrete material was 1/3, and the acceleration-similarity constant was 2.0. Then, the other parameters were determined by dimensional analysis. The specific similarity constants are shown in Table 1.

Table 1. Structure-similarity constants of the physical model.

Physical Property	Physical Parameter	Similarity Constant	Dimension of Quantity	Relation	Value
Geometrical properties	Length	S_l	$[L]$	S_l	1/8
	Area	S_A	$[L^2]$	$(S_l)^2$	1/64
	Linear displacement	S_l	$[L]$	S_l	1/8
	Angular displacement	S_u	$[1]$	S_σ/S_E	1.0
Material properties	Strain	S_ε	$[1]$	S_σ/S_E	1.0
	Elastic modulus	S_E	$[ML^{-1}T^{-2}]$	$S_E = S_\sigma$	1/3
	Stress	S_σ	$[ML^{-1}T^{-2}]$	S_σ	1/3
	Mass density	S_ρ	$[ML^{-3}]$	$S_\sigma/(S_a \cdot S_l)$	1.333
	Mass	S_m	$[M]$	$S_\sigma \cdot (S_l)^2/S_a$	2.604×10^{-3}
	Concentrated force	S_F	$[MLT^{-2}]$	$S_\sigma \cdot (S_l)^2$	5.208×10^{-3}

Table 1. *Cont.*

Physical Property	Physical Parameter	Similarity Constant	Dimension of Quantity	Relation	Value
Load properties	Line load	S_q	$[MT^{-2}]$	$S_\sigma \cdot S_l$	4.167×10^{-2}
	Surface load	S_p	$[ML^{-1}T^{-2}]$	S_σ	1/3
Dynamic properties	Moment	S_M	$[ML^2T^{-2}]$	$S_\sigma \cdot (S_l)^3$	6.510×10^{-4}
	Period	S_T	$[T]$	$(S_l)^{0.5} \cdot (S_a)^{-0.5}$	0.25
	Frequency	S_f	$[T^{-1}]$	$(S_l)^{0.5} \cdot (S_a)^{0.5}$	4
	Speed	S_v	$[LT^{-1}]$	$(S_l \cdot S_\sigma)^{0.5}$	0.5
	Acceleration	S_a	$[LT^{-2}]$	S_a	2.0
	Gravity	S_g	$[LT^{-2}]$	$S_g \cdot S_a$	1.0

2.2. Material Performance Test

2.2.1. C30 Concrete Material Performance Test

The main materials used in the physical model in this paper are as follows: steel bars, C30 concrete, micro-concrete, galvanized iron wire, etc. The main structure adopted galvanized iron wire and micro-concrete, and the base adopted HRB400 steel bars and C30 concrete.

In order to ensure the quality of the C30 concrete, a concrete cube sample with a size of 100 (mm) × 100 (mm) × 100 (mm) was made to test the cube's compressive strength [41]. The C30 concrete specimens were numbered HL1, HL2, and HL3, respectively. The average compressive strength of the three specimens was 27.20 (MPa). The test results verify the stability and reliability of C30 self-mixing concrete, and the test results are shown in Table 2.

Table 2. C30 concrete material performance test results.

Numbering	Size (mm)	Failure Load (kN)	Compressive Strength (MPa)	Average Compressive Strength (MPa)	Error (Absolute Value)
HL1	100 × 100 × 100	301.58	28.65	27.20	5.06%
HL2		264.90	25.17		8.07%
HL3		292.32	27.77		2.05%

2.2.2. Material Performance Test of Micro-Concrete

According to the similarity calculation above, for the physical model, a mix ratio of cement, yellow sand, and water of 1:7:2 [42] was selected to make micro-concrete. The material performance test of the micro-concrete was carried out. The performance tests of the micro-concrete materials mainly included a compressive strength test and an elastic modulus test. Two kinds of specimens were made in this experiment, which were cube specimens and cuboid specimens. The cube sample size was 100 (mm) × 100 (mm) × 100 (mm), which was used to test the compressive strength of the micro-concrete [41]. The cube samples were numbered WL1~WL9. The size of the cuboid specimen was 100 (mm) × 100 (mm) × 300 (mm), which was used for the elastic modulus test of the micro-concrete [41]. The cuboid specimens were numbered WC1~WC9. The test results are shown in Tables 3 and 4. The average compressive strength of the micro-concrete was 7.40 (MPa), and the average elastic modulus was 8.31 (GPa).

Table 3. Compressive strength test results of different micro-concretes.

Numbering	Size (mm)	Failure Load (kN)	Compressive Strength (MPa)	Average Compressive Strength (MPa)	Error (Absolute Value)
WL1		20.81	8.32		11.06%
WL2		18.23	7.29		1.51%
WL3		19.58	7.83		5.49%
WL4		18.44	7.38		0.27%
WL5	70.7 × 70.7 × 70.7	17.09	6.84	7.40	8.19%
WL6		15.98	6.39		15.81%
WL7		20.36	8.14		9.09%
WL8		17.44	6.98		6.02%
WL9		18.57	7.42		0.27%

Table 4. Elastic modulus test results of different micro-concretes.

Numbering	Size (mm)	Elastic Modulus (MPa)	Average Elastic Modulus (GPa)	Error (Absolute Value)
WC1		9475.3		12.30%
WC2		7045.9		17.94%
WC3		8535.6		2.64%
WC4		7756.7		7.13%
WC5	100 × 100 × 300	8232.2	8.31	0.95%
WC6		9843.1		15.58%
WC7		8417.6		1.28%
WC8		7332.7		13.33%
WC9		8174.0		1.66%

2.2.3. Performance Test of Galvanized Iron Wire Material

The constitutive relationship and elastic modulus of galvanized iron wire are similar to those of steel bars. In the physical model, the integrity of the structure was maintained, and the performance of steel bars in real buildings was simulated. Moreover, the galvanized iron wire had good toughness and plasticity, which is convenient for the production and construction of the wire cages of beam and column members. The types of galvanized iron wire used in this test were 14#, 16#, and 18#, respectively. The detailed parameters are shown in Table 5.

Table 5. Performance test results of galvanized iron wire materials.

Type	Section Diameter (mm)	Cross-Sectional Area (mm^2)	Tensile Strength (MPa)			Elastic Modulus (MPa)		
			Test Value	Average Value	Error (Absolute Value)	Test Value	Average Value	Error (Absolute Value)
			248		1.21%	331		9.37%
			264		4.92%	383		5.48%
14#	2.11	3.49	230	251	9.13%	364	362	0.55%
			266		5.64%	371		2.43%
			248		1.21%	362		0.00%
			246		4.07%	358		0.91%
			223		5.83%	314		12.79%
16#	1.60	2.01	247	236	4.45%	336	334	8.24%
			220		7.27%	324		9.97%
			242		2.48%	338		7.73%
			214		0.47%	323		4.53%
			231		6.93%	341		17.49%
18#	1.20	1.14	219	215	1.83%	301	316	13.19%
			204		5.39%	305		14.82%
			207		3.86%	310		12.71%

In order to test the tensile strength and elastic modulus of the galvanized iron wire, the iron wire samples were pre-stretched to ensure that they were upright. Different types of galvanized iron wire were cut into samples with lengths of 10 (cm), and 5 samples were prepared for each type: a total of 15 samples. All samples were not bent before the test and remained upright. The test results are shown in Table 5.

2.2.4. Carbon-Fiber Cloth Material Performance

In the physical model test, the unidirectional carbon-fiber cloth of the national standard of 200-gram grade I (10-cm wide) was used for the reinforcement of the Diaojiaolou. The physical parameters of the material are shown in Table 6.

Table 6. Physical parameter table of unidirectional carbon-fiber cloth [43].

Unidirectional Carbon-Fiber Cloth	Physical Parameter
Tensile strength standard value (MPa)	≥3400
Tensile elastic modulus (MPa)	$\geq 2.4 \times 10^5$
Elongation (%)	≥1.7
Bending strength (MPa)	≥700
Interlaminar shear strength (MPa)	≥45
The positive tensile bond strength between the fiber composite and concrete under the condition of upward bonding (MPa)	≥2.5
Mass per unit area (gram/m^2)	200
Calculation of the thickness (mm/layer)	0.111

2.3. *Structure Design of the Diaojiaolou*

2.3.1. Prototype Structure Design of the Physical Model

According to the 'Code for design of concrete structures' [44], the 'Technical specification for concrete structures of tall building' [45], the 'Code for seismic design of buildings' [46], and other regulatory requirements and the actual construction of the Diaojiaolou, the prototype structure was designed. The prototype design parameters of the Diaojiaolou structure are shown in Table 7, below.

Table 7. Prototype structure parameters.

Project	Parameter
Size	14.4 (m) × 9.6 (m)
Number of layers	Diaojiaolou layer 1 (layer), standard layer 2 (layer)
Layer height	Diaojiaolou layer: 4 (m); standard layer: 3 (m)
X-direction span number	3
Y-direction span number	2
Span length	4.8 (m)
Column section size	500 (mm) × 500 (mm)
Beam section size	300 (mm) × 500 (mm)
Thickness of floor	120 (mm)
Concrete material	C30 concrete, volumetric weight of 27 (kN/m^3)
Steel bars	HRB400 longitudinal bar, HRB400 stirrup, HPB300 slab reinforcement, volumetric weight of 27 (kN/m^3)

The seismic fortification intensity of the project was 7 (degrees). The design earthquake group was the second group. The design basic seismic acceleration was 0.15 (g). The site class was Class II. The seismic grade of the Diaojiaolou was Grade II. The characteristic period was 0.4 (s). The period reduction coefficient was 0.8, and the damping ratio was 0.05. The structure of the Diaojiaolou is shown in Figure 3.

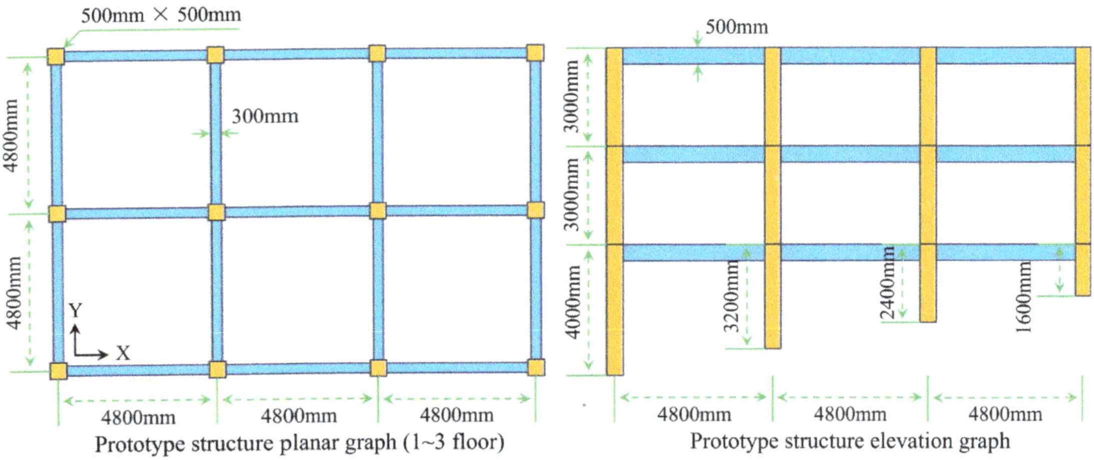

Figure 3. Prototype structure of Diaojiaolou.

2.3.2. Physical Model Design

Based on the model prototype and similarity theory, the physical model of the shaking-table test was made at a ratio of 1:8. The physical model design parameters are shown in Table 8. The base of the physical model adopted C30 concrete, and the base had four steps, numbered DZ1, DZ2, DZ3, and DZ4, respectively, as shown in Figure 4. The specific size information of the physical model is shown in Figure 4. Because the base of the physical model needed to be installed on the vibration table, a water drill was used to drill the hole after the concrete of the base was finalized, and the bolt hole of the base was aligned with the screw hole of the vibration table. The size of the shaking table was 3000 (mm) × 3000 (mm), there were 225 screw holes with diameters of 20 (mm), and the spacing of the screw holes was 200 (mm). The arrangement of the shaking table and the physical model is shown in Figure 4.

Table 8. Structural parameters of physical model.

Project	Parameter
Size	1.8 (m) × 1.2 (m)
Number of layers	Diaojiaolou layer 1 (layer), standard layer 2 (layer)
Layer height	Diaojiaolou layer: 0.5 (m); standard layer: 0.375 (m)
X-direction span number	3
Y-direction span number	2
Span length	0.6 (m)
Column section size	62.5 (mm) × 62.5 (mm)
Beam section size	37.5 (mm) × 62.5 (mm)
Thickness of floor	15 (mm)
Concrete material	Micro-concrete, volumetric weight of 24 (kN/m^3)
Steel bar	14#, 16#, 18#

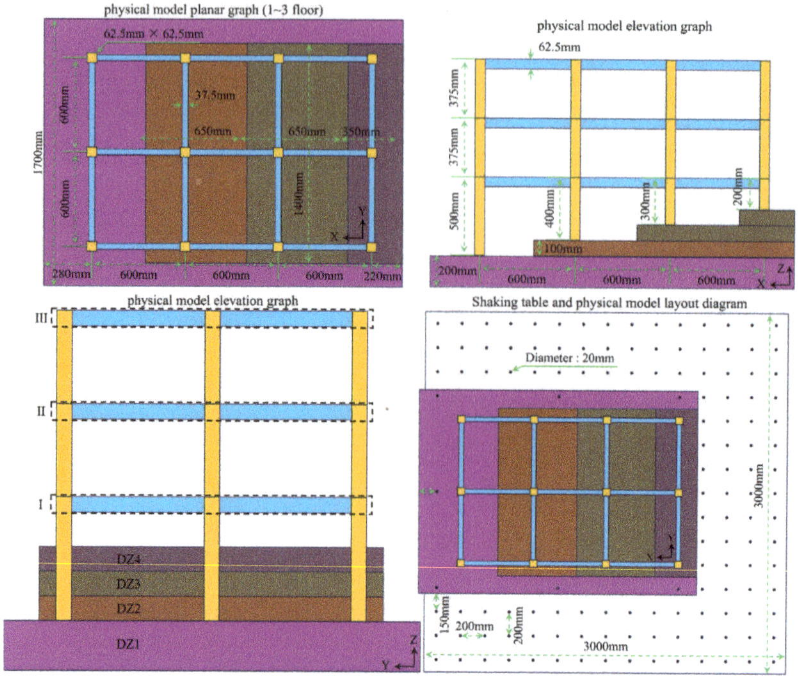

Figure 4. Size information of shaking table and physical model.

2.3.3. Physical Model Counterweight Calculation

In the experiment with the physical model, in order to ensure that the dynamic characteristics of the physical model were consistent with the model prototype, the physical model needed to be counterweighted. Based on the actual mass and similarity principle of the model prototype, concrete counterweights with specific sizes and weights were added to each layer of the physical model and fixed with expanded foam glue to increase the mass rather than the stiffness. The parameters of the concrete counterweight block used are as follows: trapezoidal concrete counterweight block, mass of 10 (kg), bottom size of 200 (mm) × 200 (mm), top size of 150 (mm) × 150 (mm), and height of 150 (mm). The concrete cube counterweight block had a mass of 1 (kg) and a size of 75 (mm) × 75 (mm) × 75 (mm). The counterweight information for each layer of the physical model structure is shown in Table 9.

Table 9. Physical model counterweight information.

Model	Layer	Counterweight (kg)	Number of Counterweight Blocks (1 (kg))	Number of Counterweight Blocks (10 (kg))
DF	I	326	6	32
	II	327	7	32
	III	151	1	15
CDF	I	326	6	32
	II	327	7	32
	III	151	1	15

2.4. Physical Model-Making Process and Monitoring Point Layout

The specific making process of the physical model is shown in Figure 5.

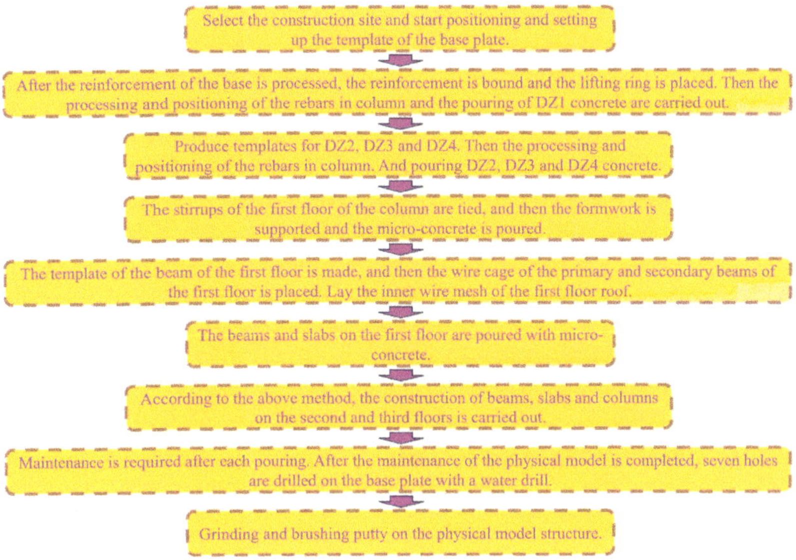

Figure 5. Physical model-making process.

The number of wrapped carbon-fiber cloths was one layer, and the wrapping direction was counterclockwise. The coverage area of a single column was 25,000 (mm^2), and the total coverage area of three columns was 75,000 (mm^2). The reinforcement position of the carbon-fiber cloth in the physical model is shown in Figure 6.

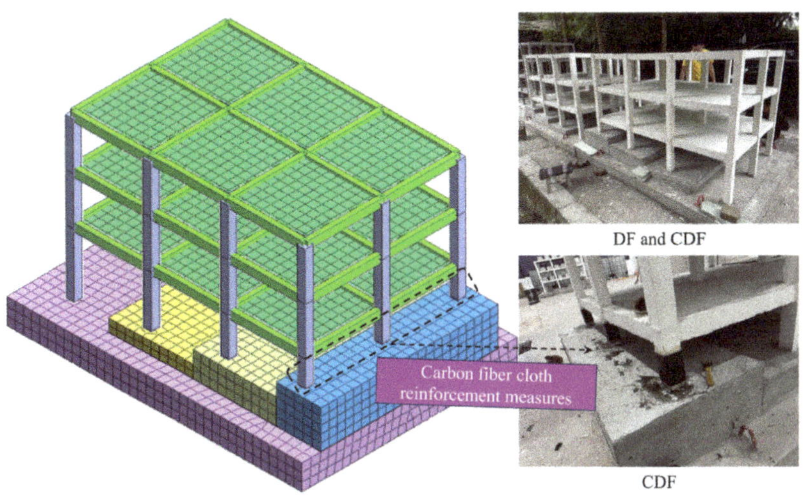

Figure 6. Reinforcement position of carbon-fiber cloth.

In the vibration test, a total of 16 YX-1210 IEPE acceleration sensors, eight KS15 cable displacement sensors, and one INV3065N2 dynamic data-acquisition system were used. The INV3065N2 dynamic data-acquisition system has a strong data-acquisition capability and can better ensure the reliability of data. Since the seismic wave only considered the X-direction, the sensors were arranged along the X-direction. Additionally, INV3065N2 has only 16 data interfaces, so each group of vibration tests used four KS15 cable displacement sensors and eight YX-1210 IEPE acceleration sensors. The displacement sensor was arranged on the outermost middle column of the physical model in the X-direction, which could greatly reduce the influence of the model on the displacement data of the cable. At the same time, in order to monitor the acceleration data of the frame structure and the base more comprehensively, the acceleration sensor was arranged diagonally. The acceleration sensors of DF were numbered AT-B-1, AT-B-2, AT-1-1, AT-1-2, AT-2-1, AT-2-2, AT-3-1, and AT-3-2. The displacement sensors of DF were numbered DT-B-1, DT-1-1, D-2-1, and DT-3-1. The acceleration sensor numbers of the CDF were RAT-B-1, RAT-B-2, RAT-1-1, RAT-1-2, RAT-2-1, RAT-2-2, RAT-3-1, and RAT-3-2. The displacement sensors of CDF were numbered RDT-B-1, RDT-1-1, RDT-2-1, and RDT-3-1. The distribution of the sensors is shown in Figure 7.

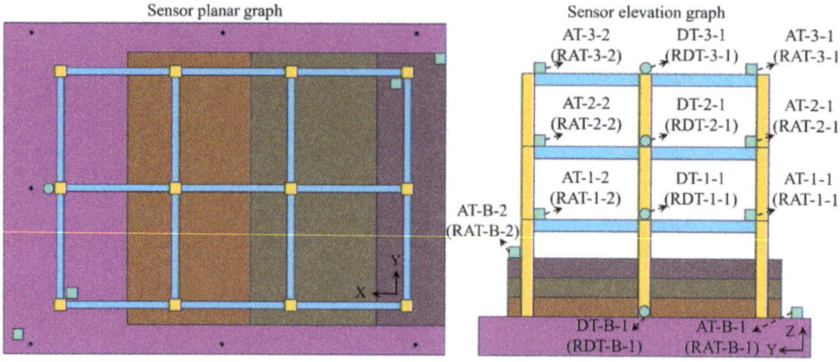

Figure 7. Sensor distribution diagram.

2.5. Loading Method of Seismic Wave

According to the dynamic characteristics of the prototype structure of the physical model, the site conditions, and other factors, five-seismic-wave information was selected as the test seismic waves. The five seismic waves were ACC (artificial seismic wave), El Centro, Kobe, Taft, and Wenchuan. The peak accelerations of the seismic waves of the prototype structure are shown in Table 10.

Table 10. Seismic peak accelerations of prototype structure (cm/s^2).

Seismic Intensity	7-Degree	8-Degree	9-Degree
Frequent earthquakes	35	70	140
Design-basis earthquake	100	200	400
Maximum considered earthquake	220	400	620

In order to adapt to the proportions of the physical model and the test requirements, the time similarity constant, $S_T = 0.25$, and the acceleration-similarity constant, $S_a = 2.0$, were used to adjust the relevant parameters of the five seismic waves to ensure the rationality and effectiveness of the test. The adjusted values of the durations of the seismic waves are shown in Table 11. The peak acceleration adjustment coefficients of the seismic waves are shown in Table 12.

Table 11. Adjustment of seismic wave duration.

Seismic Wave	Absolute Value of PGA (cm/s^2)	Time Interval (s)	Duration (s)	Adjusted Duration (s)
ACC	100.00	0.01000	30	7.5
El Centro	210.00	0.02000	30	7.5
Kobe	216.37	0.01000	20	5.0
Taft	176.80	0.01000	54	13.5
Wenchuan	1046.59	0.00125	70	17.5

Table 12. Adjustment of seismic wave peak acceleration.

Seismic Intensity	Absolute Value of PGA (cm/s^2)	Adjustment Coefficient				
		ACC	El Centro	Kobe	Taft	Wenchuan
7-degree frequent earthquakes (PGA = 0.07 (g))	70	0.7000	0.3333	0.3235	0.3959	0.0669
7-degree design-basis earthquake (PGA = 0.20 (g))	200	-	0.9524	-	-	-
7-degree maximum considered earthquake (PGA = 0.45 (g))	440	-	2.0952	-	-	-
8-degree maximum considered earthquake (PGA = 0.82 (g))	800	-	3.8095	-	-	-
9-degree maximum considered earthquake (PGA = 1.26 (g))	1240	-	5.9048	-	-	-

The adjusted ACC seismic wave, Kobe seismic wave, Taft seismic wave, Wenchuan seismic wave, and El Centro seismic wave are shown in Figures 8 and 9.

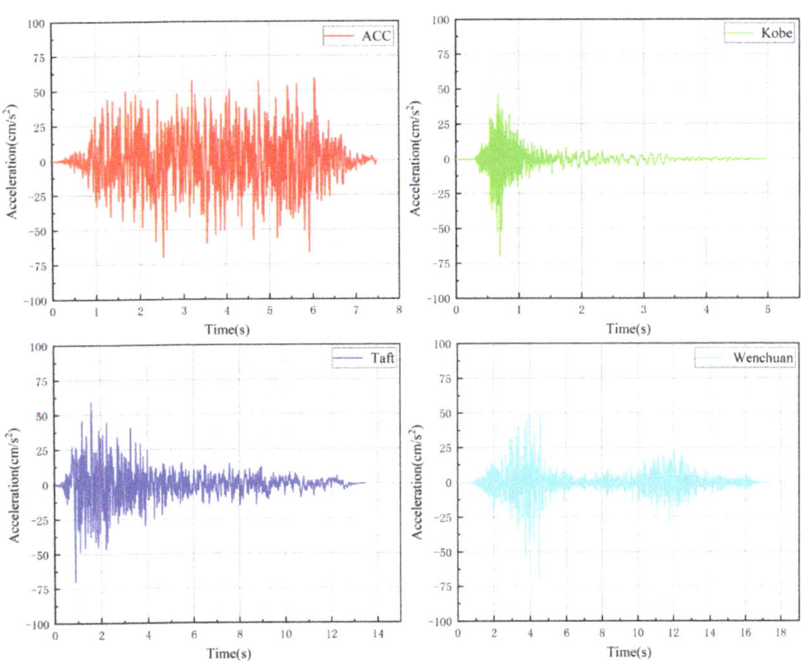

Figure 8. Acceleration time–history curves of ACC, Kobe, Taft, and Wenchuan seismic waves (PGA = 0.07 (g)).

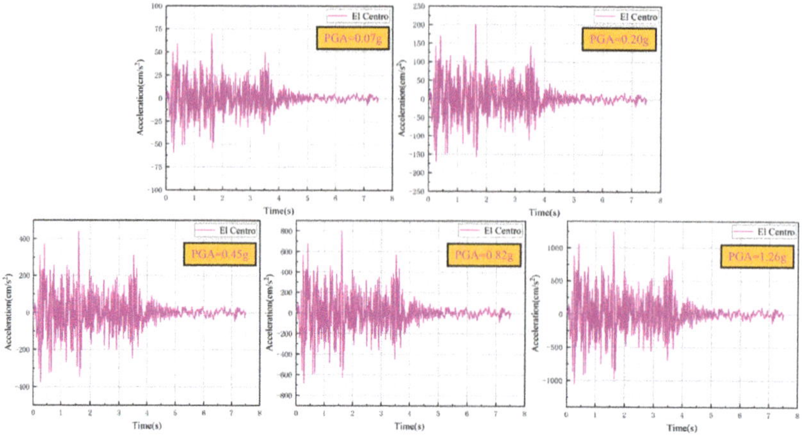

Figure 9. Acceleration time–history curves of El Centro seismic waves.

The acceleration peaks of the five seismic waves were adjusted to 0.1 (g), and then the response spectrum analysis with a damping of 5% was performed, respectively. The response spectra of the five seismic waves were compared with the design response spectrum of the 'Code for seismic design of buildings' [46], as shown in Figure 10. It can be seen from Figure 10 that the errors between Kobe, Taft, and Wenchuan and the standard response spectrum are large, so only these three waves were used for the analysis of the 7-degree frequent earthquake conditions. The response spectrum errors of ACC and El Centro were smaller than that of the standard, but the shaking-table test mainly explores the influence of seismic waves on structures in nature, so the El Centro wave was selected for the subsequent tests.

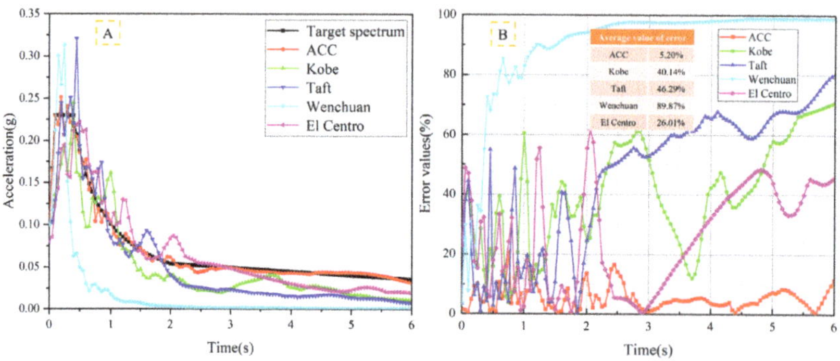

Figure 10. Response spectrum comparison diagram: (**A**) seismic wave response spectra and (**B**) response-spectrum error values.

At the beginning of the test, the dynamic characteristics, such as the natural frequency of the model, were evaluated by inputting random white-noise waves with a frequency range of 0.1~50 (Hz). A white-noise sweep lasting 120 s was inserted between seismic waves of different intensities to further monitor the dynamic characteristics of the two physical models after experiencing different seismic waves.

After the completion of the test of the 9-degree maximum considered earthquake condition, a new seismic condition was introduced to carry out the failure test, and the peak acceleration was increased until the model was destroyed. The test steps of the physical model are shown in Table 13. W1~W8 represent the white-noise sweep test. F7,

B7, R7, R8, and R9 represent the 7-degree frequent earthquake, 7-degree design-basis earthquake, 7-degree maximum considered earthquake, 8-degree maximum considered earthquake, and 9-degree maximum considered earthquake, respectively. AX, EX, KX, TX, and WX represent the ACC wave X-direction input, El Centro wave X-direction input, Kobe wave X-direction input, Taft wave X-direction input, and Wenchuan wave X-direction input, respectively.

Table 13. Test condition information.

Condition		Seismic Wave	Peak Acceleration (g)	Direction
1	W1	White noise		
2	F7KX	ACC		
3	F7EX	El Centro		
4	F7TX	Kobe	0.07	X
5	F7AX	Taft		
6	F7WX	Wenchuan		
7	W2	White noise		
8	B7EX	El Centro	0.20	X
9	W3	White noise		
10	R7EX	El Centro	0.45	X
11	W4	White noise		
12	R8EX	El Centro	0.82	X
13	W5	White noise		
14	R9EX	El Centro	1.26	X
15	W6	White noise		
Supplementary working conditions	160%R9EX	El Centro	2.02	X
	W7	White noise		
	240%R9EX	El Centro	3.02	X
	W8	White noise		

3. Analysis of Model Test Results

This part analyzes the above test results. The damage development for DF during the test is shown in Figure 11, and the damage development for CDF during the test is shown in Figure 12. When the PGA was 0.07 (g), the vibration amplitudes of DF and CDF were not obvious, and no cracks were found. Therefore, Figures 11 and 12 are the test results of the El Centro seismic wave.

It can be seen from Figures 11 and 12 that when the PGA is 0.20 (g), there are no cracks in DF and CDF, and only the putty of some nodes has slight peeling. When the PGA is 0.45 (g), small cracks begin to appear in both DF and CDF. When the PGA is 0.82 (g), the cracks of DF and CDF expand rapidly, and the joints of the beams and columns begin to loosen. When the PGA is 1.26 (g), both DF and CDF have large cracks. When the PGA is 2.02 (g), DF has collapsed, and the plastic deformation of CDF continues to intensify, but CDF still has not collapsed. When the PGA is 3.02 (g), CDF has collapsed. This shows that after carbon-fiber reinforcement, the Diaojiaolou can withstand greater earthquake action and significantly enhance the seismic capacity of the structure. The PGA when CDF collapsed increased by 49.5% compared with the PGA when DF collapsed. At the same time, the collapse of DF was mainly due to the failure of the connection between the column base and the base and the failure of the beam–column joints, and the failure points of the beam–column joints were mainly concentrated on the first and second layers. The reason for the collapse of CDF is similar to that of DF, but the failure points of the beam–column joints of CDF were mainly concentrated in the first layer. It can be seen that the failure modes of DF and CDF are mainly node failure, which is the same as the typical failure mode of RC structures under earthquakes. Additionally, after the collapse of CDF, there is no obvious shedding and cracking of the carbon-fiber cloth. This also shows that it is reasonable to use epoxy resin adhesive as adhesive.

Figure 11. The macroscopic damage-development process of DF.

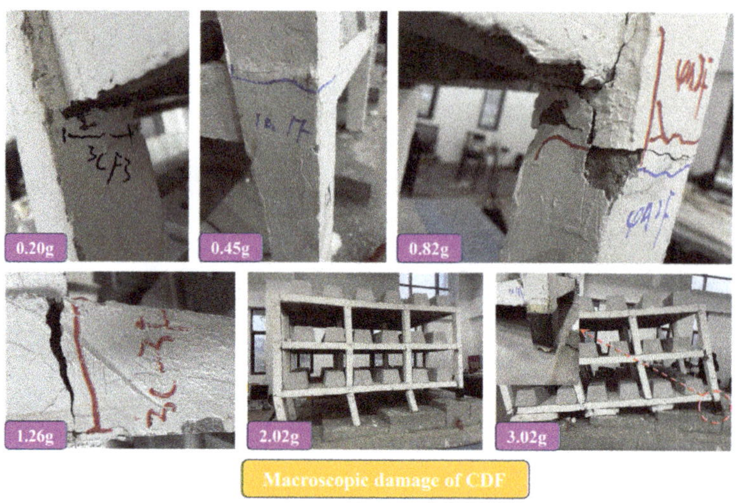

Figure 12. The macroscopic damage-development process of CDF.

3.1. Dynamic Characteristic Analysis

The variations in DF and CDF's natural frequencies were obtained by a white-noise random-wave sweep-model structure, as shown in Table 14.

Table 14. Natural vibration period and natural vibration period growth rate.

Condition	DF				CDF			
	Natural Frequency	Natural Vibration Period	Growth Rate	Relative Growth Rate	Natural Frequency	Natural Vibration Period	Growth Rate	Relative Growth Rate
W1	8.34375	0.120	0	0	12.7187	0.078	0	0
W2	8.25	0.121	0.83%	0.83%	12.2917	0.081	3.85%	3.85%
W3	7.5625	0.132	10.00%	9.09%	10.8125	0.092	17.95%	13.58%
W4	4.625	0.216	80.00%	63.64%	6.125	0.165	111.54%	78.35%
W5	3.90625	0.256	113.33%	18.52%	5.6875	0.176	125.64%	6.67%
W6	2.09375	0.478	298.33%	86.72%	3.8125	0.262	235.90%	48.86%
W7	-	-	-	-	2.25	0.44	464.10%	67.94%
W8	-	-	-	-	-	-	-	-

It can be seen from Table 14 that the natural vibration period of DF is 0.12 (s) at W1. After being reinforced by carbon-fiber cloth, the natural vibration period of CDF is shortened to 0.078 (s). This shows that carbon-fiber cloth reinforcement measures can effectively improve the stiffness of structures, resulting in the reduction in the natural vibration period. This improved the bending and shear resistance of CDF. At W2, the natural vibration periods of DF and CDF increase by 0.83% and 3.85%, respectively, and the natural vibration periods of both begin to increase slightly. No cracks were found during the crack and damage inspection. At W3, the natural vibration periods of DF and CDF increase by 10.00% and 17.95%, respectively. During the crack and damage inspection, it was found that no obvious cracks were found on the components of DF and CDF, but the putty at some beam–column joints was slightly peeled off. At this time, under the earthquake action, slight damage began to appear inside the structure, and the structural rigidity began to decrease. At W4, the natural vibration periods of DF and CDF increase by 80.00% and 111.54%, respectively. During the crack and damage inspection, it was found that many horizontal cracks appeared at the beam–column joints of the structure. This shows that the internal damage of DF and CDF began to increase significantly and the structural stiffness decreased significantly. At W5, the natural vibration periods of DF and CDF increase by 113.33% and 125.64%, respectively. The sloshing amplitudes of the DF and CDF model structures are very large. During the crack and damage inspection, the existing cracks continued to develop, new cracks began to appear on the top and foot of the column in the third layer, and the foot of the column began to be damaged. At W6, the natural vibration periods of DF and CDF increase by 298.33% and 235.90%, respectively. At this time, both the DF and CDF structures showed obvious plastic deformation. DF collapsed when the PGA = 2.02 (g). At W7, the natural vibration period of CDF increases by 464.10%. CDF collapsed when the PGA = 3.02 (g).

Through Figure 13, it can be seen that the natural vibration period of CDF is lower than that of DF and that CDF can withstand a greater PGA during the whole test process. This confirms the inference that carbon-fiber cloth can effectively improve the stiffness of these structures. At the same time, based on the stiffness-enhancement effect, it was found that after the Diaojiaolous were reinforced by the carbon-fiber cloth, the local stiffnesses of the Diaojiaolous increased sharply, which changed the shape of the structural vibration mode and shortened the natural vibration period. Additionally, with an increase in the PGA, the difference between the natural vibration periods of DF and CDF gradually increased. At 0~0.45 (g), the natural vibration periods of DF and CDF show a concave trend. At 0.45~0.82 (g), the natural vibration period curves of DF and CDF are relatively flat, showing a horizontal development trend. After 0.82 (g), the natural vibration period curves of DF and CDF begin to increase rapidly, showing a linear development trend. Therefore, according to the development trend of the natural vibration period of the structure, the damage process of the structure is divided into three stages. When the PGA is between 0 (g) and 0.45 (g), the structure is in the damage-development stage (DS). When the PGA is between 0.45 (g) and 0.82 (g), the structure is in the yield stage (YS). When the PGA is greater than 0.82 (g), the structure is in the plastic-hardening stage (PS). Through the fitting results, it can be seen that the development law of the natural vibration periods of DF and CDF conforms to the development law of the exponential function (see expression (1)). The fitting function of DF is $N = 0.0526 + 0.02051\exp((P + 0.81994)/0.68779)$, and its coefficient of determination (R^2) is 0.98262. The fitting function of CDF is $N = -0.17778 + 0.00984\exp((P + 7.5196)/2.30586)$, and its coefficient of determination (R^2) is 0.98772. According to the coefficient of determination, DF and CDF have achieved good fitting results:

$$N = \delta + A\exp((P + P1)/T1) \tag{1}$$

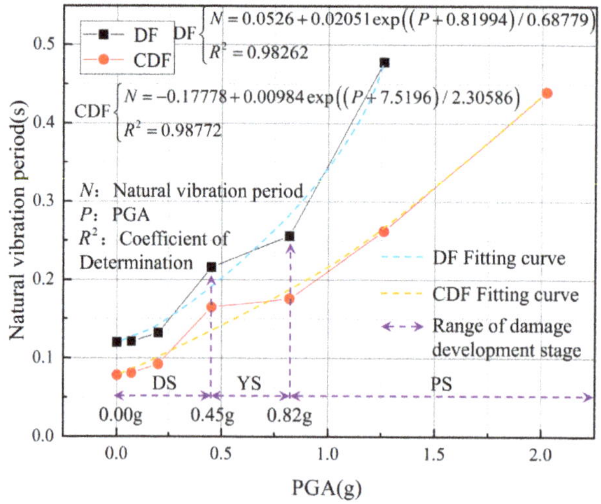

Figure 13. The development trends of the DF and CDF natural vibration periods.

In the above expression, δ, A, $P1$ and $T1$ are all fitting coefficients.

3.2. Analysis of Acceleration Characteristics

In order to better reflect the variation in acceleration with structure height, this paper introduces the acceleration-increment coefficient (ζ) to carry out the research. The calculation method of the acceleration-increment coefficient is shown in expression (2):

$$\zeta = L/DL \tag{2}$$

In this expression, ζ is the acceleration-increment coefficient; L is the maximum acceleration of each floor; and DL is the maximum acceleration of DZ1.

Under the influences of different seismic waves, the acceleration-increment coefficients of DF and CDF are shown in Tables 15 and 16, respectively. The height of AT-B-1 (RAT-B-1) on DZ1 is set to 0 (m).

Table 15. The X-direction maximum accelerations (m/s^2) and acceleration-increment coefficients of DF were obtained (PGA = 0.07 (g)).

Layer	ACC		Kobe		Taft		Wenchuan		El Centro	
	L	ζ	L	ζ	L	ζ	L	ζ	L	ζ
DZ1	0.9579	1	0.5202	1	0.4386	1	3.1556	1	0.3550	1
DZ4	0.9845	1.0278	0.5322	1.0230	0.4425	1.0088	3.1893	1.0107	0.3589	1.0109
I	1.2664	1.3220	0.6165	1.1850	0.4097	0.9340	3.223	1.0213	0.4435	1.2494
II	0.9725	1.0152	0.8219	1.5798	0.3189	0.7272	2.093	0.6632	0.7621	2.1471
III	1.4083	1.4703	1.0638	2.0448	0.4782	1.0903	3.6563	1.1587	0.8529	2.4028

Table 16. The X-direction maximum accelerations (m/s^2) and acceleration-increment coefficients of CDF were obtained (PGA = 0.07 (g)).

Layer	ACC		Kobe		Taft		Wenchuan		El Centro	
	L	ζ	L	ζ	L	ζ	L	ζ	L	ζ
DZ1	0.9445	1	0.7417	1	0.7535	1	0.5104	1	0.6382	1
DZ4	0.9449	1.0005	0.7513	1.0129	0.7554	1.0025	0.5208	1.0204	0.6465	1.0130
I	0.9901	1.0483	0.8023	1.0816	0.7479	0.9915	0.5404	1.0587	0.6613	1.0362
II	1.3325	1.4109	0.8671	1.1691	1.0235	1.3584	0.6572	1.2875	0.5360	0.8399
III	1.7997	1.9055	1.1879	1.6016	1.6194	2.1493	1.0151	1.9888	0.7935	1.2434

The variations in the acceleration-increment coefficients of DF and CDF with the height of the Diaojiaolou are shown in Figure 12.

Through Figure 14, it can be seen that the acceleration-increment coefficient of DF shows high discreteness after the height is greater than 0.3 (m). However, the overall trend of the acceleration-increment coefficient is increasing with the increase in height. At this time, the function expression of the acceleration-increment coefficient with the height is $\zeta = 0.96647 + 0.00101\exp((H + 1.62163)/0.44299)$ by fitting. However, in DF, R^2 only reaches 0.3021. After being reinforced by the carbon-fiber cloth, the acceleration-increment coefficient of CDF begins to present discrete characteristics after the height is greater than 0.5 (m). After fitting, the function expression of the acceleration-increment coefficient of CDF with the height is $\zeta = 0.98113 + 0.00331\exp((H + 0.39931)/0.30081)$ and R^2 increases to 0.74896. This shows that after strengthening with carbon-fiber cloth, under the influence of different seismic waves, the development law of the acceleration-increment coefficient of the Diaojiaolou gradually tends to be consistent and gradually conforms to the development law of exponential function (see expression (3)).

$$\zeta = \delta' + A'\exp((H + P2)/T2) \tag{3}$$

In this expression, δ', A', $P2$ and $T2$ are all fitting coefficients.

Under the influence of the El Centro seismic wave, the acceleration-increment coefficients of DF and CDF are shown in Tables 17 and 18, respectively.

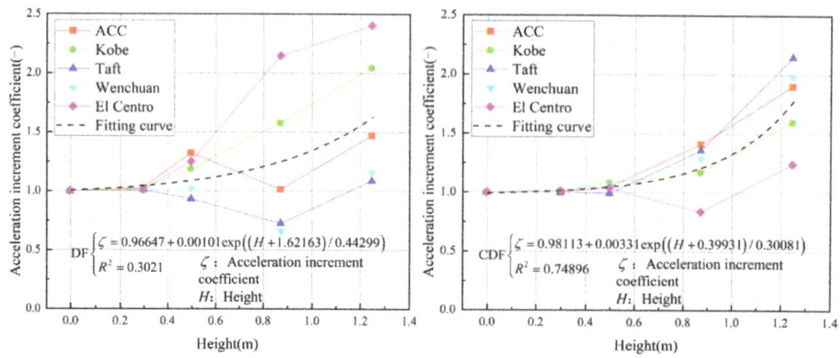

Figure 14. The variation law of the acceleration-increment coefficients of DF and CDF (PGA = 0.07 (g)).

Table 17. The X-direction acceleration-increment coefficients of DF (El Centro).

Layer	DZ1	Error (Absolute Value)	DZ4	Error (Absolute Value)	I	Error (Absolute Value)	II	Error (Absolute Value)	III	Error (Absolute Value)
0.07 (g)	1.0000	0.00%	1.0109	9.71%	1.2494	25.02%	2.1471	45.93%	2.4028	42.23%
0.20 (g)	1.0000	0.00%	1.0325	7.42%	1.2210	23.28%	1.6133	28.04%	1.8791	26.12%
0.45 (g)	1.0000	0.00%	1.0101	9.80%	1.2683	26.14%	1.3991	17.02%	1.7662	21.40%
0.82 (g)	1.0000	0.00%	0.8457	31.15%	1.0558	11.27%	0.9473	22.56%	1.0240	35.57%
1.26 (g)	1.0000	0.00%	2.5270	56.11%	0.5953	57.37%	0.7046	64.77%	0.9645	43.93%
2.02 (g)	1.0000	0.00%	0.2284	385.60%	0.2309	305.72%	0.1544	651.94%	0.2924	374.76%
Average value	1.0000	-	1.1091	-	0.9368	-	1.1610	-	1.3882	-

Table 18. The X-direction acceleration-increment coefficients of CDF (El Centro).

Layer	DZ1	Error (Absolute Value)	DZ4	Error (Absolute Value)	I	Error (Absolute Value)	II	Error (Absolute Value)	III	Error (Absolute Value)
0.07 (g)	1.0000	0.00%	1.0130	2.34%	1.0362	23.90%	0.8399	6.04%	1.2434	8.72%
0.20 (g)	1.0000	0.00%	1.0459	5.41%	1.2191	35.32%	1.7764	49.86%	2.4275	53.24%
0.45 (g)	1.0000	0.00%	0.7833	26.30%	1.0028	21.37%	1.4063	36.67%	1.6382	30.72%
0.82 (g)	1.0000	0.00%	0.8240	20.06%	1.0565	25.37%	0.9233	3.54%	1.1425	0.66%
1.26 (g)	1.0000	0.00%	2.0241	51.12%	0.8399	6.12%	1.0248	13.10%	1.1296	0.48%
2.02 (g)	1.0000	0.00%	0.2577	283.90%	0.0983	702.14%	0.0815	992.76%	0.1161	877.61%
3.02 (g)	1.0000	0.00%	0.9773	1.23%	0.2666	195.76%	0.1819	389.61%	0.2475	358.59%
Average value	1.0000	-	0.9893	-	0.7885	-	0.8906	-	1.1350	-

It can be seen from Tables 17 and 18 that DF collapsed after the PGA reached 2.02 (g). After the PGA reached 3.02 (g), CDF collapsed. This shows that carbon-fiber cloth effectively increases the anti-collapse performance of Diaojiaolou structures. It can be seen from Figure 15 that the average value of the acceleration-increment coefficient of CDF is generally smaller than the average value of the acceleration-increment coefficient of DF. This shows that reinforcement with carbon-fiber cloth can effectively reduce the transmission effect of acceleration in a Diaojiaolou. Additionally, in DF, the average value of the acceleration-increment coefficient shows an N-shaped development law with the height. In the CDF, the average value of the acceleration-increment coefficient shows a V-shaped development law with the height. When the height is 0.3 (m), the average value of the acceleration-increment coefficient of CDF decreases. This shows that the carbon-fiber cloth effectively changes the development law of the average value of the acceleration-increment coefficient, making it change from N-shaped to V-shaped. The average value of the acceleration-increment coefficients of DF and CDF reaches the minimum at a height of 0.5 (m). This shows that the Diaojiaolou structure is least affected by acceleration at I.

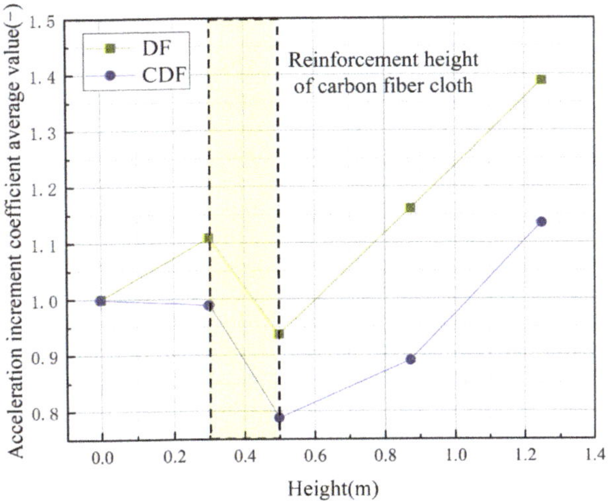

Figure 15. Under the influence of the El Centro seismic wave, the development law of the average value of the X-direction acceleration-increment coefficients of DF and CDF is obtained.

3.3. Analysis of Displacement Characteristics

In this test, the pull-line displacement meters were arranged in the X-directions of DZ1, I, II, and III of DF and CDF, respectively. Through these displacement sensors, the displacement values of the structure during the test could be directly measured and output. The height of DT-B-1 (RDT-B-1) on DZ1 was set to *0* (m). When the PGA was 0.07, the displacement value and displacement-increment coefficients (ζ') of DF and CDF affected by different seismic waves were as shown in Tables 19 and 20, respectively. The calculation method of the displacement-increment coefficient is shown in expression (4).

$$\zeta' = L'/DL' \tag{4}$$

Table 19. The maximum displacement values (mm) and displacement-increment coefficients of DF were obtained (PGA = 0.07 (g)).

Layer	ACC		Kobe		Taft		Wenchuan		El Centro	
	L'	ζ'	L'	ζ'	L'	ζ'	L'	ζ'	L'	ζ'
DZ1	1.1749	1.0000	0.8022	1.0000	0.1629	1.0000	1.4084	1.0000	0.7031	1.0000
I	1.1971	1.0189	0.9809	1.2228	0.1852	1.1369	1.5102	1.0723	0.8567	1.2185
II	1.4214	1.2098	1.2553	1.5648	0.2523	1.5488	2.1007	1.4916	1.1887	1.6907
III	1.4889	1.2673	1.3448	1.6764	0.2864	1.7581	2.2306	1.5838	1.2674	1.8026

In this expression, ζ' is the displacement-increment coefficient; L' is the maximum displacement of each floor; and DL' is the maximum displacement of DZ1.

The variations in the displacement-increment coefficients of DF and CDF with the heights of the Diaojiaolous is shown in Figure 14.

Table 20. The maximum displacement values (mm) and displacement-increment coefficients of CDF were obtained (PGA = 0.07 (g)).

Layer	ACC		Kobe		Taft		Wenchuan		El Centro	
	L'	ζ'	L'	ζ'	L'	ζ'	L'	ζ'	L'	ζ'
DZ1	1.2855	1.0000	1.0512	1.0000	0.3281	1.0000	0.7306	1.0000	0.9165	1.0000
I	1.3724	1.0676	1.0985	1.0450	0.3513	1.0707	0.7564	1.0353	0.9535	1.0404
II	1.4279	1.1108	1.1443	1.0886	0.3732	1.1375	0.7680	1.0512	0.9824	1.0719
III	1.4967	1.1643	1.1792	1.1218	0.3924	1.1960	0.7860	1.0758	1.0050	1.0966

It can be seen from Figure 16 that the displacement-increment coefficient of DF has high discreteness as a whole, but the displacement-increment coefficient shows an increasing trend with the increase in height. At this time, the function expression of the displacement-increment coefficient with the height is $\zeta' = -0.42779 + 0.17449\exp((H+6.70712)/3.21531)$ by fitting. R^2 is 0.74607. After being reinforced by the carbon-fiber cloth, the development trend of the displacement-increment coefficient of CDF shows a low discreteness. After fitting, the function expression of the displacement-increment coefficient of CDF with the height is $\zeta' = -7.30092 + 3.72704\exp((H+63.85585)/79.74574)$ and R^2 increases to 0.75586. This shows that after strengthening with the carbon-fiber cloth, the development law of the displacement-increment coefficient is more consistent under the influences of different seismic waves. The development trend of the displacement-increment coefficients of DF and CDF with the heights conforms to the development law of exponential function (see expression (5)).

$$\zeta' = \delta'' + A''\exp((H+P3)/T3) \tag{5}$$

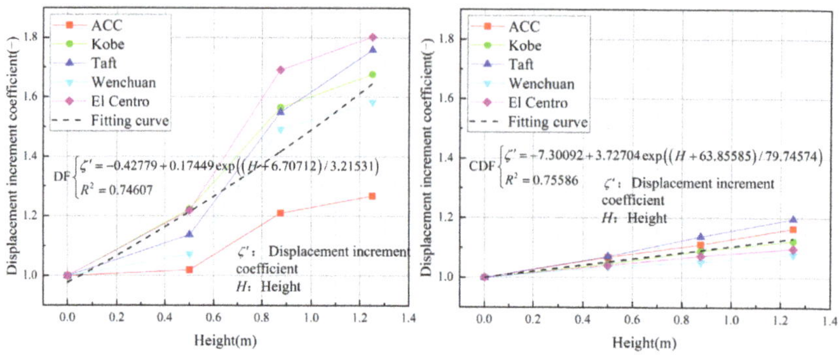

Figure 16. The variation law of the displacement-increment coefficients of DF and CDF (PGA = 0.07 (g)).

In this expression, δ'', A'', $P3$ and $T3$ are all fitting coefficients.

Under the influence of the El Centro seismic wave, the displacement-increment coefficients of DF and CDF were as shown in Tables 21 and 22, respectively.

Table 21. The displacement-increment coefficients of DF (El Centro).

Layer	DZ1	Error (Absolute Value)	I	Error (Absolute Value)	II	Error (Absolute Value)	III	Error (Absolute Value)
0.07 (g)	1.0000	0.00%	1.2185	16.42%	1.6907	30.32%	1.8026	26.27%
0.20 (g)	1.0000	0.00%	1.0398	2.06%	1.2177	3.26%	1.2777	4.02%
0.45 (g)	1.0000	0.00%	1.1592	12.15%	1.4346	17.89%	1.5414	13.77%
0.82 (g)	1.0000	0.00%	1.2014	15.23%	1.2837	8.23%	1.5077	11.85%
1.26 (g)	1.0000	0.00%	1.0961	7.09%	1.2493	5.71%	1.3543	1.86%
2.02 (g)	1.0000	0.00%	0.5953	71.07%	0.7046	67.19%	0.9645	37.80%
Average value	1.0000	-	1.0184	-	1.1780	-	1.3291	-

Table 22. The displacement-increment coefficients of CDF (El Centro).

Layer	DZ1	Error (Absolute Value)	I	Error (Absolute Value)	II	Error (Absolute Value)	III	Error (Absolute Value)
0.07 (g)	1.0000	0.00%	1.0404	14.85%	1.0719	6.54%	1.0966	8.44%
0.20 (g)	1.0000	0.00%	1.0412	14.92%	1.0729	6.63%	1.0964	8.45%
0.45 (g)	1.0000	0.00%	1.1872	25.38%	1.4217	29.54%	1.5023	20.85%
0.82 (g)	1.0000	0.00%	1.2047	26.46%	1.4417	30.51%	1.5526	23.41%
1.26 (g)	1.0000	0.00%	1.2000	26.18%	1.3397	25.22%	1.7053	30.27%
2.02 (g)	1.0000	0.00%	0.3442	157.38%	0.4292	133.41%	0.7231	64.44%
3.02 (g)	1.0000	0.00%	0.3379	162.18%	0.3055	227.92%	0.5551	114.21%
Average value	1.0000	-	0.8859	-	1.0018	-	1.1891	-

It can be seen from Tables 21 and 22 that DF collapsed after the PGA reached 2.02 (g). After the PGA reached 3.02 (g), CDF collapsed. This also shows that the carbon-fiber cloth effectively increased the anti-collapse performance of the Diaojiaolou structure. It can be seen from Figure 17 that the average value of the displacement-increment coefficient of CDF is smaller than the average value of the displacement-increment coefficient of DF. This shows that after reinforcement with carbon-fiber cloth, the displacement change value in a Diaojiaolou can be effectively reduced. Additionally, in DF, the average value of the displacement-increment coefficient shows a concave shape with the development of the height. In CDF, the average value of the displacement-increment coefficient shows a V-shaped pattern with the development of the height. When the height is less than 0.5 (m), the average value of the displacement-increment coefficient of CDF decreases. This shows that the carbon-fiber cloth effectively changed the development law of the average value of the displacement-increment coefficient and made it change from concave to V-shaped. The average value of the displacement-increment coefficient of DF has its minimum value when the height is 0 (m). The average value of the displacement-increment coefficient of the CDF reaches its minimum at a height of 0.5 (m). This shows that when a Diaojiaolou structure is not reinforced by carbon-fiber cloth, DZ1 is the least affected by seismic waves. After the Diaojiaolou structure was reinforced by carbon-fiber cloth, I was the least affected by the seismic waves.

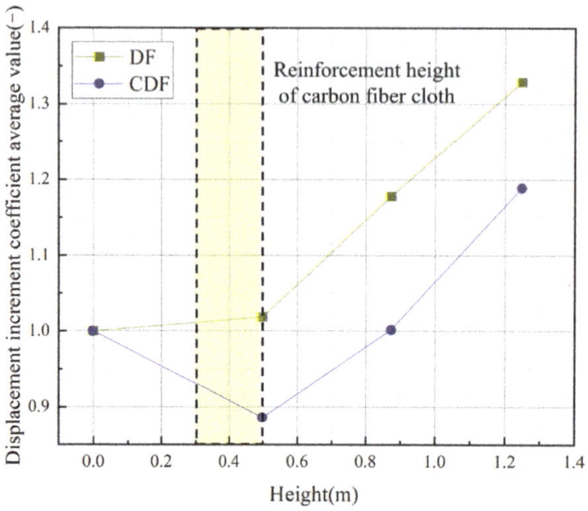

Figure 17. Under the influence of the El Centro seismic wave, the development law of the average values of the displacement-increment coefficients of DF and CDF is obtained.

4. Conclusions

In this paper, the response characteristics of RC Diaojiaolou structures under different seismic intensities and the reinforcement of carbon-fiber cloth were studied using similar physical models. The specific research results are as follows:

(1) Based on the analysis results of the dynamic characteristics, it was found that carbon-fiber cloth can effectively improve the stiffness of these structures. According to the development trend of the natural vibration period of the structures, the damage process of the structures is divided into three stages. The PGA between 0 (g) and 0.45 (g) was divided into the damage-development stage (DS). The PGA between 0.45 (g) and 0.82 (g) was divided into the yield stage (YS). The range of PGA greater than 0.82 (g) was divided into the plastic-hardening stage (PS). It was concluded that the development law of the natural vibration period of DF and CDF conformed to the exponential function development law. The function expression of DF is $N = 0.0526 + 0.02051\exp((P + 0.81994)/0.68779)$. The function expression of CDF is $N = -0.17778 + 0.00984\exp((P + 7.5196)/2.30586)$.

(2) Based on the analysis results of the acceleration characteristics, it was found that under the influence of different seismic waves, the development law of the acceleration-increment coefficients of Diaojiaolous tends to be consistent and gradually conforms to the development law of the exponential functions after reinforcement with carbon-fiber cloth. The function expression of the acceleration-increment coefficient of DF with the height is $\zeta = 0.96647 + 0.00101\exp((H + 1.62163)/0.44299)$. The function expression of the acceleration-increment coefficient of CDF with the height is $\zeta = 0.98113 + 0.00331\exp((H + 0.39931)/0.30081)$. After the reinforcement with the carbon-fiber cloth, the transmission effect of the acceleration in the Diaojiaolous could be effectively reduced. Additionally, it effectively changes the development law of the average value of the acceleration-increment coefficient, making it change from N-shaped to V-shaped. At the same time, it was also found that the Diaojiaolou structures were least affected by acceleration at I.

(3) Based on the analysis results of the displacement characteristics, it was found that the development law of the displacement-increment coefficients in DF and CDF conforms to the development law of the exponential functions. The development trend of the displacement-increment coefficient of CDF showed lower discreteness. The

function expression of the displacement-increment coefficient of DF with the height is $\zeta' = -0.42779 + 0.17449\exp((H + 6.70712)/3.21531)$. The function expression of the displacement-increment coefficient of CDF with the height is $\zeta' = -7.30092 + 3.72704\exp((H + 63.85585)/79.74574)$. It was also found that carbon-fiber cloth can effectively change the development law of the average value of the displacement-increment coefficient so that it changes from concave to V-shaped. At the same time, it was also found that the displacement deformation of DF was least affected by the seismic waves at DZ1. The displacement deformation of CDF was least affected by the seismic waves at I.

(4) In this paper, the structural layer most affected by the seismic waves could be found according to the variation in the acceleration-increment coefficient and the displacement-increment coefficient with the height. Then, during the seismic design of the Diaojiaolou, the structural layer that was most affected by the seismic waves was strengthened to enhance the seismic performance of the structure. Moreover, carbon-fiber cloth and epoxy resin adhesive are not only low-cost but also simple to operate, which makes carbon-fiber reinforcement measures widely used in rural mountainous areas. However, the reinforcement method adopted in this paper is relatively simple, and only one-way seismic input was considered. This has some shortcomings with the response modes of full-scale structures under multi-directional earthquakes. In a follow-up study, more accurate function expressions of the acceleration-increment coefficient and displacement-increment coefficient need to be studied. The mathematical model proposed in this paper was compared with other results. At the same time, the influence of ground-motion characteristics on other types of buildings of the ethnic minorities in the mountainous areas will also continue to be explored. The response characteristics of a variety of reinforcement methods and full-scale structures under multi-directional earthquakes will be considered in order to improve the seismic design requirements of the buildings in the above areas and enhance their emergency response capacities.

Author Contributions: All authors contributed to the study conception and design. Formal analysis, Z.S., W.F., X.Z., X.M. and Z.F.; investigation, Z.S., W.F., X.Z., X.M. and Z.F.; validation, Z.S., W.F., X.Z., X.M. and Z.F.; writing—original draft, W.Z.; and all authors commented on previous versions of the manuscript. All authors have read and agreed to the published version of the manuscript.

Funding: This work was supported by the Chengdu Municipal Bureau of Science and Technology (grant no. RZ2400002931).

Data Availability Statement: The original contributions presented in this study are included in this article, and further inquiries can be directed to the corresponding author.

Conflicts of Interest: We, the authors, declare that there are no conflicts of interest associated with this manuscript. We confirm that this research was conducted independently and free from any commercial or financial relationships that could be construed as potential conflicts of interest.

References

1. The State Council of the People's Republic of China. Available online: https://www.gov.cn/guoqing/2005-06/24/content_2582624.htm (accessed on 22 April 2025).
2. China Earthquake Networks Center. Available online: https://news.ceic.ac.cn/index.html (accessed on 22 April 2025).
3. Central People's Government of the People's Republic of China. Available online: https://www.gov.cn/gongbao/content/2008/content_1120416.htm (accessed on 22 April 2025).
4. Heydarpour, K.; Tehrani, P. Influence of Abutment Stiffness and Strength on the Seismic Response of Horizontally Curved RC Bridges in Comparison with Equivalent Straight Bridges at Different Seismic Intensity Levels. *Shock Vib.* **2022**, *2022*, 3532331. [CrossRef]
5. Massumi, A.; Gholami, F. The influence of seismic intensity parameters on structural damage of RC buildings using principal components analysis. *Appl. Math. Model.* **2016**, *40*, 2161–2176. [CrossRef]

6. Pachla, F.; Tatara, T.; Aldabbik, W. Impact of Various High Intensity Earthquake Characteristics on the Inelastic Seismic Response of Irregular Medium-Rise Buildings. *Appl. Sci.* **2024**, *14*, 9740. [CrossRef]
7. Chester, D.K. The 1755 Lisbon earthquake. *Prog. Phys. Geogr.* **2001**, *25*, 363–383. [CrossRef]
8. Griffin, J.D.; Stirling, M.W.; Wang, T. Periodicity and Clustering in the Long-Term Earthquake Record. *Geophys. Res. Lett.* **2020**, *47*, e2020GL089272. [CrossRef]
9. Omar, K.A.; Mangira, O. A Stress Release Model in Egypt. *Geotectonics* **2020**, *54*, 106–112. [CrossRef]
10. Zoller, G. Recurrence Times of Large Earthquakes: Assimilating the Effect of Seismic Coupling into a Renewal Model. *Bull. Seismol. Soc. Am.* **2024**, *114*, 1754–1761. [CrossRef]
11. Greco, R.; Vanzi, I.; Lavorato, D.; Briseghella, B. Seismic duration effect on damping reduction factor using random vibration theory. *Eng. Struct.* **2019**, *179*, 296–309. [CrossRef]
12. Kolli, M.K.; Bora, S.S. On the use of duration in random vibration theory (RVT) based ground motion prediction: A comparative study. *Bull. Earthq. Eng.* **2021**, *19*, 1687–1707. [CrossRef]
13. Stanko, D.; Gulerce, Z.; Markusic, S.; Salic, R. Evaluation of the site amplification factors estimated by equivalent linear site response analysis using time series and random vibration theory based approaches. *Soil Dyn. Earthq. Eng.* **2019**, *117*, 16–29. [CrossRef]
14. Zhang, H.; Zhao, Y.-g. Damping modification factor based on random vibration theory using a source-based ground-motion model. *Soil Dyn. Earthq. Eng.* **2020**, *136*, 106225. [CrossRef]
15. Caglar, N.; Sert, S.; Serdar, A.H. Basement-Storey Effect on the Seismic Response of RC Buildings on Soft Surface Soil. *Arab. J. Sci. Eng.* **2021**, *46*, 11291–11302. [CrossRef]
16. Oinam, R.M.; Sahoo, D.R. Numerical evaluation of seismic response of soft-story RC frames retrofitted with passive devices. *Bull. Earthq. Eng.* **2018**, *16*, 983–1006. [CrossRef]
17. Qiu, W.; Wang, K.; Guo, W. Analysis of impact factors on seismic response of the upper building structure of RC beam-type covered bridge. *Structures* **2024**, *66*, 106889. [CrossRef]
18. Katsimpini, P.S.; Papagiannopoulos, G.A.; Askouni, P.K.; Karabalis, D.L. Seismic response of low-rise 3-D steel structures equipped with the seesaw system. *Soil Dyn. Earthq. Eng.* **2020**, *128*, 105877. [CrossRef]
19. Papagiannopoulos, G.A. On the seismic behaviour of tension-only concentrically braced steel structures. *Soil Dyn. Earthq. Eng.* **2018**, *115*, 27–35. [CrossRef]
20. Xiang, Y.; Luo, Y.-f.; Zhu, Z.-c.; Shen, Z.-y. Estimating the response of steel structures subjected to vertical seismic excitation: Idealized model and inelastic displacement ratio. *Eng. Struct.* **2017**, *148*, 225–238. [CrossRef]
21. Jahangiri, M.; Hadianfard, M.A.; Shojaei, S. Evaluation of the frequency, damping, and vulnerability variations of masonry structures during the construction process using seismic response acquisition. *Measurement* **2024**, *237*, 115201. [CrossRef]
22. Zhang, H.; Sun, T.; Hou, S.-W.; Gao, Q.-M.; Li, X. Effect of Aftershocks on Seismic Fragilities of Single-Story Masonry Structures. *Front. Phys.* **2021**, *9*, 695111. [CrossRef]
23. Zhang, Y.; Wang, Z.; Jiang, L.; Skalomenos, K.; Zhang, D. Seismic analysis method of unreinforced masonry structures subjected to mainshock-aftershock sequences. *Bull. Earthq. Eng.* **2022**, *20*, 2619–2641. [CrossRef]
24. Altunisik, A.C.; Sunca, F.; Sevim, B. Modal parameter identification and seismic assessment of historical timber structures under near-fault and far-fault ground motions. *Structures* **2023**, *47*, 1624–1651. [CrossRef]
25. Cui, L.; Zhang, X.; Hu, W.; Meng, Z.; Chen, J.; Qiu, Z. Seismic Performance Analysis of Ancient Palace-Style Timber Structure Considering Column Rocking Effect. *J. Earthq. Eng.* **2024**, *28*, 3175–3190. [CrossRef]
26. Ren, Y.; Ma, D.; Wang, W.; Guo, X.; Wang, Z.; Chen, D.; Wu, J.; Fei, Z.; Zhao, X. Application of SPO2IDA in Seismic Performance Evaluation of Ancient Timber Structures. *Int. J. Archit. Herit.* **2024**, *18*, 1308–1322. [CrossRef]
27. Liu, L.; Zhong, R.; Yu, Y.; Li, R. Study on seismic performance of reinforced concrete stilted building with steel braces in mountainous areas. *Structures* **2023**, *55*, 2199–2216. [CrossRef]
28. Li, R.; Liu, L.; Li, Y.; Wang, G. Experimental Investigate on Seismic Performance of RC Hill-Side Stilted Buildings Affected by Vertical Stiffness Irregularity. *J. Earthq. Eng.* **2024**, *28*, 2273–2291. [CrossRef]
29. Kumar, V.; Anand, P. Study of Tunnelling-Induced Response of Building Frame with Shake Table Test. *Arab. J. Sci. Eng.* **2024**, *49*, 5159–5169. [CrossRef]
30. Zhou, Y.; Chen, P.; Zhang, L.; Wang, C.; Lu, Z. Shaking table tests and numerical analysis of an over-track multi-tower building. *Struct. Infrastruct. Eng.* **2019**, *15*, 230–243. [CrossRef]
31. Beroza, G.C.; Segou, M.; Mostafa Mousavi, S. Machine learning and earthquake forecasting-next steps. *Nat. Commun.* **2021**, *12*, 4761. [CrossRef]
32. Mousavi, S.M.; Beroza, G.C. Machine Learning in Earthquake Seismology. *Annu. Rev. Earth Planet. Sci.* **2023**, *51*, 105–129. [CrossRef]
33. Yavas, C.E.; Chen, L.; Kadlec, C.; Ji, Y. Improving earthquake prediction accuracy in Los Angeles with machine learning. *Sci. Rep.* **2024**, *14*, 24440. [CrossRef]

34. Habib, A.; Habib, M.; Bashir, B.; Bachir, H. Exploring the Sustainability Benefits of Digital Twin Technology in Achieving Resilient Smart Cities During Strong Earthquake Events. *Arab. J. Sci. Eng.* **2025**. [CrossRef]
35. Yu, S.; Lei, Q.; Liu, C.; Zhang, N.; Shan, S.; Zeng, X. Application research on digital twins of urban earthquake disasters. *Geomat. Nat. Hazards Risk* **2023**, *14*, 2278274. [CrossRef]
36. Charles, S.; Natarajan, J. Identification of Key Gene Modules and Novel Transcription Factors in Tetralogy of Fallot Using Machine Learning and Network Topological Features. *Medinformatics* **2023**, *1*, 27–34. [CrossRef]
37. Chattopadhyay, S. Decoding Medical Diagnosis with Machine Learning Classifiers. *Medinformatics* **2024**. [CrossRef]
38. Kazemi, F.; Asgarkhani, N.; Jankowski, R. Machine learning-based seismic response and performance assessment of reinforced concrete buildings. *Arch. Civ. Mech. Eng.* **2023**, *23*, 94. [CrossRef]
39. Wood, D.A. Spatio-Temporal Attributes of Varicella-Zoster Case Number Trends Assist with Optimizing Machine Learning Predictions. *Medinformatics* **2023**, *1*, 43–53. [CrossRef]
40. Zhou, Y.; Lu, X.L. *Method and Technology for Shaking Table Model Test of Building Structures*; Science Press: Beijing, China, 2016.
41. *GB/T 50081-2019*; Standard for Test Methods of Concrete Physical and Mechanical Properties. China Architecture & Building Press: Beijing, China, 2019.
42. Yu, H.X. Shaking Table Test on the Seismic Performance of Frame Structure and Stilted Frame Structure. Master's Thesis, Chongqing University, Chongqing, China, 2020.
43. *GB/T 3354-2014*; Test Method for Tensile Properties of Orientation Fiber Reinforcedpolymer Matrix Composite Materials. Standards Press of China: Beijing, China, 2014.
44. *GB 50010-2010*; Code for Design of Concrete Structures. China Architecture & Building Press: Beijing, China, 2010.
45. *JGJ3-2010*; Technical Specification for Concrete Structures of Tall Building. China Architecture & Building Press: Beijing, China, 2010.
46. *GB 50011-2010*; Code for Seismic Design of Buildings. China Architecture & Building Press: Beijing, China, 2010.

buildings

MDPI

Article

In-Plane Stability of Circular Arch Under Uniform Vertical Load Based on the Asymptotic Method

Jing Jin [1] and Mingzhou Su [1,2,*]

1 School of Civil Engineering, Xi'an University of Architecture and Technology, Xi'an 710055, China; jinjing1005@xauat.edu.cn
2 Key Lab of Structural Engineering and Earthquake Resistance of the Ministry of Education, Xi'an University of Architecture and Technology, Xi'an 710055, China
* Correspondence: sumingzhou@xauat.edu.cn

Abstract: Conventional analyses often simplify vertical loads as uniform radial loads while neglecting axial force effects in the buckling analyses of arches, leading to discrepancies between theoretical predictions and actual loading conditions. To address this issue, this research proposes a nonlinear analytical approach based on asymptotic methods, include the parameter perturbation method and the Wentzel–Kramers–Brillouin (WKB) method. The results show the following: (1) The parameter perturbation method is effective for the snap-buckling of a shallow arch, and the fifth-order solution is sufficiently accurate. (2) For shallow arches with a large modified slenderness ratio, the influence of the axial load component cannot be neglected. (3) Regardless of the rise-to-span ratio of the arch, the nonlinear bending moment is significantly larger than the linear bending moment. (4) In the anti-symmetric buckling analysis, the eigenvalue obtained using the second-order WKB method is smaller than that obtained using the third-order WKB method; therefore, the second-order solution can be used as the critical load. (5) For shallow arches with a small rise-to-span ratio, the critical load for anti-symmetric buckling closely matches the classical solution, and the results from arches subjected to a uniformly distributed radial load are reliable. For deep arches with a large rise-to-span ratio, the influence of the axial load component cannot be ignored.

Keywords: arch; snap-through buckling; thin-walled curved beam; nonlinear analysis; perturbation techniques; WKB method

1. Introduction

Arches are classified as shallow or deep based on the rise-to-span ratio, with a ratio of less than 0.2 typically considered shallow [1]. Under in-plane loading, shallow arches may exhibit either anti-symmetric bifurcation or snap-through buckling, while deep arches typically experience only anti-symmetric buckling [1–6]. Classical buckling theory is used to determine the elastic anti-symmetric buckling load of arches subjected to uniform radial loads, assuming linear pre-buckling behavior [5–8]. Analytical solutions for the arch stability problem can be obtained under specific load conditions, such as uniform radial and central concentrated loads; however, solutions are generally not available for other loading scenarios. In the analysis of arches subjected to uniform vertical loads, the vertical load is typically simplified to a uniform radial load [3,9], neglecting the influence of the axial load component. Y.-L. Pi et al. studied the nonlinear buckling behavior of circular shallow arches subjected to uniform radial loads and provided formulas for both snap-through instability and anti-symmetric buckling of the arches [2–5]. G. Tong et al. [1]

investigated the anti-symmetric buckling of pin-ended circular arches subjected to uniform vertical and radial loads, considering the effects of transverse and shear stresses. The critical load of arches was obtained using the Galerkin method, though their study focused solely on deep arches. Eslami and Babaei [10–12] conducted a series of theoretical studies on the nonlinear snap-through buckling of shallow arches using the two-step perturbation method and the Galerkin method. Novel parametric studies have also been conducted to show the effects of nonlocal and length scale parameters, and the geometrical parameters of the shell. Y. Guo et al. [13] conducted experimental research on the in-plane stability and bearing capacity of pin-ended circular arches with I-beam cross-sections under combined compression and bending. They validated theoretical analysis results and proposed design formulas for the in-plane stability and bearing capacity of compressed circular arches. However, their study did not further investigate the differences between uniform vertical loads and uniform radial loads. Caporale et al. [14] illustrated the application of an original approach to the case of masonry arches to limit the analysis of structural models based upon the intersection properties of convex sets. Micaela et al. [15] investigated the nonlinear fracture of masonry arch structures under vertical loads using the Lattice Discrete Particle Model (LDPM) method and analyzed the size effect. Obviously, in the stability analysis of the arches subjected to vertical load, numerical methods such as the finite element method, Lattice Discrete Particle Model, and Galerkin method are conventionally employed. However, solutions obtained through purely numerical methods are often isolated and do not easily provide a systematic understanding of the physical phenomena. In contrast, solutions derived from asymptotic methods possess a clear analytical structure, providing semi-analytic solutions that are more advantageous for theoretical analysis. To investigate the stability of circular arches under uniform vertical loads, this study employs the asymptotic method, an approximate analytical technique for solving nonlinear structural problems [16,17]. The method of parameter perturbation is a classical approach in the large deflection and buckling analysis of plates [11,12], while the WKB method is widely used in the eigenvalue stability analysis of tubular and spherical shells and solid structures [18–20].

In this study, a specific analysis of the stability of a pin-ended circular arch subjected to a uniform vertical load was conducted. First, the nonlinear equilibrium equations of the arch were derived. Second, a nonlinear analysis of arches subjected to uniformly distributed radial and vertical loads was performed using the parameter perturbation method, and the results were compared with classical solutions. Third, the buckling equilibrium equations were established, and the asymptotic solution of the arch subjected to uniformly distributed vertical loads was obtained using the WKB method. The third-order and second-order solutions were compared along with the critical loads obtained using linear and nonlinear axial forces.

2. Nonlinear Equilibrium Equations

Figure 1 shows a pin-ended circular arch subjected to a uniformly distributed vertical load, with a radius of curvature R and a central angle of 2α.

The following strain–displacement relations for the arch are obtained [2]:

$$\begin{gathered} \varepsilon = \varepsilon_\theta^0 + z\kappa_\theta \\ \varepsilon_\theta^0 = \tfrac{1}{R}\left(\overline{u}' + \overline{\omega}\right) + \tfrac{1}{2R^2}\overline{\omega}'^2,\ \kappa_\theta = -\tfrac{1}{R^2}\left(\overline{\omega}'' + \overline{\omega}\right), \end{gathered} \tag{1}$$

where $()' = d()/d\theta$ and $\overline{u}$ and $\overline{\omega}$ are the mid-axis axial and radial displacements, respectively. At equilibrium, the total virtual potential energy of the system is null, leading to the following form:

$$R\int_{-\alpha}^{\alpha}\overline{N}\delta\left[\overline{u}'+\overline{\omega}+\frac{1}{2R}(\overline{\omega}')^2\right]d\theta-\int_{-\alpha}^{\alpha}\overline{M}\delta(\overline{\omega}''+\overline{\omega})d\theta-R^2\int_{-\alpha}^{\alpha}(\overline{q}_{\omega}\delta\overline{\omega}+\overline{q}_{u}\delta\overline{u})d\theta=0 \quad (2)$$

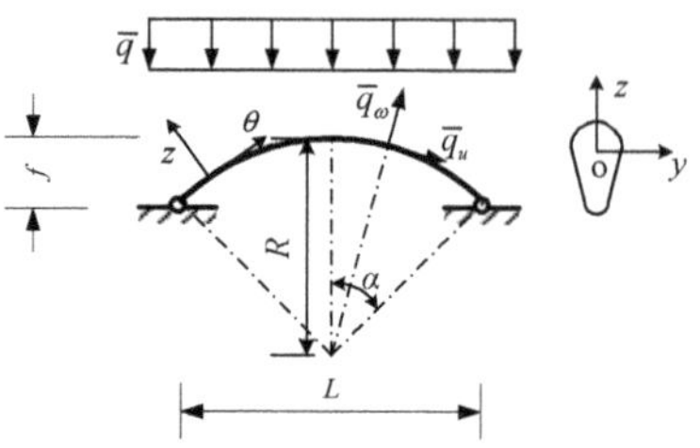

Figure 1. Calculation mode of the arch.

The axial force and bending moment are calculated as follows:

$$\overline{N}=\int_A\sigma_\theta dA=\frac{EA}{R}(\overline{u}'+\overline{\omega})+\frac{EA}{2R^2}\overline{\omega}'^2,\ \overline{M}=\int_A z\sigma_\theta dA=-\frac{EI}{R^2}(\overline{\omega}''+\overline{\omega}), \quad (3)$$

Nonlinear equilibrium equations and boundary conditions can be obtained through the application of the fundamental lemma of the calculus of variations:

$$\begin{aligned}\overline{N}'&=-\overline{q}Rsin\theta cos\theta\\ R\overline{N}-\left(\overline{N}\overline{\omega}'\right)'-\overline{M}''-\overline{M}&=-\overline{q}R^2cos\theta cos\theta\end{aligned} \quad (4)$$

The boundary conditions of the arch are obtained as follows:

$$\theta=\pm\alpha:\ \overline{u}=0,\overline{\omega}=0,\overline{M}=0 \quad (5)$$

To maintain generality, the following dimensionless quantities are defined:

$$\omega=\frac{\overline{\omega}}{R},\ u=\frac{\overline{u}}{R},q=\frac{\overline{q}R}{EA},\lambda=\frac{R}{r}, \quad (6)$$

where r represents the radius of gyration of the section. Based on Equations (3) and (6), the nonlinear equilibrium equations can then be revised as follows:

$$\begin{aligned}u''+\omega'+\omega'\omega''&=-qsin\theta cos\theta\\ \tfrac{1}{\lambda^2}(\omega''''+2\omega''+\omega)+(u'+\omega)+\tfrac{\omega'^2}{2}-\omega'(u''+\omega'+\omega'\omega'')-\omega''\left(u'+\omega+\tfrac{1}{2}\omega'^2\right)&=-qcos^2\theta\end{aligned} \quad (7)$$

The boundary conditions are as follows:

$$\theta=\pm\alpha:\ u=0,\omega=0,\omega''=0 \quad (8)$$

3. Snap-Through Buckling of Shallow Arch

Equation (8) represents the nonlinear dimensionless equilibrium equation of the circular arch, which is difficult to solve analytically. Therefore, the parameter perturbation method is employed to solve it.

3.1. Solution Procedure

For the arch, the dimensionless deflection at the crown is undoubtedly a very small quantity (<<1). Therefore, the dimensionless vertical displacement at the crown is introduced as a small parameter, namely,

$$\varepsilon = \omega_{\mathrm{m}} = (\omega)_{\theta=0} \tag{9}$$

Expanding q, ω, and u in an ascending power series yields

$$\omega = \sum_{j=1} \varepsilon^j \omega_j(\theta), u = \sum_{j=1} \varepsilon^j u_j(\theta), q = \sum_{j=1} \varepsilon^j q_j(\theta) \tag{10}$$

Substituting Equation (10) into Equations (7) and (8), the perturbation equations and boundary conditions for each order can be obtained. Equation (11) presents the recurrence relations for the perturbation term:

$$\begin{gathered} O(\varepsilon^n) : u_n'' + \omega_n' + \sum_{k=1}^{n-1} \omega_k' \omega_k'' - q_n sin\theta cos\theta \\ \frac{\omega_n'''' + 2\omega_n'' + \omega_n}{\lambda^2} + u_n' + \omega_n + \frac{1}{2} \sum_{k=1}^{n-1} \omega_k' \omega_{n-k}' - \sum_{k=1}^{n-1} \left[\omega_k' (u_{n-k}'' + \omega_{n-k}) + \omega_k'' (u_{n-k}' + \omega_{n-k})\right] \\ - \sum_{i=1}^{n-2} \sum_{j=1}^{n-i-1} \left(\omega_i' \omega_j' \omega_{n-i-j}'' + \frac{1}{2} \omega_i'' \omega_j' \omega_{n-i-j}'\right) = -q_n cos^2\theta \end{gathered} \tag{11}$$

The boundary conditions for each order are

$$u_n = 0,\ \omega_n = 0,\ \omega_n'' = 0 \tag{12}$$

Additionally, at $\theta = 0$, the following conditions hold:

$$\begin{gathered} \omega_1 = 1 \\ \omega_n = 0,\ n = 2, 3, 4 \ldots \end{gathered} \tag{13}$$

The first-order equations are linear, and their solutions can be easily obtained. By applying the boundary conditions and Equation (13), all the undetermined parameters and q_1 in the solutions can be determined. Substituting the first-order solutions into the second-order equations and boundary conditions allows for the determination of the undetermined coefficients and q_2 in the second-order solutions. Similarly, by substituting the first- and second-order solutions, the solution for the third-order equations can be obtained. This process can be continued to obtain solutions for any arbitrary n-th order. Finally, the dimensionless deformation–load relationship at the crown can be expressed in the following form:

$$q = q_1 \omega_m + q_2 {\omega_m}^2 + \cdots \tag{14}$$

Equation (14) can be further written in the following form:

$$\frac{q}{q_{\mathrm{p}}} = \frac{q_1}{q_{\mathrm{p}}} \frac{\overline{\omega}_m}{f} \frac{f}{R} + \frac{q_2}{q_{\mathrm{p}}} \left(\frac{\overline{\omega}_m}{f}\right)^2 \left(\frac{f}{R}\right)^2 + \cdots, \tag{15}$$

where q_{p} is the dimensionless classical critical load of the pin-ended circular arch subjected to a uniform radial load:

$$q_{\mathrm{p}} = \frac{\overline{q}_{\mathrm{p}} R}{EA} = \left(\frac{\pi^2}{\alpha^2} - 1\right) \frac{1}{\lambda^2} \tag{16}$$

3.2. Circular Shallow Arches Subjected to Uniform Radial Load

To verify the effectiveness of the parameter perturbation method, the perturbation solutions for a circular shallow arch subjected to a uniform radial load were first calculated using Wolfram Mathematica at different orders. These results were then compared with the analytical solution [2]. The cross-section of the arch is a doubly symmetric I-beam section, H100 × 68 × 4.5 × 7.6, with the section parameters listed in Table 1.

Table 1. Properties of the cross-section.

Area A (mm^2)	Moment of Inertia I (mm^4)	Radius of Gyration r (mm)	Elastic Modulus E (GPa)	Poisson's Ratio μ
1433	245 × 10^4	41.4	206	0.3

Figure 2 illustrates the relationship between $\overline{\omega}_m/f$ and q/q_p, where the modified slenderness is defined as

$$\lambda_s = \frac{R\alpha^2}{r} \tag{17}$$

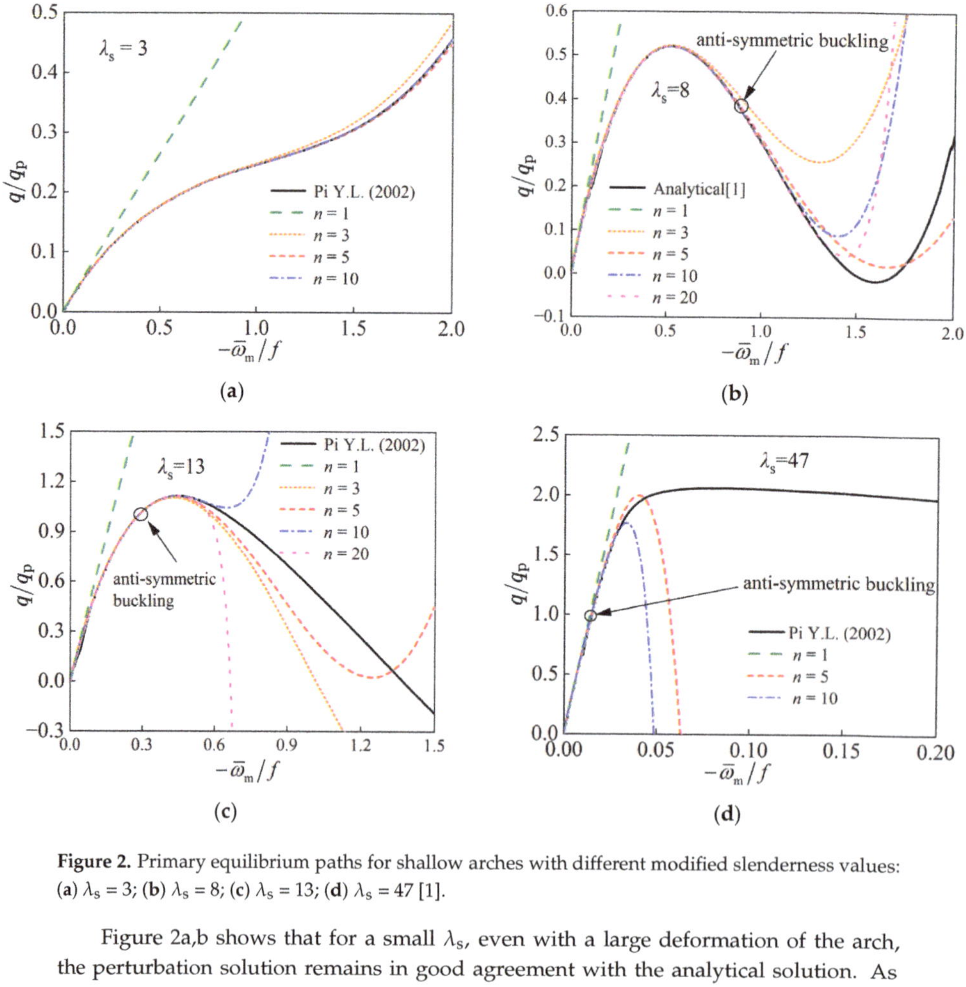

Figure 2. Primary equilibrium paths for shallow arches with different modified slenderness values: (**a**) $\lambda_s = 3$; (**b**) $\lambda_s = 8$; (**c**) $\lambda_s = 13$; (**d**) $\lambda_s = 47$ [1].

Figure 2a,b shows that for a small λ_s, even with a large deformation of the arch, the perturbation solution remains in good agreement with the analytical solution. As the perturbation order increases, the perturbation solutions progressively approach the analytical solutions. In Figure 2a, the non-dimensional load–deformation curves do not exhibit the extreme point, indicating that the shallow arches will never experience snap-through buckling when $\lambda_s = 3$. When the perturbation order $n = 3$, the perturbation solution agrees reasonably well with the analytical solution, and by the time the perturbation order reaches $n = 10$, the solutions are almost identical. In the case of $\lambda_s = 8$, as shown in Figure 2b, the arch buckles in the snap-through mode first. The higher-order perturbation solutions (for $n = 5, 10, 20$) show good agreement before the occurrence of anti-symmetric buckling.

When λ_s is relatively large, as shown in Figure 2c, for $\lambda_s = 13$, snap-through buckling occurs after anti-symmetric buckling. The higher-order perturbation solutions provide a good approximation of the path following snap-through buckling.

When λ_s is very large, as shown in Figure 2d, for $\lambda_s = 47$, the critical load for anti-symmetric buckling of the arch is very small, significantly lower than the critical load at the extreme point. It can be observed that after the extreme point, the perturbation solution diverges completely from the analytical solution, and increasing the perturbation order does not improve the accuracy of the solution. However, the perturbation method still maintains high accuracy in calculating pre-buckling nonlinear deformations and does not require high perturbation orders; a fifth-order perturbation solution can effectively approximate the load–deformation curve.

The condition for the perturbation method to effectively approximate the true solution is that each subsequent term must be a small correction to the preceding one. For this problem, this implies that starting from the second term, the following condition must be met:

$$\delta_n = \frac{q_{n-1}}{q_n}\frac{R}{f} = O\left(\frac{\overline{\omega}_m}{f}\right) \tag{18}$$

In Equation (18), δ_n and $\overline{\omega}_m/f$ should be of the same order of magnitude. Table 2 shows the values of δ_n corresponding to the first 10 perturbation terms for different λ_s values. As λ_s increases, the magnitude of δ_n gradually decreases. The perturbation solution remains valid only when the relative deformation $\overline{\omega}_m/f$ is within the same order of magnitude as δ_n, which is consistent with the trend shown in Figure 2, where the curve maintains high accuracy within the corresponding magnitude. When λ_s is further increased to 80, the perturbation solution remains valid within the order of 0.01. At this point, the corresponding value of $\overline{\omega}_m/f$ for the classical buckling load is approximately 0.005. Clearly, the perturbation method is effective for pre-buckling calculations, and the fifth-order solution is sufficiently accurate.

Table 2. δ_n with different λ_s values.

n	$\lambda_s = 3$	$\lambda_s = 8$	$\lambda_s = 13$	$\lambda_s = 47$	$\lambda_s = 80$
			δ_n		
2	1.27	0.75	0.68	0.36	0.19
3	3.04	2.76	3.01	−0.08	−0.02
4	50.78	149.53	−32.29	−0.04	−0.01
5	12.58	−0.24	−0.66	−0.04	−0.01
6	−1.29	−1.59	−0.75	−0.04	−0.01
7	−34.31	−10.03	−0.60	−0.04	−0.02
8	0.66	0.35	−0.47	−0.06	−0.02
9	−9.33	−1.36	−0.45	−0.10	−0.03
10	3.21	−6.97	−0.50	−0.04	0.02

3.3. Circular Shallow Arches Subjected to Uniform Vertical Load

Based on the computational and analytical results of circular shallow arches under uniform radial loads, the perturbation method proves highly effective in calculating nonlinear pre-buckling deformations and snap-through buckling loads, with the fifth-order solution converging to an accurate result. Therefore, the fifth-order perturbation solution is applied to analyze the pre-buckling nonlinear deformations and snap-through buckling of shallow arches under uniform vertical loads using the same arch cross-section as in Section 3.1.

Figure 3 shows the non-dimensional load–displacement curves for circular shallow arches with the same λ_s under different rise-to-span ratios f/L. Existing studies indicate that for circular shallow arches subjected to uniform radial loads, the rise-to-span ratio has a slight effect on the non-dimensional load–displacement relationship when λ_s is constant. However, for circular shallow arches under uniform vertical loads, the rise-to-span ratio clearly affects the non-dimensional load–displacement relationship.

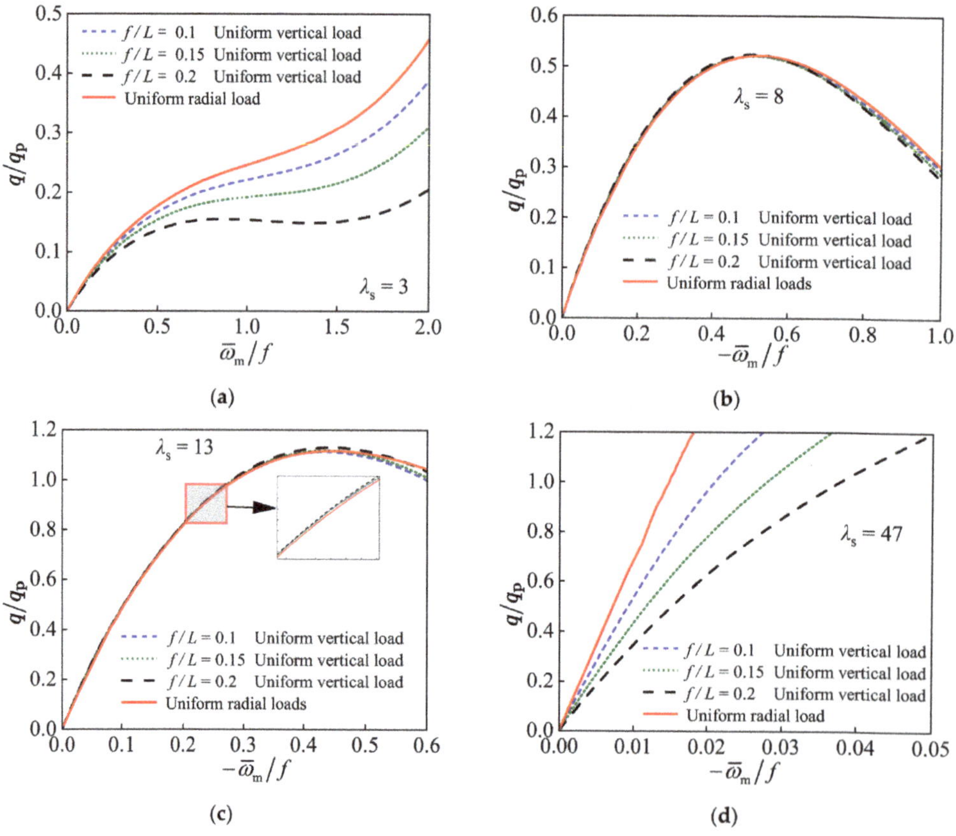

Figure 3. Dimensionless load–displacement curves of arches under uniform vertical loads with different modified slenderness values: (**a**) $\lambda_s = 3$; (**b**) $\lambda_s = 8$; (**c**) $\lambda_s = 13$; (**d**) $\lambda_s = 47$.

Figure 3a shows that when $\lambda_s = 3$, the rise-to-span ratio has a prominent effect on the non-dimensional load–deformation curve before the relative deformation $-\overline{\omega}_m/f$ reaches 1. This is because when the slenderness ratio is low, the structure behaves more like a short column than a thin-walled beam, and the axial component has a significant influence. Therefore, the axial component significantly reduces the load-bearing capacity of the structure.

When $\lambda_s = 8$ and $\lambda_s = 13$, as the rise-to-span ratio increases, the critical load for snap-through buckling also increases, indicating enhanced bearing capacity of the arch, as shown in Figure 3b,c. The bearing capacity of shallow arches subjected to uniform vertical loads exceeds that of those subjected to uniform radial loads, suggesting that the axial load component contributes to improving the structural bearing capacity.

When $\lambda_s = 47$, the load-bearing behavior of shallow arches with different rise-to-span ratios varies significantly, as shown in Figure 3d. Circular shallow arches with a larger rise-to-span ratio exhibit a lower bearing capacity. Therefore, for shallow arches with a large modified slenderness ratio (λ_s), the influence of the axial load component cannot be neglected.

3.4. Internal Force Distribution of Arches Subjected to Uniform Vertical Load

Figure 4 shows the axial force distribution of the arch under a uniformly distributed vertical load. From Figure 4, it can be observed that as the rise-to-span ratio increases, the axial force distribution along the arch axis becomes more variable. The distribution patterns of the nonlinear and linear results are completely consistent, with the linear axial force being slightly smaller than the nonlinear result. In fact, the absolute difference between the nonlinear and linear axial forces is very small, with a discrepancy of about 1%.

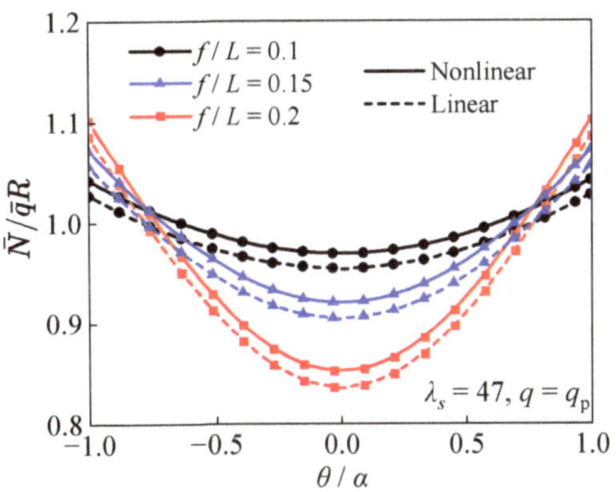

Figure 4. The axial force distribution.

Figure 5 shows the bending moment distribution of the arch under a uniform vertical load, where $\overline{M}_0 = \bar{q}L^2/8$ represents the mid-span bending moment of a simply supported beam with the same span as the arch. Figure 5 shows that regardless of the rise-to-span ratio of the arch, the nonlinear bending moment is significantly larger than the linear bending moment. This highlights the importance of considering the nonlinear effects of bending moments in the analysis of shallow arches. In addition, compared to the simply supported beam with the same span, the bending moment of the arch is much smaller, further emphasizing the advantages of arch structures in load-bearing performance.

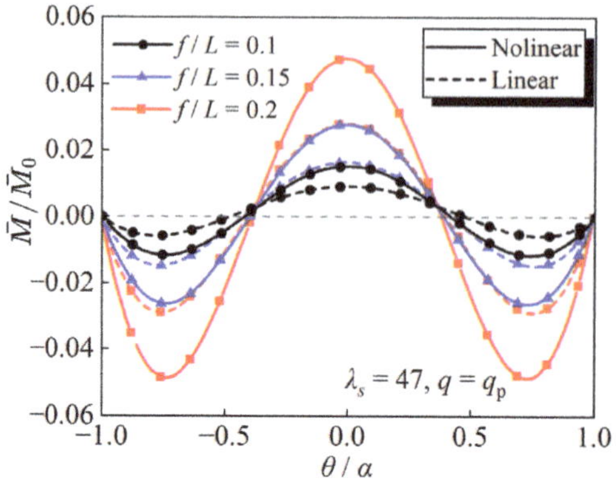

Figure 5. The bending moment distribution.

4. Anti-Symmetric Bifurcation Buckling

4.1. Differential Equations for Buckling Analysis

According to the adjacent equilibrium criterion [2–4], the governing equilibrium equations of the bifurcation stage can be obtained. For this purpose, a pre-buckling state is considered with the pre-buckling displacement components $\overline{\omega}_b$ and $\overline{u}_b$. The disturbed displacements and internal forces can then be expressed as

$$\overline{\omega}^* = \overline{\omega} + \overline{\omega}_b,\ \overline{u}^* = \overline{u} + \overline{u}_b,\ \overline{N}^* = \overline{N} + \overline{N}_b,\ \overline{M}^* = \overline{M} + \overline{M}_b \tag{19}$$

The equilibrium equations of the structure can be rewritten as

$$\begin{gathered} \overline{N}^{*\prime} + R\overline{q}_u = 0 \\ R\overline{N}^* - \left(\overline{N}^*\overline{\omega}^{*\prime}\right)' - \overline{M}^{*\prime\prime} - \overline{M}^* - R^2\overline{q}_\omega = 0 \end{gathered} \tag{20}$$

Since the structure is in a state of equilibrium before the disturbance, substituting Equation (4) into Equation (20) yields the differential equations for buckling analysis:

$$\begin{gathered} \overline{N}_b{}' = 0 \\ R\overline{N}_b - (\overline{N}\overline{\omega}_b{}')' - \overline{N}_b(\overline{\omega} + \overline{\omega}_b)'' - \overline{M}_b{}'' - \overline{M}_b = 0 \end{gathered} \tag{21}$$

The following dimensionless quantities are defined:

$$\omega_b = \frac{\overline{\omega}_b}{R},\ u = \frac{\overline{u}_b}{R},\ \widetilde{N}_b = \frac{\overline{N}_b R^2}{EI},\ \widetilde{N} = \frac{\overline{N}R^2}{EI} \tag{22}$$

The buckling equilibrium equations can be revised as follows:

$$\begin{gathered} \overline{N}_b{}' = 0 \\ \omega_b{}''' + \widetilde{N}_b - \left(\widetilde{N}\omega_b{}'\right)' - \widetilde{N}_b(\omega'' + \omega_b{}'') = 0 \end{gathered} \tag{23}$$

Similarly, the dimensionless boundary conditions can be obtained:

$$\theta = \pm\alpha:\ u_b = 0,\ \omega_b = 0,\ \omega_b{}'' = 0 \tag{24}$$

The first equation in Equation (23) shows that the buckling axial force is constant. Therefore, by neglecting higher-order small terms, the buckling axial force can be written as [2]

$$\widetilde{N}_b = \frac{\lambda^2}{2\alpha}\int_{-\alpha}^{\alpha} (u_b{}' + \omega_b + \omega'\omega_b{}')d\theta \tag{25}$$

For the anti-symmetric bifurcation buckling, the buckling radial displacement ω_b is anti-symmetric, while the pre-buckling radial displacement ω is symmetric. Therefore, the term ω_b + $\omega'\omega_b{}'$ in Equation (25) is anti-symmetric. The boundary conditions also require that u_b = 0 at $\theta = \pm\,\alpha$. Therefore, the buckling axial force $\overline{N}_b$ is zero. Alternatively, as pointed out by Eslami [6], the differential equations for buckling analysis must be homogeneous to ensure a nontrivial solution. The buckling equation can be written as

$$\omega_b{}'''' + 2\omega_b{}'' + \omega_b - \widetilde{N}\omega_b{}'' - \widetilde{N}'\omega_b{}' = 0 \tag{26}$$

4.2. WKB Theory

Obviously, when the arch is subjected to uniform radial loads, the axial force before buckling remains constant, making Equation (26) a constant-coefficient differential equation with a closed-form solution. However, when the arch is subjected to loads with axial components, such as uniform vertical loads, Equation (26) becomes a variable-coefficient differential equation, making it difficult to obtain a closed-form solution. The WKB theory is a powerful tool for studying the eigenvalue problems of variable-coefficient differential equations, with the first two-order solutions, i.e., the physical optics approximation, offering high accuracy [16].

For an arch subjected to a uniform vertical load, the expression for $\widetilde{N}$ is given by

$$\widetilde{N} = q\left(\frac{1}{4}\lambda^2 cos2\theta - N_0\right) = qN(\theta) \tag{27}$$

As shown in Section 2, the pre-buckling axial force $\widetilde{N}$ is negative across the entire span. The eigenvalue multiplicity of the differential equation remains constant within the domain, meaning that there are no turning points [21].

First, suppose that the solution to Equation (27) takes the following form:

$$\omega_b = Y(\theta)exp\left(\xi\int S(\theta)d\theta\right), \tag{28}$$

where ξ is an undetermined parameter. Substituting Equation (28) into Equation (26) leads to Equation (29).

$$\begin{aligned}\xi^4 S^4 Y + \xi^3(6S'S^2Y + 4S^3Y') + \xi^2(3S'^2Y + 2S^2Y + 12SS'Y' + 4SS''Y + 6S^2Y'')\\ +\xi(4S''Y' + 6S'Y'' + S'''Y + 4SY''' + 2S'Y + 4SY') + Y''''\\ -q\left[\xi^2NS^2Y + \xi(N'SY + NYS' + 2NSY') - NY'' - N'Y'\right] = 0\end{aligned} \tag{29}$$

Applying the principle of dominant balance yields $\xi = \sqrt{q}$. Substituting $\xi = \sqrt{q}$ into Equation (29) gives the dominant term as

$$\left(S^4 - NS^2\right)Y = 0 \tag{30}$$

Assume that $Y(\theta) \neq 0$, and let the i-th solution of $S(\theta)$ be denoted as $S(i)$. Then,

$$S^{(1)} = S^{(2)} = 0,\ S^{(3)} = i\sqrt{-N},\ S^{(4)} = -i\sqrt{-N} \tag{31}$$

$S(\theta)$ possesses a double root, which precludes the linear independence of first-order solutions. In cases involving repeated roots, the modified WKB method proposed by Fu (2002) is employed for the calculation [22]. We look for an asymptotic solution of the form

$$Y(\theta) = Y_0 + \frac{1}{\xi}Y_1 + \frac{1}{\xi^2}Y_2 + \cdots \tag{32}$$

Substituting Equation (32) into Equation (29) leads to equations at each order of the WKB theory. From these equations, the expression for $Y(\theta)$ can be obtained. Similarly, denote the i-th solution of $Y(\theta)$ as $Y(i)$. At order $q^{3/2}$, we obtain

$$Y_0^{(3)} = Y_0^{(4)} = (-N)^{\frac{-3}{4}} \tag{33}$$

At order q, we obtain

$$\begin{aligned} Y_0^{(1)} &= 1,\ Y_0^{(2)} = \int_0^\theta \frac{1}{N(t)}dt \\ Y_0^{(3)} &= i(-N)^{\frac{-3}{4}} \int_0^\theta \frac{41N'^2 - 28NN'' + 32N^2}{32(-N)^{5/2}}dt \\ Y_0^{(4)} &= -i(-N)^{\frac{-3}{4}} \int_0^\theta \frac{41N'^2 - 28NN'' + 32N^2}{32(-N)^{5/2}}dt \end{aligned} \tag{34}$$

At order $q^{1/2}$, we obtain

$$Y_1^{(3)} = Y_1^{(4)} = 0 \tag{35}$$

Finally, the approximate solution for the pre-buckling radial displacement can be obtained from the above results:

$$\begin{aligned} \omega_b^{(1)} &= 1, \omega_b^{(2)} = \int_0^\theta \frac{1}{N(t)}dt \\ \omega_b^{(3)} &= (-N)^{\frac{-3}{4}}\left[cos\left(\sqrt{q}\int_0^\theta \sqrt{-N}dt\right) - sin\left(\sqrt{q}\int_0^\theta \sqrt{-N}dt\right)\int_0^\theta \frac{41N'^2 - 28NN'' + 32N^2}{32(-N)^{5/2}}dt\right] \\ \omega_b^{(4)} &= (-N)^{\frac{-3}{4}}\left[sin\left(\sqrt{q}\int_0^\theta \sqrt{-N}dt\right) + cos\left(\sqrt{q}\int_0^\theta \sqrt{-N}dt\right)\int_0^\theta \frac{41N'^2 - 28NN'' + 32N^2}{32(-N)^{5/2}}dt\right] \end{aligned} \tag{36}$$

Obviously, when the arch is subjected to a uniform radial load, Equation (36) degenerates into the solution for a shallow arch subjected to a uniform radial load. First, it can be obtained based on the symmetry of the boundary conditions:

$$K_1 = K_3 = 0 \tag{37}$$

Substituting Equations (36) and (37) into the boundary conditions at $\theta = \alpha$ yields a matrix equation of the following form:

$$\begin{pmatrix} \omega_b^{(2)}(\alpha) & \omega_b^{(4)}(\alpha) \\ \omega_b^{(2)''}(\alpha) & \omega_b^{(4)''}(\alpha) \end{pmatrix}\begin{pmatrix} K_2 \\ K_4 \end{pmatrix} = 0, \tag{38}$$

from which the characteristic equation is obtained as follows:

$$y(q) = \begin{vmatrix} \omega_b^{(2)}(\alpha) & \omega_b^{(4)}(\alpha) \\ \omega_b^{(2)''}(\alpha) & \omega_b^{(4)''}(\alpha) \end{vmatrix} = 0 \tag{39}$$

The modified WKB method transforms the eigenvalue problem of a differential equation with variable coefficients into a matrix eigenvalue problem. Figure 6 shows part of the images of the second- and third-order characteristic polynomials, and for anti-symmetric buckling, the flexural half-wave number is 2. From Figure 6, it can be seen that the third-order eigenvalues are all larger than the second-order eigenvalues, which is due to the fact that the third-order WKB method takes into account the axial component of the load more fully. From the construction of the WKB solution, the second-order WKB solution practically ignores the effect of the term $\tilde{N}' \omega_b'$ in the buckling equation, which results in a small value of the critical load of the structure.

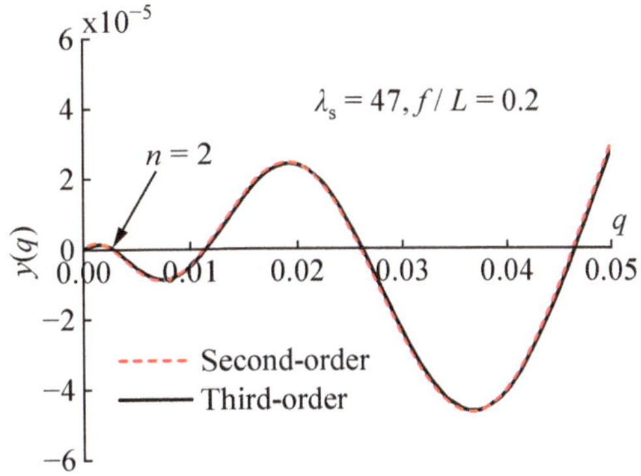

Figure 6. Characteristic polynomial $y(q)$.

Figure 7 illustrates the ratio of the critical load q_{cr} to the classical critical load q_p calculated using the WKB method under different rise-to-span ratio conditions. The figure also compares the computational results of the WKB solutions at various orders and highlights the differences in pre-buckling axial forces, depending on whether a nonlinear or linear axial force analysis is employed.

Figure 7 demonstrates the following:

- The larger the rise-to-span ratio, the larger the q_{cr}/q_p for the same modified slenderness λ_s, which is due to the effect of the axial component of the load. Obviously, the smaller the rise-to-span ratio of an arch subjected to a uniform vertical load, the closer the axial force distribution of the arch is to that of an arch subjected to a uniform radial load, and therefore the closer its critical load is to that of an arch subjected to a uniform radial load.
- The eigenvalue obtained using the second-order WKB method is smaller than that obtained using the third-order WKB method, consistent with the conclusions drawn from Figure 6 and their underlying reasons.
- The variation in q_{cr}/q_p with λ_s calculated using nonlinear axial force differs from that obtained with a linear axial force. When a linear axial force is used, the critical load decreases as λ_s increases, whereas with a nonlinear axial force, the critical load increases as λ_s increases. This difference arises because, for small values of λ_s, the nonlinear effects before buckling are significant, leading to lower critical loads when a nonlinear axial force is used. In contrast, the results based on a linear axial force are influenced solely by the axial component of the load, which results in higher critical loads compared to those under uniformly distributed radial loads. As λ_s becomes large, the difference between the critical loads calculated with nonlinear and linear axial forces diminishes.

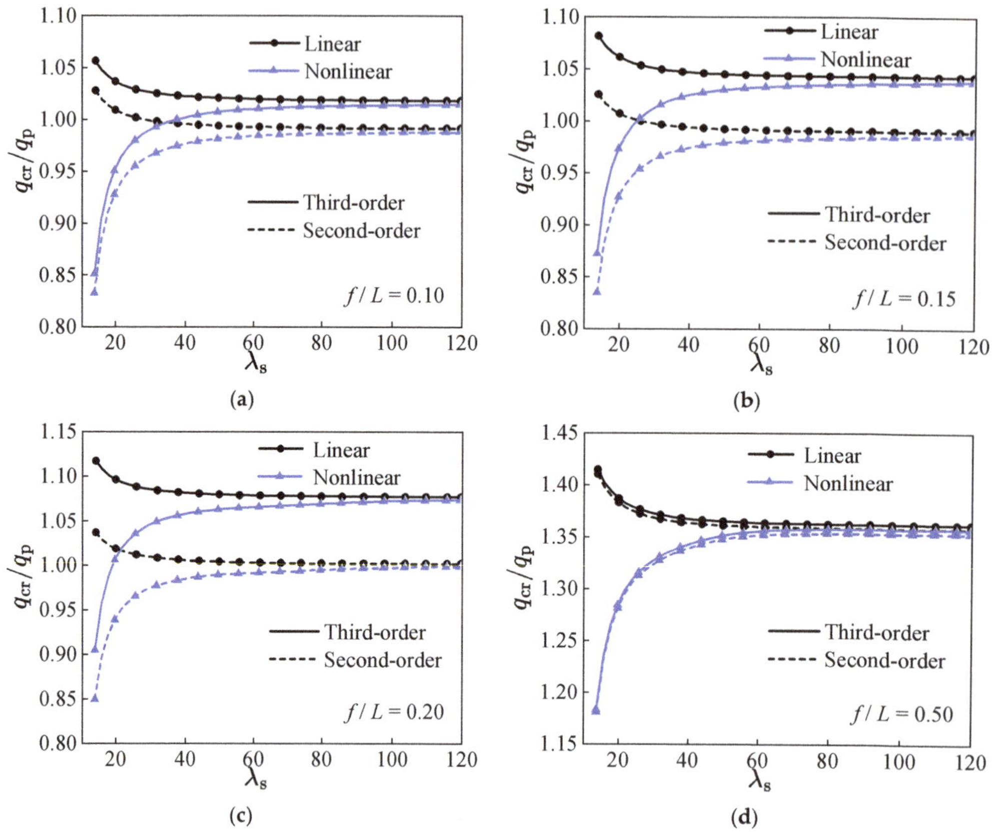

Figure 7. q_{cr}/q_p with different rise-to-span ratios: (**a**) $f/L = 0.1$; (**b**) $f/L = 0.15$; (**c**) $f/L = 0.12$; (**d**) $f/L = 0.5$.

The above analysis shows that for shallow arches with a small rise-to-span ratio, the critical load for anti-symmetric buckling closely matches the classical solution, and the results from arches subjected to a uniformly distributed radial load are reliable. For deep arches with a large rise-to-span ratio, the influence of the axial load component cannot be ignored. Since the eigenvalue obtained using the second-order WKB method is smaller than that obtained with the third-order WKB method, one can consider the second-order solution as an approximation in practical calculations. Therefore, the pre-buckling radial displacement can be simplified as follows:

$$\omega_b = K_2 \int_0^{\theta} \frac{1}{N} dt + K_4 \frac{sin\left(\int_0^{\theta} \sqrt{-N} dt\right)}{(-N)^{\frac{3}{4}}} \tag{40}$$

4.3. WKB Results with Other Analytical Studies

To validate the WKB method, we performed comparative analyses with the numerical solutions obtained using the Galerkin method by Cheng [23]. A comparative analysis between the WKB solutions and numerical solutions under varying modified semi-span slenderness ratios $\lambda_f = \alpha R/r$ and central angles is presented in Table 3.

Table 3. $q_{cr}R^3/EI$ with different modified semi-span slenderness ratios.

α (°)	$\lambda_f = 20$		$\lambda_f = 50$		$\lambda_f = 100$		$\lambda_f = 200$	
	Galerkin	**WKB**	**Galerkin**	**WKB**	**Galerkin**	**WKB**	**Galerkin**	**WKB**
90	4.1525	5.4697	4.1299	5.434	4.1270	5.4289	4.1466	5.4276
78.54	5.9379	6.3614	5.8887	6.3094	5.8814	6.3020	5.8795	6.3002
67.08	8.2705	7.9866	8.1942	7.9002	8.1817	7.8880	8.1785	7.8849
55.62	11.8221	11.0052	11.6925	10.8360	11.6693	10.8118	11.6634	10.8057
44.16	18.2921	17.0966	17.9946	16.6837	17.9391	16.6246	17.9248	16.6098
32.7	32.9204	31.6189	31.9171	30.2386	31.7297	30.0409	31.6813	29.9915

The comparison shows that when the central angle α of the arch is small, the buckling load obtained using the WKB method is lower than that obtained using the Galerkin method, while when the central angle is large, it is higher than the buckling load from the Galerkin method. The reason for this difference lies in the fact that Cheng [23] did not neglect the transverse stresses of the arch in his analysis. Overall, when the arch is relatively shallow, the WKB method demonstrates effectiveness, indicating that the semi-analytical solution obtained using the WKB method can be used as the anti-symmetric buckling solution for shallow arches ($\alpha < 70°$).

5. Conclusions

This research proposes a nonlinear analytical approach based on asymptotic methods, including the parameter perturbation method and the WKB method. The results show that

- The parameter perturbation method is effective for snap-buckling of shallow arches, and the fifth-order solution is sufficiently accurate. For shallow arches with a large modified slenderness ratio, the influence of the axial load component cannot be neglected.
- In the anti-symmetric buckling analysis, the eigenvalue obtained using the second-order WKB method is smaller than that obtained using the third-order WKB method; therefore, the second-order solution can be used as the critical load.
- The asymptotic solution for the buckling deformation of the two-hinged circular arch is provided. For shallow arches with a small rise-to-span ratio, the critical load for anti-symmetric buckling closely matches the classical solution. For deep arches with a large rise-to-span ratio, the influence of the axial load component cannot be ignored.

Author Contributions: Conceptualization, J.J. and M.S.; methodology, J.J.; software, J.J.; validation, J.J.; formal analysis, J.J.; investigation, M.S.; resources, M.S.; data curation, J.J.; writing—original draft preparation, J.J.; writing—review and editing, M.S.; visualization, J.J.; supervision, M.S.; project administration, J.J.; funding acquisition, M.S. All authors have read and agreed to the published version of the manuscript.

Funding: This work was supported by the National Natural Science Foundation of China (grant number 51978564).

Data Availability Statement: The original contributions presented in the study are included in the article; further inquiries can be directed to the corresponding author.

Conflicts of Interest: The authors declare that they have no known competing financial interests or personal relationships that could have appeared to influence the work reported in this paper.

References

1. Tong, G.; Pi, Y.L.; Bradford, M.A.; Tin-Loi, F. In-Plane Nonlinear Buckling Analysis of Deep Circular Arches Incorporating Transverse Stresses. *J. Eng. Mech.* **2008**, *134*, 362–373.

2. Pi, Y.L.; Bradford, M.A.; Uy, B. In-Plane Stability of Arches. *Int. J. Solids Struct.* **2002**, *39*, 105–125.
3. Pi, Y.L.; Trahair, N.S. In-Plane Buckling and Design of Steel Arches. *J. Struct. Eng.* **1999**, *125*, 1291–1298.
4. Pi, Y.L.; Trahair, N.S. Nonlinear Buckling and Post-Buckling of Elastic Arches. *Eng. Struct.* **1998**, *20*, 571–579.
5. Bradford, M.A.; Wang, T.; Pi, Y.L.; Gilbert, R.I. In-Plane Stability of Parabolic Arches with Horizontal Spring Supports. I: Theory. *J. Constr. Steel Res.* **2020**, *164*, 1057.
6. Eslami, M.R. Buckling and Post-Buckling of Curved Beams and Rings. In *Buckling and Postbuckling of Beams, Plates, and Shells*; Springer: Cham, Switzerland, 2018.
7. Kang, Y.J.; Yoo, C.H. Thin-Walled Curved Beams. II: Analytical Solutions for Buckling of Arches. *J. Eng. Mech.* **1994**, *120*, 2102–2125. [CrossRef]
8. Reddy, J.N. *Theory and Analysis of Elastic Plates and Shells*; CRC Press: Boca Raton, FL, USA, 2006.
9. Hu, C.; Li, Z.; Hu, Q. On Non-Linear Behavior and Buckling of Arch-Beam Structures. *Eng. Struct.* **2021**, *239*, 112214. [CrossRef]
10. Babaei, H.; Kiani, Y.; Eslami, M.R. Large amplitude free vibration analysis of shear deformable FGM shallow arches on nonlinear elastic foundation. *Thin-Walled Struct.* **2019**, *144*, 106237. [CrossRef]
11. Babaei, H.; Eslami, M.R. On Nonlinear Vibration and Snap-Through Stability of Porous FG Curved Micro-Tubes Using Two-Step Perturbation Technique. *Compos. Struct.* **2020**, *247*, 112447.
12. Babaei, H.; Eslami, M.R. On nonlinear vibration and snap-through buckling of long FG porous cylindrical panels using nonlocal strain gradient theory. *Compos. Struct.* **2021**, *256*, 113125.
13. Guo, Y.; Lin, B.; Guo, Y. Theoretical and Experimental Studies of In-Plane Stability Design of Circular Arches Subjected to Axial Force and Moment. *China Civ. Eng. J.* **2011**, *44*, 8–15. (In Chinese)
14. Caporale, A.; Luciano, R.; Rosati, L. Limit analysis of masonry arches with externally bonded FRP reinforcements. *Comput. Methods Appl. Mech. Eng.* **2006**, *196*, 247–260.
15. Mercuri, M.; Pathirage, M.; Gregori, A.; Cusatis, G. Masonry vaulted structures under spreading supports: Analyses of fracturing behavior and size effect. *J. Build. Eng.* **2006**, *45*, 103396.
16. Bender, C.M.; Orszag, S.A. *Advanced Mathematical Methods for Scientists and Engineers I: Asymptotic Methods and Perturbation Theory*; Springer: New York, NY, USA, 1999.
17. Nayfeh, A.H. *Introduction to Perturbation Techniques*; John Wiley & Sons: Hoboken, NJ, USA, 2011.
18. Jin, L.; Liu, Y.; Cai, Z. Asymptotic Solutions on the Circumferential Wrinkling of Growing Tubular Tissues. *Int. J. Eng. Sci.* **2018**, *128*, 31–43.
19. Fu, Y. Some Asymptotic Results Concerning the Buckling of a Spherical Shell of Arbitrary Thickness. *Int. J. Non-Linear Mech.* **1998**, *33*, 1111–1122.
20. Pour, M.S. WKB analysis of the buckling of a neo-Hookean cylindrical shell of arbitrary thickness subject to an external pressure. *Int. J. Appl. Mech.* **2010**, *2*, 857–870. [CrossRef]
21. Wasow, W. *Linear Turning Point Theory*; Springer: New York, NY, USA, 1985.
22. Fu, Y.; Pour, M.S. WKB Method with Repeated Roots and Its Application to the Buckling Analysis of an Everted Cylindrical Tube. *SIAM J. Appl. Math.* **2002**, *65*, 1856–1871.
23. Cheng, P. In-Plane Nonlinear Theory and Elastic-Plastic Stability of Pined Circular Arches. Ph.D. Thesis, Zhejiang University, Hangzhou, China, 2005.

buildings

MDPI

Article

Cyclic Behavior of Joints Assembled Using Prefabricated Beams and Columns with High-Ductility Recycled Powder Concrete

Xiuling Li [1,2,*], Haodong Sun [2], Kezhen Chen [2], Tianfeng Yuan [2], Long Wen [2], Xiaowei Zheng [3] and Tongxing Bu [2]

1 School of Civil Engineering, Shandong Jiaotong University, Jinan 250357, China
2 Key Laboratory of Building Structural Retrofitting and Underground Space Engineering, Shandong Jianzhu University, Ministry of Education, Jinan 250101, China; horizonsun77@163.com (H.S.)
3 School of Mechanics and Civil Engineering, China University of Mining and Technology, Xuzhou 221116, China
* Correspondence: dlutiem@163.com

Abstract: The integration of recycled powder (RP) as a partial cement replacement in concrete, combined with fiber reinforcement, facilitates the development of high-ductility recycled powder concrete (HDRPC) with enhanced mechanical properties. This approach holds significant potential for effectively recycling construction waste and reducing carbon emissions. To improve the seismic performance of prefabricated joints in industrial prefabricated building production, experimental tests under low-cycle reversed cyclic loading were conducted on four HDRPC prefabricated joints, one HDRPC cast-in-place joint, and one normal prefabricated concrete joint. The study systematically analyzed damage patterns, deformation ductility, stiffness degradation, hysteresis energy dissipation, and other performance characteristics. The results demonstrate that HDRPC effectively mitigates crack width and shear deformation in the joint core area, achieving a 17.8% increase in joint-bearing capacity and a 33.3% improvement in displacement ductility. Moreover, HDRPC improves specimen damage characteristics, enhances joint shear capacity and flexibility, and reduces the demand for hoop reinforcement in the joint core area due to its exceptional shear ductility. Based on the softened tension–compression bar model, a crack-resistance-bearing capacity equation for HDRPC joints was derived, which aligns closely with shear test results when cracks develop in the joint core area.

Keywords: high-ductility recycled powder concrete; prefabricated beam–column joints; cyclic behavior; hysteresis curve; joint shear capacity

1. Introduction

The construction industry significantly impacts global sustainability due to inefficient resource utilization, environmental degradation, and reliance on traditional production methods. Notably, construction activities account for approximately 40% of global energy consumption and 36% of carbon emissions [1,2]. As a major contributor to climate change, the industry faces growing pressure to adopt low-carbon, eco-friendly solutions [3]. Among these, prefabricated structures and recycled concrete have emerged as promising alternatives, leveraging construction and demolition waste to mitigate environmental impacts and conserve resources [4,5].

Prefabricated structures, while offering advantages in construction efficiency and sustainability, often exhibit weak characteristics under seismic loading. Previous studies have identified several critical issues, including brittle failure at beam–column joints, inadequate energy-dissipation capacity, and limited ductility due to insufficient reinforcement detailing and poor connection integrity [6,7]. For instance, post-earthquake investigations of precast reinforced concrete structures have revealed that joint failures are a common cause of structural collapse, particularly in regions with high seismic activity [8–10]. These findings underscore the need for improved joint designs that enhance seismic resilience while maintaining the benefits of prefabrication.

The incorporation of recycled concrete in construction materials offers significant environmental and social benefits by reducing carbon emissions and promoting efficient use of solid waste resources [11]. Notably, construction waste generates a large volume of fine particles under 150 μm, commonly referred to as recycled powder (RP), which constitutes approximately 20% of the raw material used [12,13]. Extensive research has explored the potential of RP in improving engineering cement-based composite materials. The feasibility of using recycled powder to produce high-performance concrete has been validated through studies on particle size distribution [14,15], chemical composition [16,17], fiber interactions [18–21], and empirical investigations [22].

High-ductility recycled powder concrete (HDRPC) is an eco-friendly material that offers substantial improvements over normal concrete in terms of ductility, crack resistance, and durability [23,24]. HDRPC, which typically incorporates a high replacement ratio of recycled powder and hybrid fibers such as basalt fibers (BFs) and polyvinyl alcohol (PVA) fibers, demonstrates enhanced fracture resistance and improved ductility due to better matrix–fiber interaction [25–28]. Yu et al. [29] developed a carbon-emission-calculation model for prefabricated buildings based on life cycle assessment (LCA) theory, evaluated the environmental impact of green building materials, and quantitatively calculated the carbon emissions of HDRPC during the materialization phase, showing that HDRPC has better environmental benefits than C30 concrete, with a 31.86 $kgCO_2e$ reduction in emissions per structure. This makes HDRPC an ideal candidate for strengthening and repairing reinforced concrete (RC) structures, including beam–column joints under both static and seismic loading conditions [30]. However, while the properties of HDRPC have been explored for general concrete applications, its performance in the joint areas of prefabricated RC structures remains under-researched.

Prefabricated buildings have gained significant attention due to their efficiency, sustainability, and potential for reducing construction time and costs. Among the various components in a prefabricated structure, the beam–column joint is particularly critical in ensuring the overall performance and safety of the structure. Current connection techniques for prefabricated elements include bolting, welding, riveting, and grouting, with grouting sleeves being one of the most widely adopted methods for connecting precast components [31]. These sleeves help join prefabricated elements either horizontally [32–34] or vertically [35–38], transferring internal forces and meeting seismic design requirements. Grouting sleeves enhance the load-bearing capacity and rigidity of joints, forming a "rigid zone" that helps prevent crushing or damage under seismic loads [39]. Joints connect elements, transfer internal loads, and maintain structural integrity in assembled concrete frame structures. Important stress point affects the seismic performance of structures. The joint must be strong, stiff, and ductile for earthquake bearing and deformation [40,41]. The strength, stiffness, and ductility of joints are crucial for earthquake resistance, and several studies have focused on improving the mechanical behavior of prefabricated beam–column joints to increase their seismic performance [42–44]. Ding et al. [42] proposed a prefabricated concrete beam–column joint with a bolt connection and investigated its seismic performance through full-scale experiments and finite element simulations, with experimental results showing that joints connected by grade 8.8 bolts exhibit higher yield and ultimate bearing capacities, initial stiffness, and energy consumption compared to those connected by grade 5.6 bolts, while simulations further explore the impact of bolt strength and axial compression ratio on the seismic behavior of the joint. Zhang et al. [43] present cyclic tests of four interior precast UHPC (ultra-high-performance concrete)/RC composite beam–column joints to evaluate the impact of replacing normal concrete with UHPC in the joint zone, showing that non-steam-cured UHPC exhibits high compressive and tensile strength, eliminates the need for stirrups, and allows for significantly reduced anchorage lengths, thus simplifying the fabrication process. Several approaches [45–47] have been proposed to increase load-bearing, ductility, and energy dissipation of prefabricated frame beam–column connections. Huang et al. [45] propose a novel dry precast concrete beam–column connection with multi-slit devices (MSDs) and investigate its seismic performance through cyclic loading tests on three full-scale specimens, showing that the precast joints outperform cast-in-place joints in terms of load-carrying capacity, stiffness, energy dissipation, and deformation capacity, making them suitable for high-risk earthquake regions. Ye et al. [46] propose a new fully precast replaceable energy-dissipation rocking beam–column joint (RBCJ) with energy-dissipating steel plates, and investigate its seismic performance through quasi-static tests on two scaled specimens, demonstrating that the joint has small residual displacement, concentrated plastic deformation in replaceable parts, and excellent repairable performance, with the short-lap specimen outperforming the long-lap one in seismic performance and recovery.

Recent research has proposed various strategies to enhance the seismic resilience of prefabricated frame joints, such as introducing new connection forms [48–51] and utilizing high-performance concretes [52–55]. Notably, some studies have examined the use of UHPC to improve the shear strength and cyclic performance of prefabricated joints [56,57]. Lin et al. [57] propose a novel precast beam–column connection with a UHPC core-shell and investigate its seismic performance through quasi-static tests on three precast and one monolithic specimen, showing that the UHPC shell effectively limits shear behavior, delays cracking, enhances ductility, and improves energy dissipation, while allowing for fewer stirrups than required by code. Similarly, high-ductility concrete (HDC) has been investigated for improving the shear-bearing capacity of prefabricated beam–column connections [58]. Deng et al. [58] investigate the use of HDC in five precast beam–column connections under cyclic loading, showing that HDC significantly improves shear and damage-tolerance capacities, shifts failure from joint shear to beam end and can replace stirrups in confining joints, though column strengthening is required to meet "strong column" criteria. Despite the advancements in prefabricated structures, there remains a significant research gap in the application of HDRPC for enhancing the seismic performance of beam–column joints. Previous studies have primarily focused on normal concrete or UHPC, with limited exploration of HDRPC in prefabricated joints.

This study seeks to address the existing gap in knowledge by evaluating the cyclic behavior of HDRPC joints under varying axial loads and stirrup configurations. The objectives of this research are as follows: (1) to investigate the seismic performance of HDRPC joints in comparison to normal concrete joints, (2) to develop design guidelines for HDRPC joints based on both experimental and theoretical analyses, and (3) to explore the potential of HDRPC in reducing the demand for hoop reinforcement, thus simplifying construction processes. The experimental results will provide valuable insights into the fracture resistance and shear-carrying capacity of HDRPC-prefabricated joints, contributing essential data for the development of design formulas. Ultimately, the findings of this study will support the advancement of prefabricated structures using HDRPC, laying a solid foundation for their application in earthquake-prone regions.

2. Materials and Mixture Proportions

2.1. Normal Concrete

To decrease the amount of cement used and enhance the efficiency of recycled powder, the non-critical stress areas of the nodal specimens in this experiment were made with normal recycled powder concrete (NRPC) [4]. Table 1 displays the two common concrete mixtures used in the PC30 joints. Table 2 shows the parameters of the regenerated powder used in the experiment. A cubic compressive strength test was conducted, and the amount of water-reducing agent was determined based on the ease of mixing. The results of the test can be found in Table 3 and Figure 1. Type I normal Portland cement and the normal river sand of fine aggregate were adopted. The coarse aggregate was used crushed aggregate with a maximum size of 16 mm in diameter. The compressive strengths were tested using materials from the same batch as the beam–column joint specimens.

Table 1. Mix proportions of concrete.

Type of Concrete	RCP Replacement Percentage (%)	Water-to-Binder Ratio	Sand-to-Binder Ratio	HRWR [3] (%)	PVA Fiber [4] (vol.%)	Basalt Fiber (vol.%)
C30	-	0.41	0.42	0.21	-	-
C35	-	0.38	0.42	0.26	-	-
NRPC [1]	50	0.41	0.42	0.26	-	-
HDRPC [2]	45	0.32	0.40	0.17	1.7	0.2

[1] "NRPC" is normal recycled concrete; [2] "HDRPC" is high-ductility recycled powder concrete; [3] "HRWR" is a high-range water reducer used to improve the flow ability of concrete; [4] "PVA fiber" is polyvinyl alcohol fiber.

Table 2. Parameters of recycled powder.

Density ($g \cdot cm^{-3}$)	Moisture Content (%)		Heat Loss (%)	Characteristic Parameters of Particle Size (μm)		
				D10	D50	D90
2.63	2.1		25.7	0.99	4.51	16.18
Major oxides content (%)	CaO 44.5	SiO_2 23.05	Al_2O_3 6.41	MgO 2.6	Fe_2O_3 2.48	K_2O 1.3

Table 3. Compressive strength of NRPC.

Replacement Rate (%)	Water-Reducing Agent ($kg \cdot m^{-3}$)	7 d Water Curing (MPa)	28 d Water Curing (MPa)	28 d Natural Curing (MPa)
0	4.40	48.4	58.2	53.6
10	4.40	40.5	51.8	46.4
20	4.62	40.8	52.8	49.2
30	4.84	35.3	46	41.4
40	5.72	35.7	46.7	41.9
50	6.60	27.6	36.8	34.5

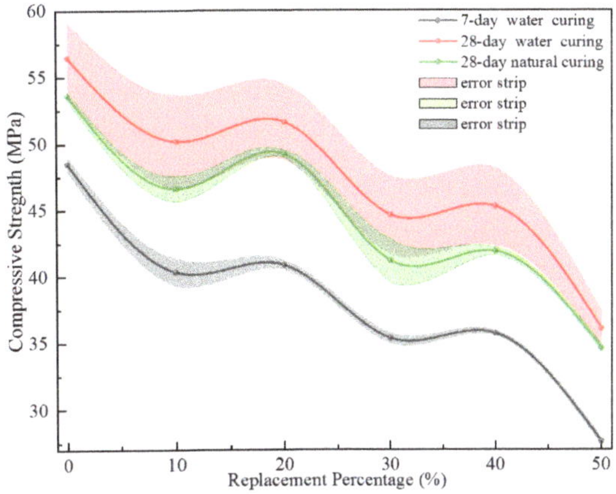

Figure 1. Compressive strength of NRPC.

2.2. High-Ductility Recycled Powder Concrete

High-ductility recycled powder concrete (HDRPC) was formulated using cement, fly ash, recycled powder, quartz sand, water, a high-range water reducer, basalt fibers (BFs), and polyvinyl alcohol (PVA) fibers. Table 4 shows the parameters of different fibers used in the experiment. Detailed mix proportions are presented in Table 1. To evaluate tensile behavior, four dog-bone-shaped specimens were tested, yielding an average tensile strength of 5.15 MPa and an ultimate strain of 3.63%, meeting the requirements for engineered cementitious composites (ECC) (Figure 2).

Table 4. Parameters of fiber.

Type of Fiber	Density (kg·m^{-3})	Tensile Strength (MPa)	Tensile Limit (%)	Length (mm)	Modulus (GPa)	Diameter (μm)
PVA	1300	1620	6.0–8.0	12	42.8	39
Basalt fiber	2650	2320	2.6	18	84.5	15

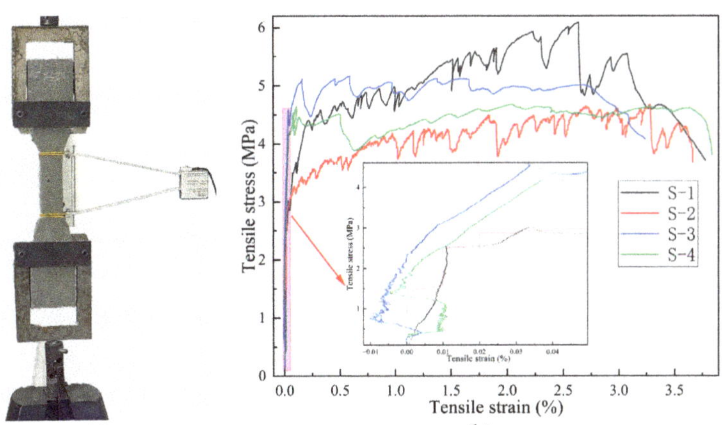

Figure 2. Tensile test of the HDRPC: (**a**) specimen (unit: mm); (**b**) stress–strain curves.

2.3. Grouting Sleeves and Grouts

According to the "Technical Specification for Splicing of Reinforced Grouting Sleeves" JGJ 355-2015 [59], the selected cast iron fully grouted sleeve GTZQ4 16 (Figure 3) was tested. Three fully grouted sleeves were randomly selected for testing to verify their uniaxial tensile performance, and the test results showed that the steel bars were broken and ultimately destroyed. The test results are listed in Table 5. The results showed that the fully grouted sleeve used in this experiment met the specified requirements and achieved an efficient and reliable connection of steel bars between prefabricated beams and columns.

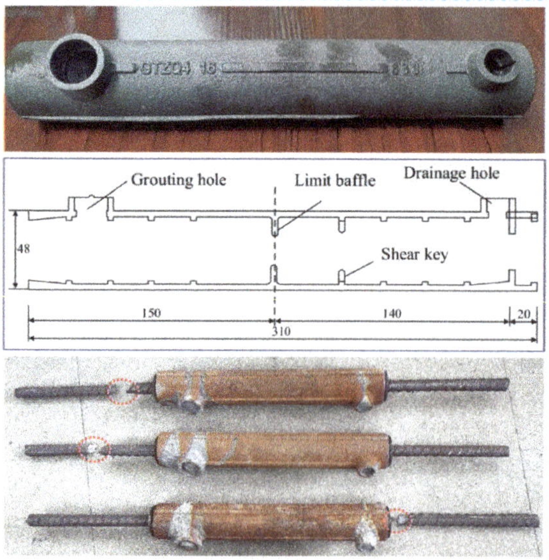

(a)

(b)

Figure 3. Cast iron fully grouted sleeve: (**a**) grouting sleeve and failure mode; (**b**) test setup.

Table 5. Uniaxial tensile properties of grouted sleeve steel bar connections.

Connector Number	1	2	3	Regulatory Requirement
Tensile strength (MPa)	635	612	637	$\geq f_{stk}$ [1]
Total elongation (%)	12.2	11.4	12.5	$\geq$6.0

[1] f_{stk} is the standard value of the ultimate strength of steel bars.

3. Mechanical Properties

3.1. Design of Test Specimens

To evaluate the connection behavior of the RC frame with beam and column, six full-scale specimens were prepared in this study. The beam cross-sections had dimensions of 250 mm × 300 mm, while the column cross-sections had dimensions of 300 mm × 300 mm. The precast beams and columns are joined together using grout sleeves to extend the whole depth of the joint beam contact. During seismic loading, the core area of the joint and the neighboring beam and column ends in the back-cast zone are anticipated to endure significant shear forces and bending moments. Consequently, the areas of high stress were reinforced using HDRPC material, whereas the rest of the specimen was made with NRPC. The axial load ratios (0.3, 0.5, 0.7) were selected to represent a range of axial compression conditions typically encountered in prefabricated structures. These values were chosen based on previous studies and design practices, where 0.3 simulates moderate axial loading, while 0.5 and 0.7 represent higher axial compression scenarios, such as those found in multi-story buildings or structures subjected to significant gravity loads [60].

The reinforcement detailing and stirrup spacing were designed by the Chinese Code for Design of Concrete Structures (GB 50010-2010) [61]. Stirrup spacings of 100 mm and 200 mm were selected to evaluate the effect of hoop-reinforcement density on joint performance. A spacing of 100 mm represents a dense reinforcement configuration, while 200 mm represents a reduced reinforcement scenario, allowing us to assess whether HDRPC can compensate for reduced stirrup density. The specific characteristics of the specimen, including its specifications and construction details, are provided in Table 6 and Figure 4a–d. The compressive strength of normal concrete and HDRPC was tested at 28 days, yielding values of 38.1 MPa and 32.3 MPa, respectively, which exceed the minimum requirements of GB 50010-2010 [61]. The yield strength of the steel reinforcement was 420 MPa, consistent with the design value specified in the code.

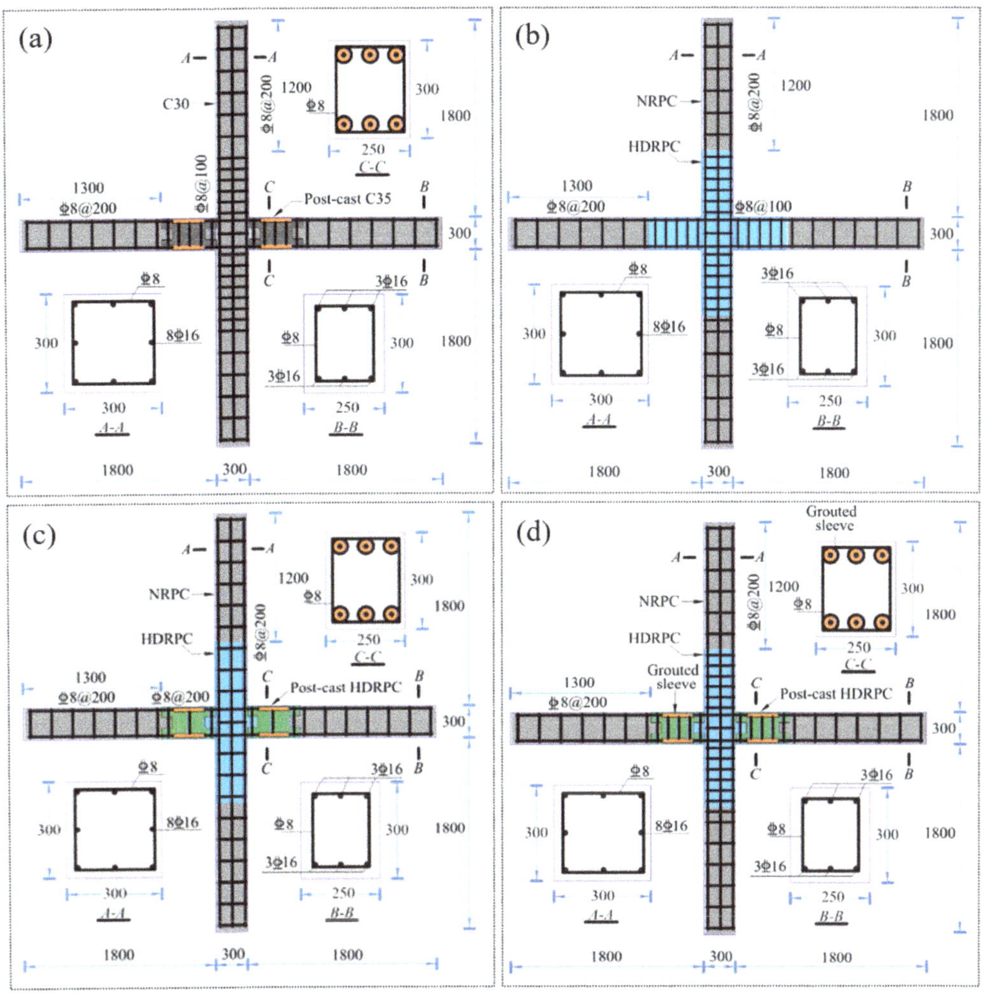

Figure 4. Dimension and details of reinforcements for specimens: (a) PC30; (b) HDJ; (c) PHDJ1; (d) PHDJ2-4.

Table 6. Summary of specimens.

Specimens	Construction Method	Materials	Axial Pressure Ratio	Shaft Pressure (kN)	Hoop Arrangement
PC30	precast beam and column assembly	C30	0.3	386.1	⏀ 8@100/200
HDJ	cast-in-place	HDRPC/C30	0.3	386.1	⏀ 8@100/200
PHDJ1	precast beam and column assembly	HDRPC/NRPC	0.3	386.1	⏀ 8@200
PHDJ2		HDRPC/NRPC	0.3	386.1	⏀ 8@100/200
PHDJ3		HDRPC/NRPC	0.5	643.5	⏀ 8@100/200
PHDJ4		HDRPC/NRPC	0.7	900.9	⏀ 8@100/200

For the manufacturing procedure of the assembled joint as shown in Figure 5, firstly the precast parts undergo fabrication and the longitudinal reinforcement spacing is controlled using a limiting template to ensure precise docking of the outstretched reinforcement bars. Additionally, a bubble film is applied to the inner side of the limiting template, which enhances interface roughness and improves the bond strength and flexural performance of the old and new cementitious materials [62]. Then, the sleeve connection is executed. The outstretched reinforcement bars of the prefabricated beams and columns are connected and grouted through the grouting process. Finally, the secondary pouring was conducted to securely attach the formwork for the "I"-shaped back-cast area, remove the bubble film at the keyway interface, and pour the HDRPC.

Figure 5. Construction sequence of HDRPC prefabricated beam–column joints.

3.2. Test Setup, Instrumentation, and Loading Protocol

Figure 6 illustrates the experimental setup. The setup consisted of a rigid steel frame anchored to the laboratory floor, which provided the reactive force during loading. The bottom of the column was supported by a 25 mm thick steel plate fixed to a hinge, while the top was constrained by two steel frames to limit lateral displacement. To ensure uniform load distribution and prevent localized deformation, the system incorporated a well-designed load-distribution mechanism. Additionally, hinge points were designed with low friction and zero play to minimize unnecessary deformation. Two MTS actuators, each with a capacity of 250 kN, were connected to the beam ends to apply cyclic loads, while a third actuator, with a capacity of 1000 kN, applied a constant vertical axial load of 386 kN to the top of the column. The system provided high-precision testing capabilities, facilitating a comprehensive evaluation of the mechanical properties of material in a controlled environment. The data-acquisition system ensured real-time synchronization of load, deformation, and force measurements, guaranteeing the accuracy and reliability of the experimental data. Vertical displacement at the beam ends was measured using Linear Variable Differential Transformers (LVDTs), and strain gauges were installed on the longitudinal reinforcement and concrete surface to monitor strain development. The constant axial load, equivalent to 30% of the axial load capacity of the column, was chosen to simulate severe loading conditions typically encountered during seismic events, thereby providing a basis for comparing joint behavior under varying stress levels. This axial load also simulated the effect of gravity load on the column support layer.

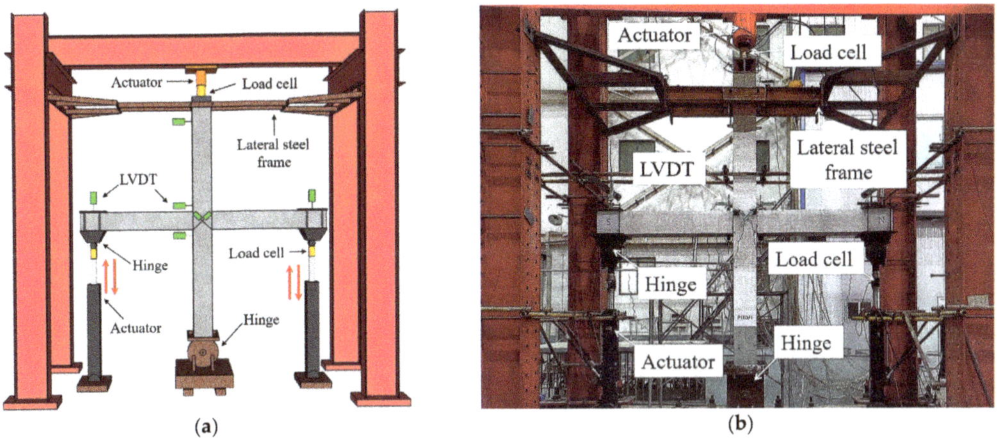

Figure 6. Test setup: (**a**) schematic diagram of loading device; (**b**) loading on-site layout plan.

As shown in Figure 7, strain gauges are placed on both the longitudinal and transverse reinforcement at key locations near the grouting sleeves to measure strain during loading. This configuration enables precise monitoring of the reinforcement response, providing valuable data on joint performance under stress. As depicted in Figure 8, an amplitude-increasing loading scheme is applied to each sample under displacement control. In this scheme, vertical load at the beam end is gradually applied: 4 mm step increments are used before joint yielding, cycled once, while 8 mm step increments are applied after joint yielding, cycled three times. The actuators for the two beams are synchronized: when one actuator pushes a beam upward, the other pulls the opposite beam downward. The test is terminated when the load on either or both actuators drops to 85% of the peak load [60].

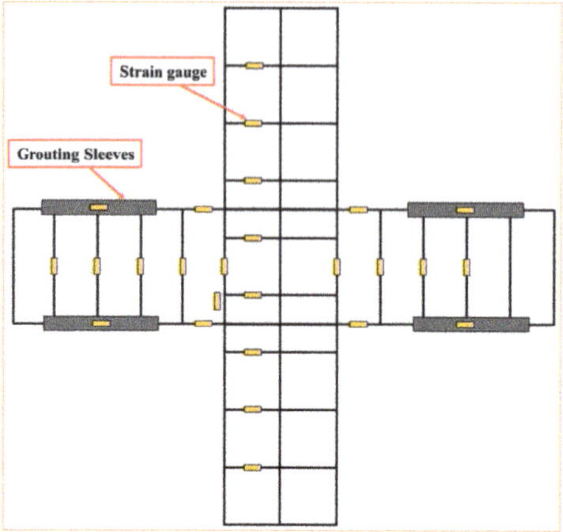

Figure 7. Arrangement of strain gauges.

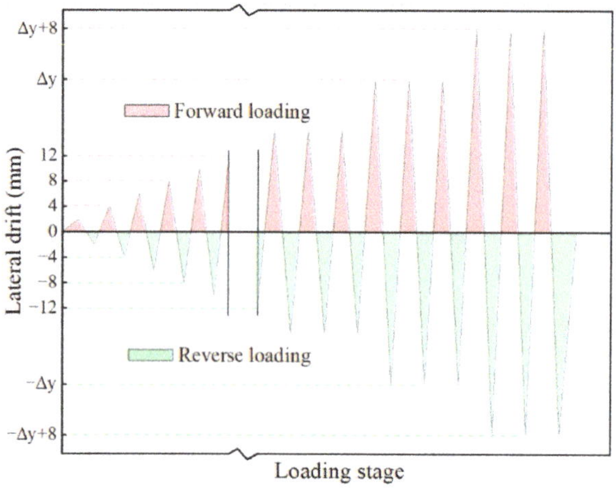

Figure 8. Loading protocol.

4. Main Test Results and Analysis

4.1. Test Phenomena and Failure Characteristics

Figure 9a shows the crack pattern of specimen PC30 made of normal concrete. The failure mode of PC30 is characterized by severe damage, with a displacement of 4 mm near the material contact, causing cracks at the column edge (600 mm from the beam root) and a rapidly propagating 50 mm diagonal core crack. As the displacement increases to 36 mm, transverse cracks appear at the column end, and diagonal cracks (0.2 mm wide) form at the core. Ultimately, the failure is attributed to shear failure, with concrete peeling off at the joints. This type of failure, while observed in the experiment, is typically undesirable in practical structural design. The intention behind designing specimen PC30 in this manner was to investigate the shear failure mechanism in prefabricated concrete joints, which could provide insights into improving joint design for real-world applications.

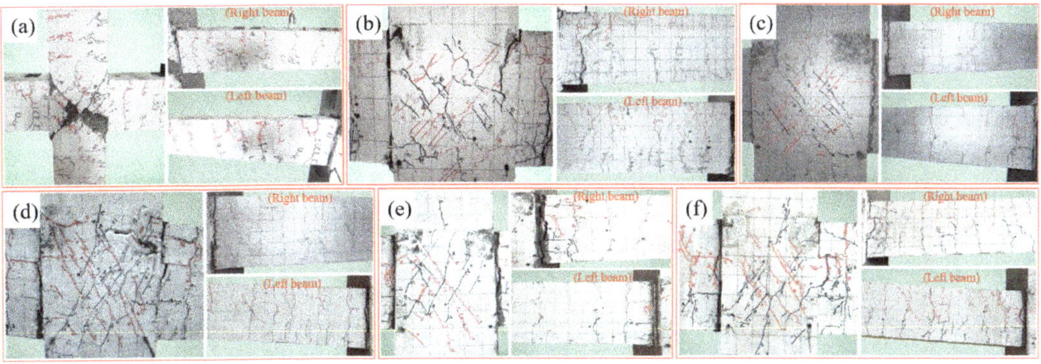

Figure 9. Component failure diagram under cyclic loading: (**a**) crack pattern of PC30 under cyclic loading; (**b**) crack pattern of HDJ under cyclic loading; (**c**) crack pattern of PHDJ1 under cyclic loading; (**d**) crack pattern of PHDJ2 under cyclic loading; (**e**) crack pattern of PHDJ3 under cyclic loading; (**f**) crack pattern of PHDJ4 under cyclic loading.

Figure 9b presents the crack pattern of specimen HDJ made of HDRPC. The crack pattern and hysteresis curve indicate that cracking occurs at all precast joints (PHDJ2-4) in the secondary pouring area, where the precast beams meet. Despite these cracks, the rough surface of the bubble film enhances interfacial adhesion and bending resistance, preventing significant damage. The maximum crack width observed is 0.5 mm, and vertical bending cracks are concentrated approximately 1000 mm from the root to the edge of the beam and column. This suggests that the steel bar joint grouting sleeve method is effective in ensuring sufficient bearing capacity at the connection between prefabricated beams and columns, thereby enhancing mechanical interlocking and delaying premature failure. The initial fracture in HDJ occurs away from the base or joint surface, making it more durable than PHDJ2.

As shown in Figure 9c, increasing the stirrup spacing in HDRPC at the core area from 100 mm to 200 mm results in cracking along the keyway of the I-beam in specimen PHDJ1, which lacks ring reinforcement. The cracks appear in the core region, extending from the upper and lower columns to the central area, with transverse cracks in the keyway reaching 0.5 mm. As displacement increases, concrete bulging and peeling occur. The crack at the bottom of the beam widens significantly to 20 mm under a 128 mm load. The failure mode of PHDJ1 changes from bending to shear failure, attributed to the absence of hoop reinforcement, which leads to plastic deformation at the column end under reciprocating loads. In contrast, specimen PHDJ2, which includes additional longitudinal steel bars, exhibits improved bending strength, thus preventing bending failure before shear failure.

Figure 9d shows that specimen PHDJ2, an HDRPC assembled joint with the same steel reinforcement and structure as PC30, initially cracks at the material interface between the left and right beams, 500 mm from the column edge, when the displacement amplitude reaches 4 mm. As the load increases, diagonal microcracks form in the core, and vertical cracks develop in the post-casting area. When the displacement amplitude reaches 32 mm, the load-displacement curve steepens, indicating joint yielding. At this stage, the crack width peaks at 0.13 mm in the core and 1 mm at the root of the beam. Vertical cracks are evident in the prefabricated beam area, and the base fracture becomes the main crack, widening as the beam end moves. The core crack grows to 0.5 mm, and surface cracks appear on the beam. At a displacement amplitude of 128 mm, the width of the main crack at the root of the beam reaches 17 mm, leading to the collapse of the material in the compression zone and subsequent joint failure. In comparison, specimen PC30 experienced core shear failure accompanied by concrete detachment, while PHDJ2, although exhibiting minor cracks (with a width of 450 μm), showed no concrete peeling due to the properties of HDRPC.

Finally, comparing Figure 9d–f, the failure modes of three prefabricated joints (PHDJ2-4) with identical steel reinforcement and materials are discussed. As axial compressive force increases, vertical compressive stress in the core area and the angle of diagonal cracks increase. The ultimate load induces the first fracture near the beam foundation at the connection. Subsequent load variations cause the main beam to rotate around both ends, with the grouting sleeve reinforcing the load-bearing capacity at the beam ends.

4.2. Hysteresis Curves

Figure 10 shows the relationship between the applied load on six specimens and the vertical displacement measured from LVDT deployed at the end of the beam. When two actuators load the beam upward and downward relative to the ground, the positive load and negative load (and displacements) correspond to the upward and downward directions. For each specimen, overall, the load-displacement curves of the two beams show the hysteresis curve and the magnitude of the load and deformation.

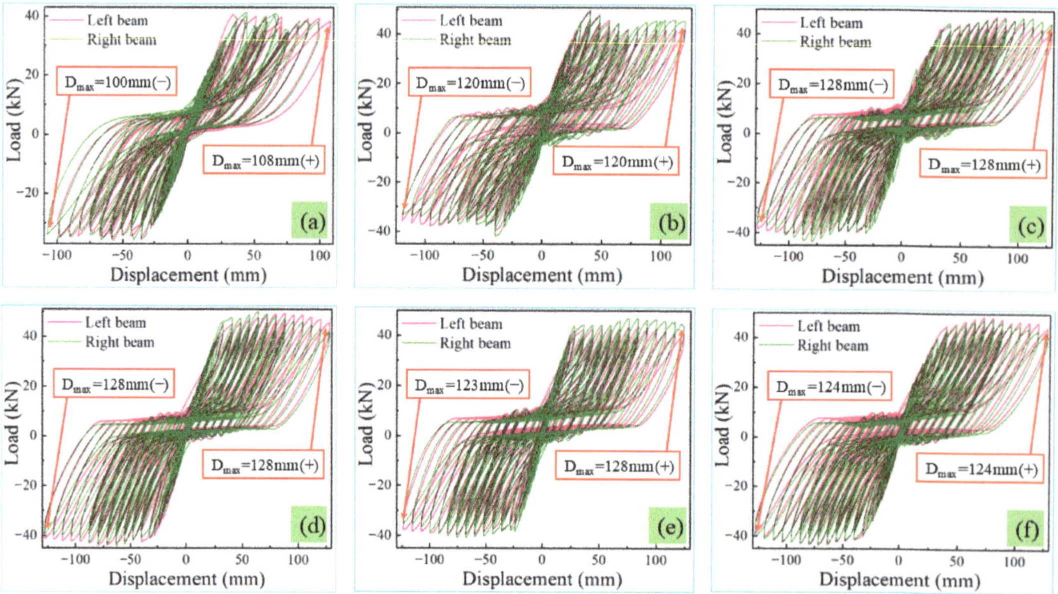

Figure 10. Hysteretic curves of different components: (**a**) P30; (**b**) HDJ; (**c**) PHDJ1; (**d**) PHDJ2; (**e**) PHDJ3; and (**f**) PHDJ4.

Figure 10a shows the load-displacement results of the beams of specimen PC30, respectively. The results indicate that when the displacement amplitude is less than 25 mm, the load during the loading and unloading stages increases approximately linearly with the displacement. When the displacement exceeds 25 mm, the specimen exhibits hysteresis during the loading–unloading phase, which may be due to internal damage such as concrete cracking and steel yield. After the concrete cracks, the magnitude of the load will still increase due to the action of the steel bars. However, the concrete has poor damage resistance and cracks develop rapidly in the core area. Overall, the hysteresis loop of the PC30 joint shows the most obvious shrinkage. In the later stage of loading, plastic hinges are formed at the ends of each joint beam, and the main crack at the root of the beam alternately opens, closes, and continues to widen under reciprocating loads. The steel bar slip phenomenon is severe, and the hysteresis curve is Z-shaped.

Figure 10b shows the load-displacement results of the beams of specimen HDJ, respectively. Compared with specimen PC30, specimen HDJ has higher bearing capacity, deformation capacity, and hysteresis performance. Before concrete cracking, the load increases approximately linearly with displacement, similar to the results of PC30. After a slight crack appeared at the root of the beam near the column, hysteresis behavior was observed due to damage in the specimen. As the displacement increases, cracks appear at beams and joints far away from the column. However, the size of the load still increases. After the load reaches its peak, it begins to decrease, and when the displacement amplitude reaches 75 mm, the load increases twice. This phenomenon is attributed to the unique tensile and crack resistance of HDRPC. After the cracking of HDRPC, due to the dispersion of mixed fibers in HDRPC, the tensile strain of HDRPC is higher, and it has better-coordinated deformation ability with steel bars. Its excellent crack resistance often hinders the development of damage in HDRPC, thereby reducing the squeezing effect.

Figure 10d shows the load-displacement results of the beams of specimen PHDJ2, respectively. Similar to the HDJ specimen, it exhibits hysteresis behavior and post-cracking resistance. However, overall, the cast-in-place HDJ joint has less shrinkage due to its superior overall integrity compared to prefabricated joints, and the development of crack width at the root of the beam is slower. When fine cracks appear at the root of the side beam of the column, PHDJ2 can still maintain the bearing capacity of the joint and has limited hoop effect, indicating that the design configuration of the prefabricated column and beam enables the assembly joint to achieve reasonable cyclic performance and achieve the same purpose as cast-in-place.

Figure 10c–f show the load-displacement results of the beams of specimens PHDJ1-4, respectively. Under the condition of reducing the reinforcement ratio of stirrups in the core area of the joint, the results of specimen PHDJ2 a high degree of overlap with specimen PHDJ1, indicating that the spacing between stirrups has little effect on the bearing capacity and deformation capacity of the joint. HDRPC itself has good shear performance and can replace some stirrups, which is conducive to simplifying the arrangement of steel bars in the joint area.

4.3. *Skeleton Curves and Drift Ratio*

The skeleton curve shows the stiffness of component and ductility by correlating force and deformation under alternating load. Individual joint skeletal curves are shown in Figure 10, which shows the joints went through elasticity, yielding, stability, and disintegration. The beam end load and displacement are linear in the elastic stage, and each specimen has a similar initial stiffness. As displacement increases, joint cracks fully develop and internal material damage accumulates, reducing specimen stiffness until yield. HDRPC assembled joint beam end material has a higher ultimate compressive strain in the middle and late stages of loading and benefits from grouting sleeve reinforcement. This keeps bearing capacity stable and improves, lengthening the horizontal segment of the skeleton curve. The bearing capacity of the specimen decreases as the joint beam end material collapses.

The drift ratio is a critical parameter for evaluating the seismic performance of structures, representing the lateral displacement relative to the height of an element. It reflects the deformation capacity of a structure under cyclic loading, indicating its ability to absorb energy and resist damage. Analyzing the drift ratio provides insights into the load-carrying capacity, stability, and overall seismic resilience of various materials and connection types, such as normal concrete and HDRPC, under seismic conditions.

The drift ratio is calculated using the following formula:

$$\text{Drift ratio} = \frac{\Delta}{h} \times 100\% \tag{1}$$

where Δ is the lateral displacement of the top of the beam, measured during cyclic loading, h is the height of the beam. The drift ratio is typically expressed as a percentage, and it helps in assessing the deformation behavior and seismic performance of structural elements, particularly their ability to resist lateral forces without collapsing or undergoing excessive damage.

Figure 11a shows that the elastic phase, HDRPC has a lower elastic modulus than normal concrete, so PC30 joints are stiffer than PHDJ2. The specimen PC30 joints have much lower bearing capacity and stiffness than PHDJ2 during later loading. This is because reciprocating loading weakens concrete, causing cracks and rapid mechanical degradation, while HDRPC has greater crack and energy-dissipation properties than normal concrete.

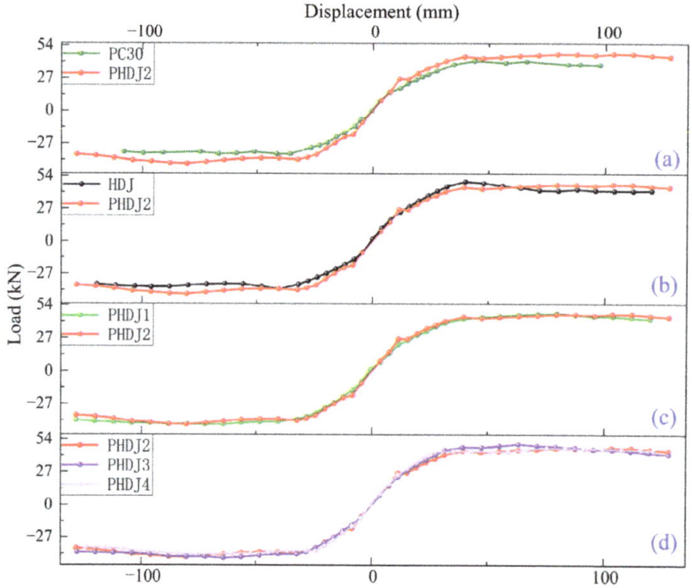

Figure 11. Comparison of skeleton curves and drift ratio: (**a**) different materials; (**b**) different fabrications; (**c**) different hoop ratios; (**d**) different axial compression ratios.

Figure 11b illustrates that the initial stiffness of the HDRPC assembled joint and cast-in-place joint are similar. However, after reaching the peak load, the bearing capacity of the HDJ specimen gradually decreases as the displacement increases. The grouting sleeve at the girder end of the PHDJ2 specimen greatly improves the sectional bending-bearing capacity, resulting in a slight improvement in its bearing capacity initially, followed by a gradual decrease with increasing girder-end displacement. The grouted sleeves with higher stiffness and strength inhibit the propagating and widening of the cracks [35]. Therefore, the PHDJ2 specimen exhibits better ductility compared to the cast-in-place joint.

As shown in Figure 11c, the skeleton curves of specimens PHDJ1 and PHDJ2 exhibit significant overlap. This indicates that the distance between the hoops has a limited influence on the load-bearing and deformation capacity of the joint. Additionally, compared to PC30, PHDJ1 can still maintain a high bearing capacity and load-carrying level while reducing the number of stirrups, HDRPC can replace some of the hoops, simplifying the reinforcing arrangement in the joint area. This approach addresses the issue of dense and complex reinforcing tying during construction.

Figure 11d indicates that the peak bearing capacity of the joint initially rises and subsequently drops as the axial compression ratio increases. This is because an appropriate axial compression force can widen the compression zone area of the column section, enhance the effectiveness of the inclined compression bar, and improve the shear strength of HDRPC with the increase in compressive stress. However, when the axial compression force becomes severe, the material in the compression zone of the joint will crack, resulting in a drop in shear strength.

The drift ratio analysis of the experimental specimens highlights the significant improvements in seismic performance achieved by HDRPC compared to normal concrete (PC30). The PC30 specimen exhibits a relatively lower deformation capacity, as it demonstrates substantial displacement at smaller drift ratios. In contrast, the HDJ specimen, made of cast-in-place HDRPC, shows a higher load-carrying capacity and offers greater resistance to deformation under cyclic loading. The prefabricated HDRPC specimens (PHDJ1, PHDJ2, PHDJ3, PHDJ4) further outperform both PC30 and HDJ, particularly PHDJ3 and PHDJ4, which exhibit the highest load capacity and maintain consistent performance at higher drift ratios.

International design codes, such as ACI 318 [63] and CEB-FIB [64], typically assume uniform material properties and do not fully account for the enhanced performance of high-ductility materials like HDRPC. For instance, ACI 318 limits the drift ratio to 2% for seismic design, which may be overly conservative for HDRPC joints exhibiting higher ductility. Similarly, the CEB-FIB Model Code lacks specific guidelines for fiber-reinforced concrete joints. These limitations underscore the need for updated design codes that incorporate the unique properties of advanced materials like HDRPC.

4.4. Load Capacity and Ductility

In seismic design, ductility refers to the capacity of structures and members to withstand deformation without a significant decrease in bearing capacity after entering the plastic phase. The ductility coefficient is commonly used to measure the quality of ductility and is calculated using the following formula:

$$\mu = \frac{\Delta_\mu}{\Delta_y} \tag{2}$$

where Δ_u is the ultimate displacement of the beam end of the joint. This refers to the vertical displacement that corresponds to the reduction of the load capacity of the member to 85% of its maximum load capacity. Δ_y, on the other hand, represents the vertical displacement of the beam end of the joint when it reaches its yield point. The yield load was calculated using the farthest point method, where the yield point is defined as the intersection of the initial elastic slope and a line parallel to the post-yield slope passing through the ultimate load point [65]. The displacement ductility coefficients for each joint are listed in Table 5.

As shown in Table 7, the experimental results demonstrate that specimens incorporating HDRPC significantly outperform the PC30 specimen, made of normal concrete, particularly in terms of ultimate load capacity. The influence of concrete strength on ultimate load was investigated by comparing the performance of HDRPC with that of normal concrete. The results reveal that HDRPC increases the ultimate load by 17.8%, underscoring the crucial role of concrete strength in joint performance. Specifically, the HDJ specimen, which uses cast-in-place HDRPC, exhibits a peak load of 49.4 kN in the forward direction and an ultimate load of 41.9 kN, significantly surpassing the PC30 specimen, which has an ultimate load of 38.4 kN in the same direction. The prefabricated HDRPC specimens (PHDJ1, PHDJ2, PHDJ3, PHDJ4) consistently outperform both the PC30 and HDJ specimens. Notably, PHDJ3 achieves the highest peak load of 50.1 kN in the forward direction and an ultimate load of 42.6 kN, indicating superior seismic performance. The PHDJ4 specimen also demonstrates the highest mean value, showing excellent consistency under reverse loading. These findings confirm that HDRPC, whether cast-in-place or prefabricated, enhances the seismic performance and load-carrying capacity of concrete joints, with concrete strength playing a significant role in improving joint performance.

Table 7. Displacement ductility coefficient.

Specimens	Load Direction	Yield Displacement (mm)	Yield Load (kN)	Peak Load (kN)	Ultimate Load (kN)	Ultimate Displacement (mm)	Mean
PC30	forward	36.0	39.2	40.0	38.4	108	3.0
	backward	28.7	32.3	35.9	34.8	100	3.5
HDJ	forward	30.3	43.9	49.4	41.9	120	4.0
	backward	30.8	35.0	39.5	35.7	120	3.9
PHDJ1	forward	31.9	39.8	47.5	41.7	128	4.0
	backward	31.9	39.5	43.7	40.0	128	4.0
PHDJ2	forward	32.0	41.3	47.1	44.7	128	4.0
	backward	28.0	38.5	43.6	37.0	128	4.3
PHDJ3	forward	31.9	45.3	50.1	42.6	123	3.9
	backward	28.1	38.9	44.3	39.3	128	4.5
PHDJ4	forward	28.0	43.9	47.0	44.8	124	4.3
	backward	24.0	38.3	41.0	35.3	124	5.1

Based on the provided table, it is evident that the displacement ductility coefficients of HDRPC joints are considerably greater than those of assembled concrete joints. This suggests that incorporating HDRPC materials in the core is beneficial for enhancing the joint damage pattern, increasing the joint load-carrying capacity, and improving the joint ductility. The ductility of HDRPC assembled joints is typically greater than that of the HDJ cast-in-place joints, indicating that the grouted sleeve connection in its assembled form can enhance the ductility of the girder cross-section. When comparing PHDJ2 with PHDJ1, it was found that the ductility of PHDJ1 joints with unencrypted hoop reinforcement was slightly decreased. However, it still reached a value of 4.0, which suggests that the excellent shear ductility of the material can serve a similar purpose as the hoop reinforcement in terms of shear and restraint.

4.5. Stiffness Degradation

The stiffness and bearing capacity of the joint gradually decrease under the reciprocating load. To measure the change in stiffness of the specimen at different loading displacements, the cut-line stiffness K_i [64] is commonly utilized. The stiffness degradation curve of the joint is determined for the right beam, as shown in Figure 11.

$$K_i = \frac{|+F_i| + |-F_i|}{|+X_i| + |-X_i|} \tag{3}$$

where K_i represents the secant stiffness of the ith loading level; $+F_i$ and $-F_i$ represent the positive and negative peak loads corresponding to the ith loading level; $+X_i$ and $-X_i$ represent the displacement toward the positive and negative peak loads corresponding to the ith loading level.

As shown in Figure 12, before reaching the yield state, the overall stiffness trend of the six specimens rapidly decreases with increasing displacement. After yielding, the rate of stiffness degradation slows down, and the curve tends to flatten as it approaches the limit state.

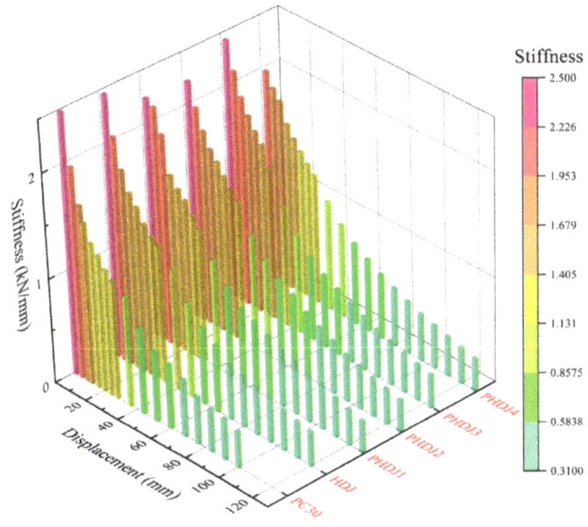

Figure 12. Results of stiffness degradation.

Compared with PHDJ2 precast joints, normal concrete precast joints PC30 have higher initial stiffness but a faster degradation rate. The main reason is that HDRPC does not contain coarse aggregates and has a lower elastic modulus than normal concrete. However, as displacement increases, concrete cracks under tensile action, rapidly develop and accumulate plastic damage. The stiffness of normal concrete joints rapidly decreases, while the tensile strength and ultimate tensile strain of HDRPC are much higher than those of concrete. Therefore, the rate of stiffness reduction is slower, resulting in a higher final stiffness.

The HDJ specimen is a cast-in-place integral joint; however, there is a weak interface caused by secondary pouring in the assembly joint specimen PHDJ2. At the initial stage of loading, there will be varying degrees of cracking at the connection between the post-cast area of specimen PHDJ2 and the prefabricated beam. Therefore, HDJ joints have higher initial stiffness. As the displacement increases, cracks in both specimen HDJ and specimen PHDJ2 fully develop, and the stiffness degradation curves of the two joints become similar. In the later stage of loading, both specimen beam ends showed significant plasticity. However, due to the larger cross-sectional area of the grouting sleeve, it can serve as a form of local reinforcement, and the ultimate stiffness of specimen PHDJ2 is slightly higher than that of the HDJ specimen.

When comparing specimens with different axial compression ratios (PHDJ2-4), it was observed that the stiffness degradation curves of PHDJ2 and PHDJ3 with lower axial compression ratios were similar to each other. However, PHDJ4 with a high axial compression ratio exhibits significantly higher initial stiffness. At the limit state, the final stiffness of all three joints is similar, indicating that the degradation rate of the joints is faster under high axial compression ratios. PHDJ1 has the lowest initial stiffness, indicating that the constraint force at the beam end is weakened and the interface is more prone to cracking when the spacing between the circumferential steel bars is wide.

4.6. Energy-Consumption Capacity

The structure will fracture and deform under seismic forces, thereby dissipating and absorbing seismic energy. Hysteretic energy dissipation is advantageous for mitigating the displacement response of a structure during seismic activity and enhancing its seismic performance. The energy-consumption capacity of each specimen was calculated by integrating the area under the load-displacement hysteresis loops. This represents the cumulative energy dissipated during cyclic loading, which is a key indicator of seismic performance. The hysteresis energy-dissipation curve is constructed by plotting the energy dissipation of the first cycle at each loading stage, as depicted in Figure 13.

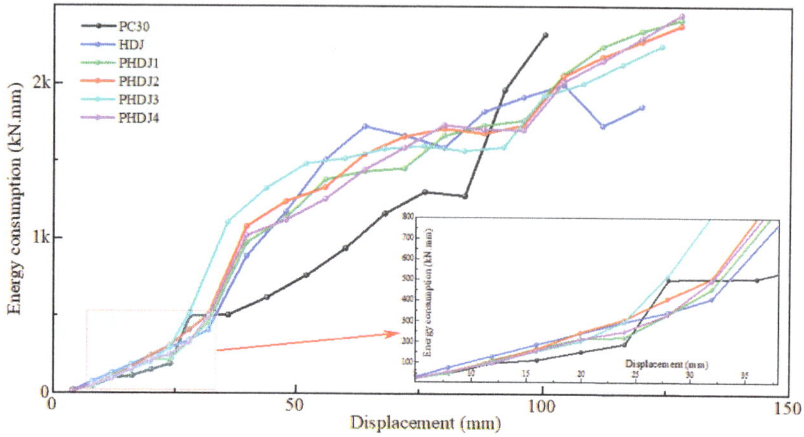

Figure 13. Energy-dissipation capacity of specimens.

During the initial loading phase, each specimen remains primarily in the elastic stage, with minimal and gradual energy dissipation, which remains relatively consistent across all specimens. However, once the yield displacement is reached, hysteresis energy dissipation at the joints increases rapidly as the load increases. Concrete exhibits limited resistance to damage, and as internal cracks develop, both its strength and stiffness decrease significantly. Consequently, the energy-dissipation capacity of PC30 joints is much lower than that of HDRPC joints after reaching the yield point. As shown in Figure 13, the energy consumption of the PC30 specimen increases sharply beyond 80 mm of displacement. This substantial rise in energy dissipation is attributed to the brittle failure behavior of normal concrete. Once cracks form, the stiffness of material rapidly decreases, leading to accelerated energy consumption. In contrast, HDRPC demonstrates superior crack resistance and ductility, resulting in more stable energy dissipation at larger displacements. During the later stages of loading, the beam end undergoes plastic deformation and forms a plastic hinge, which becomes the primary region for energy dissipation. The energy-dissipation effect of the grouting sleeve at the beam end of the assembly joint is evident, resulting in a higher level of energy dissipation compared to the cast-in-place joint. The energy-consumption patterns of PHDJ1 and PHDJ4 are similar, suggesting that increasing the hoop spacing does not reduce the energy-dissipation level of the HDRPC assembly joint.

4.7. Core Area Shear Deformation

Under reciprocating load, the core area of the joint experiences shear deformation, resulting in tensile stress along one diagonal direction and compressive stress along the other diagonal direction. Figure 13 illustrates how the core area tends to shift from a rectangle to a rhombus, with the direction of the rhombus alternating with each application of the beam end load. The extent of shear deformation in the core area can be evaluated by the shear angle. Equations (4) and (5) provide formulas for calculating the shear angle.

$$\overline{X} = \frac{|\delta_1 + \delta_1'| + |\delta_2 + \delta_2'|}{2} \tag{4}$$

$$\gamma = \frac{\sqrt{a^2 + b^2}}{a \cdot b} \cdot \overline{X} \quad (5)$$

where $\overline{X}$ is the average change in the diagonal direction of the core area, and γ represents the shear angle of the core area. The remaining variables are detailed in Figure 14, and the calculation results are presented in Figure 15.

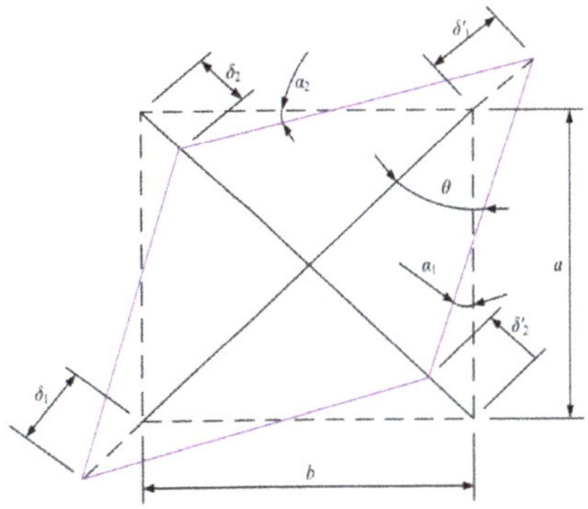

Figure 14. Shear deformation in the joint core.

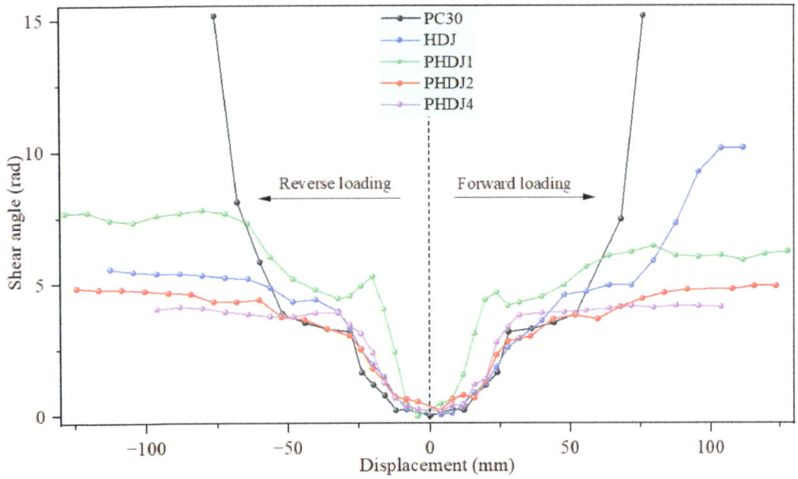

Figure 15. Shear angle of the joint core.

Figure 15 illustrates that during the initial loading phase, the shear deformation in the core region of each specimen is minimal, and the curves closely align with each other. However, as the displacement at the beam end increases, noticeable concrete cracking occurs in the core area of the PC30 joint. The growth rate and peak value of the shear angle in this joint are significantly higher compared to the other joints. The shear angle of PHDJ1, PHDJ2, and HDJ increases gradually after complete cracking in the core, reaching a maximum value that is only one-third of that observed in the PC30 joint. Because the addition of PVA fibers and basalt fibers in HDRPC helps establish bridging connections and promote stress transfer, replacing HDRPC at the joint exhibits enhanced shear ductility and maintains steady-state cracking characteristics [31]. The experimental strain readings were compared with analytical values calculated using the sectional equilibrium method. The results showed good agreement, with a maximum deviation of 5% in the joint core area. The shear angle of PHDJ1 in the core is approximately 1.5 times more than that of

PHDJ4, suggesting that the spacing between the hoops has a stronger impact on the shear deformation of the joint. However, it is still less than half of the shear angle observed in the PC30 joint, indicating that PHDJ1 performs better than the completed concrete joint.

4.8. Load Capacity of Crack Resistant

Certain reinforced concrete frame structures have special requirements for durability under normal conditions of use and need to avoid cracking. Su et al. [66] have been given the derivation process and general form of the formula for calculating the cracking capacity of joints:

$$V_{j,cr} = f_{tk} b_j h_j \sqrt{1 + \frac{N_c}{f_{tk} b_c h_c}} \tag{6}$$

where $V_{j,cr}$ is the joint shear force when the core area cracks first; b_j is the effective width of the joint core area section; h_j is the height of the joint core area section; N_c is the axial force transmitted by the column; b_c is the column section width; h_c is the column section height. In the case of HDRPC, f_{tk} is the initial cracking stress of the dumbbell-shaped pull plate in the uniaxial tensile test, with an actual test measurement of $f_{tk} = 3.5$ MPa. The tensile strength design value for concrete is approximately considered.

Practically, the bond stress between the longitudinal reinforcement of the beam and the cementitious material in the core area of the joint is unevenly distributed [60]. As a result, the horizontal shear stress transferred to the core area of the joint is also uneven. To account for this, a comprehensive influence coefficient η needs to be introduced for correction, based on theoretical formulas [35]. When considering HDRPC joints, it is important to take into account the positive impact of incorporating PVA fibers and basalt fibers in the material, which improves fracture resistance. Conversely, the negative impact of using recycled powder should also be addressed.

The initial shear force test values of joints are generally lower than the calculated values. According to the calculation results in the table, the comprehensive influence coefficient η = 0.75 can be conservatively taken in the design application, The test value of shear force at the initial cracking of the joint is calculated by the following Equation (7) [67]:

$$V_j = \frac{\sum M_b}{h_0 - a_s'} \left(1 - \frac{h_0 - a_s'}{H_c - h_b}\right) \tag{7}$$

where $\sum M_b$ is the total of the bending moments at the ends of the beam when the joint experiences cracking; h_0 is the effective height of the beam section; a_s' is the distance from the point where the longitudinal reinforcement of the beam intersects to the edge of the section; H_c is the overall height of the column; and h_b is the height of the beam section.

Utilize Equations (6) and (7) to compute the estimated value and test value, respectively, for the fracture-resistance carrying capacity of the core area of each joint. The outcomes are presented in Table 8.

According to the calculation results in Table 6, it can be seen that the tensile strength of specimen PC30 normal concrete is poor, and its initial crack load is significantly lower than other specimens. The initial shear force of specimen PHDJ2-4 increases with the increase in of axial compression ratio. This is because the increase of axial compression increases the compression zone area of the column, delaying the occurrence of diagonal tensile cracks. When the axial compression ratio is the same, the initial cracking loads of PHDJ1, PHDJ2, and HDJ are close, indicating that the prefabricated joints connected by grouting sleeves can achieve the same crack resistance as cast-in-place joints, and the lack of reinforcement has no adverse effect on the initial cracking of the joints. The initial shear force test values of joints are generally lower than the calculated values. According to the calculation results in Table 6, the comprehensive influence coefficient η = 0.75 can be conservatively taken in the design and application. It is recommended that the formula for the crack-resistance-bearing capacity of the HDRPC joint oblique section be:

$$V_{j,cr} = 0.75 f_{tk} b_j h_j \sqrt{1 + \frac{N_c}{f_{tk} b_c h_c}} \quad (8)$$

Table 8. Load capacity of crack resistance (unit: kN).

Specimens	Load Direction	Incipient Crack Load, P_{cr}	Initial Cracking Shear Test Value $V_{j,exp}$	The Calculated Value of Initial Cracking Shear $V_{j,cal}$	$V_{j,exp}/V_{j,cal}$
PC30	forward	13.9	226.4	257.4	0.88
	backward	16.8	273.6		1.06
HDJ	forward	23.3	379.5	469.9	0.81
	backward	24.3	395.8		0.84
PHDJ1	forward	22.2	364.1	469.9	0.77
	backward	25.8	423.2		0.90
PHDJ2	forward	23.7	388.9	469.9	0.83
	backward	24.5	401.9		0.86
PHDJ3	forward	28.2	462.9	549.3	0.84
	backward	31.5	516.7		0.94
PHDJ4	forward	30.2	495.5	618.9	0.80
	backward	29.3	480.6		0.78

4.9. Shear-Bearing Capacity

Hwang and Lee [68] proposed a Softened Strut-and-Tie (SST) model to calculate the shear strength of external beam–column joints. This model has a clear transmission path and mechanism, dividing the force-bearing mechanism of the joint area into three parts: oblique, vertical, and horizontal as shown in Figure 16. The calculation formula for the shear-bearing capacity of the joint is:

$$V_j = K \zeta f_c' A_{str} \cos\theta \quad (9)$$

where K is the coefficient of the softened tension rod compression rod model; ζ is the softening coefficient of compressive strength of concrete; f_c' is the compressive strength of cylindrical concrete, taken as $f_c' = 0.79 f_{cu}$; A_{str} is the effective area of the diagonal strut; θ is the angle between the diagonal strut and the horizontal direction.

$$A_{str} = a_s \times b_s \quad (10)$$

where a_s is the effective cross-sectional height of the diagonal strut, b_s is the effective cross-sectional width of the diagonal compression bar, which can be taken as the effective width of the joint b_j.

$$a_s = \sqrt{a_b^2 + a_c^2} \quad (11)$$

where a_b is the height of the compression zone of the beam section, a_c is the height of the compression zone of the column section.

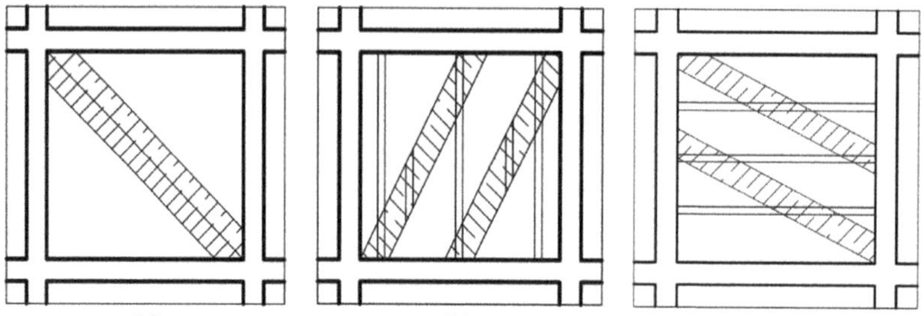
(a) (b) (c)

Figure 16. The theory of SST model: (**a**) diagonal; (**b**) vertical; (**c**) horizontal.

In this experiment, under the action of reciprocating loads, plastic hinges were formed at the ends of the beams on both sides of the core area, and the root of the beam cracked under tension. The neutral axis of the section moved upward, and the relative height of the compression zone decreased. a_b can be approximately taken as one-fifth of the height of the beam section [69]; a_c is determined based on axial pressure, and the calculation formula is:

$$a_c = \left(0.25 + 0.85\frac{N}{b_c h_c f_c'}\right)h_c \tag{12}$$

When appropriate axial pressure is applied to the top of the column, the effective cross-sectional height of the diagonal brace will increase, which is beneficial for the shear resistance of the joint. However, excessive axial pressure will exacerbate the damage to the concrete in the core area, which is unfavorable for shear resistance.

$$\theta = \tan^{-1}\left(\frac{h_b''}{h_c - 2a_c/3}\right) \tag{13}$$

where h_b'' is the distance between the centerline of the outermost longitudinal reinforcement of the beam section.

$$\zeta = 3.35/\sqrt{f_c'} \tag{14}$$

$$K = K_h + K_v - 1 \tag{15}$$

$$K_h = 1 + \frac{(\overline{K_h} - 1)A_{th}f_{yh}}{\overline{F_h}} \le \overline{K_h} \tag{16}$$

$$K_v = 1 + \frac{(\overline{K_v} - 1)A_{tv}f_{yv}}{\overline{F_v}} \le \overline{K_v} \tag{17}$$

$$\overline{K_h} = \frac{1}{1 - 0.2(\gamma_h - \gamma_h^2)} \tag{18}$$

$$\overline{K_v} = \frac{1}{1 - 0.2(\gamma_v - \gamma_v^2)} \tag{19}$$

where A_{th} is the total cross-sectional area of the longitudinal reinforcement in the column; A_{tv} is the total cross-sectional area of the hoop reinforcement along the horizontal loading direction within the core area of the column; f_{yh} is the yield strength of the longitudinal reinforcement of the column; f_{yv} is the yield strength of the hoop reinforcement in the core area of the column; γ_h represents the proportion of horizontal tie rod shear force to the total shear force of the joint; γ_v is the proportion of vertical tension rod shear force to the total shear force of the joint. $\overline{F_h}$ is the balance tension of the horizontal pull rod; $\overline{F_v}$ is the balance tension of the vertical pull rod.

$$\gamma_h = \frac{2tan\theta - 1}{3} \tag{20}$$

$$\gamma_v = \frac{2cot\theta - 1}{3} \tag{21}$$

$$\overline{F_h} = \gamma_h \overline{K_h} \zeta f_c' a_s b_j cos\theta \tag{22}$$

$$\overline{F_v} = \gamma_v \overline{K_V} \zeta f_c' a_s b_j sin\theta \tag{23}$$

The standard GB 50011-2010 [61] is founded on the truss shear mechanism and accounts for the transfer of shear force to the frame joints through the concrete diagonal compression bars. This includes the axial force of the concrete and columns between the diagonal cracks in the core area. Additionally, the horizontal hoops exert a restraining influence on the diagonal compression bars in the joint region. The calculation of this effect is determined by the following formula:

$$V_j = \frac{1}{\gamma_{RE}}\left(1.1\eta_j f_t b_j h_j + 0.05\eta_j N \frac{b_j}{b_c} + f_{yv} A_{svj} \frac{h_{b0} - a_s'}{s}\right) \tag{24}$$

where γ_{RE} is the seismic adjustment coefficient of bearing capacity and is assigned a value of 0.85. η_j is the influence coefficient of orthogonal beam constraints, assigned a value of 1.0. N is the axial pressure of the column. A_{svj} is the total cross-sectional area of the hoop reinforcement within the valid check width of the core area, in the same direction and with the same cross-section.

The test value of nodal shear is determined by measuring the beam end load at the core during cracking and then substituted into Equation (7). The theoretical value of shear capacity is obtained using Equation (24). The resulting values are presented in Table 9. The calculation results indicate that the test value of nodal capacity, obtained by applying the design formulas specified, closely aligns with the calculated value. The ratio between the two has a mean value of 1.03, with a standard deviation and coefficient of variation of 0.05.

Table 9. Comparison of test and calculated values (unit: kN).

Specimen	Load Direction	Left Beam Load	Right Beam Load	Shear Test Value V_{exp}	Normative Formula Calculated Value V_{cal}	V_{exp}/V_{cal}
PC30	forward	30.8	30.4	450.5	469.9	0.96
	backward	30.2	31.8	456.4		0.97
HDJ	forward	39.5	43.3	609.6	586.4	1.04
	backward	38.3	45.5	616.9		1.05
PHDJ1	forward	39.9	39.8	586.7	537.5	1.09
	backward	40.2	39.6	587.5		1.09
PHDJ2	forward	39.8	44.8	622.8	586.4	1.06
	backward	39.2	43.9	611.8		1.04
PHDJ3	forward	42.8	47.7	666.3	601.6	1.11
	backward	44.9	40.9	631.6		1.05
PHDJ4	forward	37.8	44.5	605.9	616.7	0.98
	backward	42.6	38.3	595.6		0.97

5. Conclusions

This study investigated the cyclic behavior of prefabricated beam–column joints using high-ductility recycled powder concrete (HDRPC) under low-cycle reversed loading. The experimental results and theoretical analyses provide valuable insights into the seismic performance of HDRPC joints, with the following key conclusions:

(1) HDRPC significantly improves the seismic performance of prefabricated joints compared to normal concrete. The bearing capacity of HDRPC joints increased by 17.8%, and the displacement ductility improved by 33.3%. This is attributed to the superior crack resistance and energy-dissipation properties of HDRPC, which mitigate damage in the joint core area and delay failure. The use of HDRPC in the joint core area resulted in multi-crack steady-state cracking, reducing crack widths and shear deformation. This behavior is critical for achieving robust joint designs in earthquake-prone regions.

(2) The grouting sleeve connection method proved highly effective in ensuring reliable force transfer between prefabricated beams and columns. It enhanced the bending capacity of the joint cross-section, allowing for better rotation of the plastic hinge zone at the beam end. The skeleton curves of HDRPC joints exhibited longer horizontal segments and gradual degradation, indicating improved ductility and energy dissipation compared to cast-in-place joints. This demonstrates that grouting sleeves are a superior connection method for prefabricated structures.

(3) Increasing the axial compression ratio improved the cracking load of HDRPC joints but accelerated stiffness degradation. The shear capacity initially increased with higher axial compression but decreased under excessive axial pressure due to concrete damage in the compression zone. Despite these variations, HDRPC joints maintained excellent ductility across the tested range of axial compression ratios (0.3 to 0.7), making them suitable for a wide range of structural applications.

(4) HDRPC joints with reduced hoop reinforcement (e.g., PHDJ1 with 200 mm spacing) exhibited superior performance compared to normal concrete joints with densified stirrups. This suggests that HDRPC can partially replace hoop reinforcement, simplifying steel reinforcement arrangements and addressing construction challenges related to dense reinforcement tying.

(5) The proposed crack-resistance-bearing capacity formula is closely aligned with experimental results, providing a reliable design tool for HDRPC joints. The Softened Strut-and-Tie Model (SST) and normative formula (GB 50011-2010) accurately predicted the shear capacity of HDRPC joints, with a mean ratio of test-to-calculated values of 1.03 and a standard deviation of 0.05.

Author Contributions: Conceptualization, X.L.; Methodology, X.L. and H.S.; Software, X.L. and K.C.; Formal analysis, X.L.; Investigation, X.L., H.S., K.C., T.Y., L.W., X.Z. and T.B.; Resources, X.L.; Data curation, X.L., H.S. and K.C.; Writing—original draft, H.S. and K.C.; Writing—review and editing, X.L., T.Y., L.W., X.Z. and T.B.; Supervision, X.L.; Project administration, X.L.; Funding acquisition, X.L. All authors have read and agreed to the published version of the manuscript.

Funding: This research was funded by the National Natural Science Foundation of China [grant No. 52278507]; the Natural Science Foundation of Shandong Province [grant No. ZR2020ME245]; and the Major Scientific & Technological Innovation Project of Shandong Province [grant No. 2021CXGC011204).

Data Availability Statement: The original contributions presented in the study are included in the article, further inquiries can be directed to the corresponding author.

Conflicts of Interest: The authors declare that we have no known competing financial interests or personal relationships that could have appeared to influence the work reported in this paper.

References

1. Hassan, Y.; Shankar, V.; Usman, S.; Alhassan, A. Carbon footprint management: A review of construction industry. *Clean. Eng. Technol.* **2022**, *9*, 100531.
2. Zhao, Y.; Wang, T.; Yi, W. Emergy-accounting-based comparison of carbon emissions of solid waste recycled concrete. *Constr. Build. Mater.* **2023**, *387*, 131674. [CrossRef]
3. Azarbayjani, M.; Thaddeus, D. *Environmental Dimensions of Climate Change: Endurance and Change in Material Culture*; Springer International Publishing: Berlin/Heidelberg, Germany, 2022; pp. 293–371.
4. Xiao, J.; Ma, Z.; Sui, T.; Akbarnezhad, A.; Duan, Z. Mechanical properties of concrete mixed with recycled powder produced from construction and demolition waste. *J. Clean. Prod.* **2018**, *188*, 720–731. [CrossRef]
5. Wu, K.; Wang, Z.; Jin, W.; Wang, W. Quantitative analysis of the sustainable cost of prefabricated concrete structures. *J. Build. Struct.* **2021**, *42*, 133–144. [CrossRef]
6. Marco, C.; Frappa, G.; Sabbà, M.; Campione Pauletta, M. Shear strength formula for interior beam-column joints with plain bars in existing buildings. *Eng. Struct.* **2023**, *293*, 116656. [CrossRef]
7. Kalogeropoulos, G.; Tsonos, A.-D.; Iakovidis, P. Hysteresis Behavior of RC Beam–Column Joints of Existing Substandard RC Structures Subjected to Seismic Loading–Experimental and Analytical Investigation. *Buildings* **2024**, *14*, 1609. [CrossRef]
8. Zhuang, M.-L.; Sun, C.; Yang, Z.; An, R.; Bai, L.; Han, Y.; Bao, G. Numerical Investigation on the Seismic Behavior of Novel Precast Beam–Column Joints with Mechanical Connections. *Buildings* **2024**, *14*, 1199. [CrossRef]
9. Pekgokgoz, R.K.; Yakut, I. Investigation of Passive Controlled Post-Tensioning System on the Structural Behaviour of Precast Reinforced Concrete Beam–Column Connections. *Buildings* **2024**, *14*, 3910. [CrossRef]
10. Feng, S.; Yang, Y.; Xue, Y.; Yu, Y. Numerical Investigation on the Hysteretic Performance of Self-Centering Precast Steel–Concrete Hybrid Frame. *Buildings* **2024**, *14*, 3202. [CrossRef]
11. Xiao, J.; Ma, X.; Liu, Q.; Zhang, H.; Duan, Z. Evolvement and research progress of concept for full recycled concrete. *J. Archit. Civ. Eng.* **2021**, *38*, 1–15. [CrossRef]
12. Tam, V.; Soomro, M.; Evangelista, A. A review of recycled aggregate in concrete applications (2000–2017). *Constr. Build. Mater.* **2018**, *172*, 272–292. [CrossRef]
13. Li, Q.; Yue, G.; Guo, Y. *Property Regulation and Mix Proportion Design of Recycled Concrete*; China Architecture & Building Press: Beijing, China, 2019; pp. 1–8.
14. Schoon, J.; De Buysser, K.; Van Driessche, I.; De Belie, N. Fines extracted from recycled concrete as alternative raw material for Portland cement clinker production. *Cem. Concr. Compos.* **2015**, *58*, 70–80. [CrossRef]
15. Lin, Y.; He, T.; Da, Y.; Yang, R.; Zheng, D. Effects of recycled powders mixing methods on the properties of recycled concrete. *J. Build. Eng.* **2023**, *80*, 107994. [CrossRef]
16. Zhang, D.; Zhang, S.; Huang, B.; Yang, Q.; Li, J. Comparison of mechanical, chemical, and thermal activation methods on the utilisation of recycled concrete powder from construction and demolition waste. *J. Build. Eng.* **2022**, *61*, 105295. [CrossRef]
17. Li, Z.; Bian, Y.; Zhao, J.; Wang, Y.; Yuan, Z. Recycled concrete fine powder (RFP) as cement partial replacement: Influences on the physical properties, hydration characteristics, and microstructure of blended cement. *J. Build. Eng.* **2022**, *62*, 105326. [CrossRef]
18. Yu, J.; Jiang, F.; Yu, K.; Dong, F.; Duan, X. Deformability enhancement of fiber-reinforced cementitious composite by incorporating recycled powder. *J. Reinf. Plast. Compos.* **2020**, *39*, 119–131. [CrossRef]
19. Yu, K.; Zhu, W.; Ding, Y.; Lu, Z.; Yu, J.; Xiao, J. Micro-structural and mechanical properties of ultra-high performance engineered cementitious composites (UHP-ECC) incorporation of recycled fine powder (RFP). *Cem. Concr. Res.* **2019**, *124*, 105813. [CrossRef]
20. Jiang, H.; Li, F. Effect of recycled fine powder content on mechanical properties of UHPC. *New Build. Mater.* **2020**, *47*, 79–81.
21. Zhang, C.; Wang, Y.; Zhang, X.; Ding, Y.; Xu, P. Mechanical properties and microstructure of basalt fiber-reinforced recycled concrete. *J. Clean. Prod.* **2021**, *278*, 123252. [CrossRef]
22. Liang, C.; Zhang, Y.; Wu, R.; Yang, D.; Ma, Z. The utilization of active recycled powder from various construction wastes in preparing ductile fiber-reinforced cementitious composites: A case study. *Case Stud. Constr. Mater.* **2021**, *15*, e00650. [CrossRef]
23. Li, X.; Lv, X.; Zhou, X.; Meng, W.; Bao, Y. Upcycling of waste concrete in eco-friendly strain-hardening cementitious composites: Mixture design, structural performance, and life-cycle assessment. *J. Clean. Prod.* **2022**, *330*, 129911. [CrossRef]
24. Song, S.; Li, X.; Wang, Z.; Wang, W. Orthogonal experimental and theoretical study on mechanical properties of fiber-reinforced recycled powder concrete. *Case Stud. Constr. Mater.* **2022**, *17*, e01546. [CrossRef]
25. Li, Y.; Zhang, J.; He, Y.; Huang, G.; Li, J.; Niu, Z. A review on durability of basalt fiber reinforced concrete. *Compos. Sci. Technol.* **2022**, *225*, 109519. [CrossRef]
26. Zheng, Y.; Zhang, Y.; Zhuo, J.; Zhang, Y.; Wan, C. A review of the mechanical properties and durability of basalt fiber-reinforced concrete. *Constr. Build. Mater.* **2022**, *359*, 129360. [CrossRef]
27. Wang, J.; Dai, Q.; Si, R.; Guo, S. Investigation of properties and performances of Polyvinyl Alcohol (PVA) fiber-reinforced rubber concrete. *Constr. Build. Mater.* **2018**, *193*, 631–642. [CrossRef]

28. Xu, H.; Shao, Z.; Wang, Z.; Cai, L.; Li, Z.; Jin, H. Experimental study on mechanical properties of fiber reinforced concrete: Effect of cellulose fiber, polyvinyl alcohol fiber and polyolefin fiber. *Constr. Build. Mater.* **2020**, *261*, 120610. [CrossRef]
29. Yu, X.; Li, X.; Ma, R.; Sun, H.; Su, Z. Study on Carbon Emission of Assembled High Ductility Recycled Powder Concrete Structure Based on LCA Theory. *Mater. Rep.* **2024**, *38*, 89–95.
30. Li, X.; Xu, Z.; Bao, Y.; Cong, Z. Post-fire seismic behavior of two-bay two-story frames with high-performance fiber-reinforced cementitious composite joints. *Eng. Struct.* **2019**, *183*, 150–159. [CrossRef]
31. Fan, W.; Su, H.; He, Y.; Sun, W.; Shao, X. Experimental investigation on UHPC-based gravity-type half grouted sleeve connections under tensile and cyclic loadings. *J. Build. Eng.* **2022**, *61*, 105284. [CrossRef]
32. Su, H.; Fan, W.; Wang, H.; Liu, B.; Sun, W.; Shao, X. Direct shear strength of UHPC-based gravity-type grouted sleeve connection members. *J. Build. Eng.* **2023**, *69*, 106266. [CrossRef]
33. Lu, Z.; Wu, B.; Sakata, H.; Huang, J.; Zhang, M. Mechanical performance of prefabricated concrete beam-column joints with double-grouted sleeve connectors: A numerical and theoretical study. *Structures* **2023**, *56*, 104870. [CrossRef]
34. Yang, X.; Jiang, Y.; Lin, F. Experimental study on direct shear resistance of horizontal grouted joints in PC shear walls under compression and in-plane bending moment. *J. Build. Eng.* **2024**, *90*, 109419. [CrossRef]
35. Liu, H.; Chen, J.; Xu, C.; Du, X. Seismic performance of precast column connected with grouted sleeve connectors. *J. Build. Eng.* **2020**, *31*, 101410. [CrossRef]
36. Qiao, D.; Xu, Y.; Zhang, X.; Pang, J.; Liu, K.; Wang, S. Seismic behaviour and size effect of column base joints with inverted exposed grouted sleeves. *J. Build. Eng.* **2022**, *51*, 104333. [CrossRef]
37. Qiao, D.; Sun, H.; Ci, M.; Men, J.; Wang, S. Plastic hinge and damage mechanism of RC assembly column base joint with post pouring area. *Eng. Struct.* **2023**, *294*, 116834. [CrossRef]
38. Wei, Y.; Li, K.; Fan, J.; Li, Y.; Li, S. Seismic performance of RC columns strengthened with HSSSWR meshes reinforced ECC under high axial compression ratio. *J. Build. Eng.* **2023**, *76*, 107350. [CrossRef]
39. Xu, L.; Pan, J.; Cai, J. Seismic performance of precast RC and RC/ECC composite columns with grouted sleeve connections. *Eng. Struct.* **2019**, *188*, 104–110. [CrossRef]
40. Wu, G.; Feng, D. Research progress on fundamental performance of precast concrete frame beam-to-column connections. *J. Build. Struct.* **2018**, *39*, 1–16. [CrossRef]
41. Gao, P.; Deng, Y.; Li, A.; Lu, F. Review on seismic performance of connection joints of prefabricated concrete frame structures. *Ind. Constr.* **2021**, *51*, 171–185+152. [CrossRef]
42. Ding, K.; Ye, Y.; Ma, W. Seismic performance of precast concrete beam-column joint based on the bolt connection. *Eng. Struct.* **2021**, *232*, 111884. [CrossRef]
43. Zhang, Z.; Ding, R.; Nie, X.; Fan, J. Seismic performance of a novel interior precast concrete beam-column joint using ultra-high performance concrete. *Eng. Struct.* **2020**, *222*, 111145. [CrossRef]
44. Xue, W.; Hu, X.; Song, J. Experimental study on seismic behavior of precast concrete beam-column joints using UHPC-based connections. *Structures* **2021**, *34*, 4867–4881. [CrossRef]
45. Huang, W.; Hu, G.; Zhang, J. Experimental study on the seismic performance of new precast concrete beam-column joints with replaceable connection. *Structures* **2022**, *35*, 856–872. [CrossRef]
46. Ye, Y.; Hou, Z.; Qiao, P.; Wang, Z.; Kang, Q. Experimental Study on Seismic Behavior of a new fully precast rocking beam-column joint. *J. Build. Eng.* **2022**, *62*, 105337. [CrossRef]
47. Zhang, J.; Li, C.; Rong, X.; Zhang, X.; Li, Y. Experimental study on the seismic performance of steel–concrete beam–column connections for prefabricated concrete frames. *J. Build. Eng.* **2021**, *43*, 103236. [CrossRef]
48. Zhang, J.; Pei, Z.; Rong, X. Experimental seismic study of an innovative precast steel–concrete composite beam–column joint. *Soil Dyn. Earthq. Eng.* **2022**, *161*, 107420. [CrossRef]
49. Xue, H.; Ashour, A.; Ge, W.; Cao, D.; Sun, C.; Cao, S. Seismic response of prestressed precast reinforced concrete beam-column joints assembled by steel sleeves. *Eng. Struct.* **2023**, *276*, 115328. [CrossRef]
50. Tong, C.; Wu, J.; Zakari, M.; Gao, Y. Hysteretic behavior of a new precast concrete interior beam-to-column joint with a controllable bending moment. *J. Build. Eng.* **2022**, *45*, 103557. [CrossRef]
51. Yang, Y.; Gao, S.; Zheng, Y.; Liu, X.; Zhang, W.; Lin, B. Seismic performance of precast column-beam joint with artificial plastic hinge. *J. Build. Eng.* **2023**, *67*, 105942. [CrossRef]
52. Zhang, X.; Wang, B.; Ju, Y.; Wang, D.; Han, Y.; Song, Y. Seismic performance of precast frame with UHPC composite beams and HSC columns under cyclic loadings. *Eng. Struct.* **2024**, *302*, 117429. [CrossRef]
53. Liu, T.; Zhang, L.; An, X.; Chen, L. Experimental and numerical investigation on collapse behavior of precast reinforced concrete beam-column sub-assemblages with cast-in-place ECC joints. *Case Stud. Constr. Mater.* **2022**, *17*, e01421. [CrossRef]
54. Xiong, X.; Wei, Z.; Zhang, D.; Liu, J.; Xie, Y.; He, L. An Experimental Study on the Seismic Performance of New Precast Prestressed Concrete Exterior Joints Based on UHPC Connection. *Buildings* **2025**, *15*, 729. [CrossRef]

55. Yuan, C.; Cai, J.; Chen, Q.; Liu, X.; Huang, H.; Zuo, Z. Experimental study on seismic behavior of precast recycled fine aggregate concrete beam-column joints with pressed sleeve connections. *J. Build. Eng.* **2022**, *58*, 104988. [CrossRef]
56. Yan, Q.; Chen, T.; Xie, Z. Seismic experimental study on a precast concrete beam-column connection with grout sleeves. *Eng. Struct.* **2018**, *155*, 330–344. [CrossRef]
57. Lin, Y.; Chen, Z.; Guan, D.; Guo, Z. Experimental study on interior precast concrete beam-column connections with UHPC core shells. *Structures* **2021**, *32*, 1103–1114. [CrossRef]
58. Deng, M.; Ma, F.; Song, S.; Lv, H.; Sun, H. Seismic performance of interior precast concrete beam-column connections with highly ductile fiber-reinforced concrete in the critical cast-in-place regions. *Eng. Struct.* **2020**, *210*, 110360. [CrossRef]
59. *JGJ 355-2015*; Ministry of Construction of the People's Republic of China: Technical Specification for Grout Sleeve Splicing of Rebars. China Architecture & Building Press: Beijing, China, 2015.
60. Li, X.; Li, Y.; Yan, M.; Meng, W.; Lu, X.; Chen, K. Cyclic behavior of joints assembled using prefabricated beams and columns with Engineered Cementitious Composite (ECC). *Eng. Struct.* **2021**, *247*, 113115. [CrossRef]
61. *GB 50011-2010*; Ministry of Construction of the People's Republic of China: Code for Seismic Design of Buildings. China Architecture & Building Press: Beijing, China, 2010.
62. Zhu, Z.; Guo, Z.; Liu, J.; Luan, L. Shear bearing performance of precast concrete component interface by bubble film. *China Civ. Eng. J.* **2020**, *53*, 21–27. [CrossRef]
63. *ACI 318-19(22)*; Building Code Requirements for Structural Concrete and Commentary. American Concrete Institute: Farmington Hills, MI, USA, 2022.
64. International Federation for Structural Concrete. *fib Model Code for Concrete Structures 2020: Final Draft*; International Federation for Structural Concrete: Lausanne, Switzerland, 2023.
65. Feng, P.; Qiang, H.; Ye, L. Discussion and definition on yield points of materials, members and structures. *Eng. Mech.* **2017**, *34*, 36–46.
66. Su, J.; Xv, S. Research on cracking behavior of a new type of UHTCC frame beam-column joints. *Ind. Constr.* **2010**, *40*, 99–102. [CrossRef]
67. Xing, G.; Wu, T.; Liu, B. Study on crack resistance of beam-column joints in reinforced concrete frame structures. *Eng. Mech.* **2011**, *28*, 163–169.
68. Hwang, S.; Lee, H. Strength Prediction for Discontinuity Regions by Softened Strut-and-Tie Model. *J. Struct. Eng.* **2002**, *128*, 1519–1526. [CrossRef]
69. Dong, J.; Yang, H.; Zhao, W. Improved diagonal strut model and parameters determination for interior joints in RC frames. *Earthq. Eng. Eng. Dyn.* **2017**, *37*, 155–164.

buildings

MDPI

Article

Collapse Fragility Analysis of RC Frame Structures Considering Capacity Uncertainty

Tailin Zeng and Yang Li *

College of Civil Engineering, Hunan University, Changsha 410082, China; tailin@hnu.edu.cn
* Correspondence: lyang@hnu.edu.cn

Abstract: To analyze the impact of capacity uncertainty on the seismic collapse fragility of reinforced concrete (RC) frame structures, a fragility analysis framework based on seismic reliability methods is proposed. First, incremental dynamic analysis (IDA) curves are plotted by IDA under a group of natural seismic waves. Subsequently, collapse points are identified based on recommendations from relevant standards, yielding the probability distribution of the maximum inter-story drift ratios (MIDRs) at collapse points. Then, the distribution of the MIDRs under various intensity measures (IMs) of artificial seismic waves is calculated by using the fractional exponential moments-based maximum entropy method (FEM-MEM). Next, the structural failure probability is determined based on the combined performance index (CPI), and a seismic collapse fragility curve is plotted using the four-parameter shifted generalized lognormal distribution (SGLD) model. The results indicate that the collapse probability is lower considering the capacity uncertainty. Compared to deterministic MIDR limits of 1/25 and 1/50, the median values of the structure's collapse resistance increased by 13.2% and 87.3%, respectively. Additionally, the failure probability obtained by considering the capacity uncertainty is lower than the results based on deterministic limits alone. These findings highlight the importance of considering capacity uncertainty in seismic risk assessments of RC frame structures.

Keywords: seismic fragility; capacity uncertainty; seismic reliability; collapse analysis

1. Introduction

Earthquakes are one of the common natural disasters, and strong earthquakes may cause structural damage or even collapse, resulting in significant loss of life and property. Such earthquakes often trigger nonlinear structural failures and collapses. Consequently, it is imperative to investigate the seismic collapse fragility under seismic loading. Fragility analysis is a crucial component of seismic risk analysis, with the main purpose of evaluating the probability that structures exceed one specific limit state under varying IMs. This analysis facilitates a probabilistic assessment of structural seismic performance, revealing the intrinsic relationship between earthquake intensity and structural damage on a macro scale.

Seismic fragility is the possibility of a certain degree of damage to engineering structures under the influence of earthquakes of different intensities [1–3]. It can quantitatively describe the seismic performance of structures from a probabilistic perspective, reflecting the relationship between earthquake intensity and structural failure probability. In the early fragility analysis, a damage probability matrix (DPM) is proposed to describe the degree of damage to building structures exposed to different intensities [4]. Another method of describing seismic fragility results is to use smooth and continuous curves to characterize the variation of structural failure probability with seismic intensity. It originated from the

https://doi.org/10.3390/buildings15050694

seismic probability risk assessment in the nuclear industry in the 1970s [2], and subsequently, fragility curves have been widely used for fragility analysis of various engineering structures. The fragility curve can be categorized as empirical, judgmental, analytical, and hybrid based on the different sources of data and methods [5]. The analytical method is widely used for seismic fragility analysis with a large number of time history analyses. The incremental dynamic analysis (IDA) method was first adopted for fragility analysis [6,7]; multiple stripe analysis [8,9] and cloud analysis [10,11] are alternative methods for time history analysis. Sainct et al. [12] proposed a method for estimating fragility curves based on a combination of support vector machines and active learning algorithms. Kazemi et al. [13] proposed a seismic risk assessment method for reinforced concrete buildings based on machine learning algorithms, providing an effective and accuracy tool for RC building seismic risk assessment. Feng et al. [14] proposed a PDEM- based non-parametric seismic fragility assessment framework without pre-defined distribution of the structural demand and performance. Liu et al. [15] predicted the seismic damage of a steel frame by training an artificial neural network. Xu et al. [16] deduced the failure probability of the structure under earthquake action using the method of dynamic reliability. Mai et al. [17] computed the fragility curves for a three-story steel frame using kernel density estimation without assuming the shape for the fragility curve. Furthermore, artificial intelligence has been applied in fragility analysis as well, presenting novel alternatives for seismic risk assessment [18–20].

IDA is a widely employed method for assessing structural collapse fragility in structural collapse fragility analysis, initially proposed by Bertero [21] in 1977 and later endorsed by the Federal Emergency Management Agency (FEMA) in the FEMA 350 [22] document in 2000. Fattahi et al. [23] used the IDA to investigate the seismic collapse fragility of steel moment frames of varying heights. Cardone et al. [24] employed IDA to study the collapse fragility functions of both an isolation system and superstructure. Meanwhile, Dabaghi et al. [25] explored the impact of different structural parameters on the seismic collapse fragility of reinforced concrete shear wall structures. The IDA derives results from the dynamic time history analyses of structures subjected to seismic excitations. It typically identifies collapse points based on fixed response limits and yields fragility curves using lognormal distributions. However, relying solely on fixed parameters such as a specific displacement to determine a structure's ultimate capacity may lead to misjudgments in assessing the collapse fragility. Moreover, the uncertainties associated with both structural capacity and engineering demand parameters (EDPs) have not yet been incorporated into fragility studies. Therefore, accurately defining the collapse limit states of structures is crucial for the seismic collapse fragility analysis of structures.

According to Chinese regulations, previous studies [16] have classified the levels of structural damage based on inter-story drift ratios. These classifications are as follows: slight damage (SD = 1/550), moderate damage (MD = 1/100), extensive damage (ED = 1/50), and complete damage (CD = 1/25). In fragility analysis focused on collapse risk, the collapse limit states are typically determined based on specific thresholds of key indicators, such as inter-story drift ratios.

However, in engineering practice, the inter-story drift ratio at the collapse point is not a fixed value. To investigate the distribution of the maximum inter-story drift ratios (MIDRs) at collapse points, this study defines collapse scenarios based on the thresholds from FEMA 350 [22] and other literature [26]: A structure is considered to be in a collapse scenario when the tangent stiffness of the IDA curve decreases to 20% or less of the initial value or when the inter-story drift ratio exceeds 0.045.

Based on the defined collapse criteria, this study proposes a fragility analysis framework grounded in seismic reliability methods, innovatively incorporating uncertainties related to the bearing capacity of RC frame structures and the EDPs. The purpose of the research is to provide a more comprehensive and accurate analytical framework for earthquake risk assessment, which helps to better understand and predict the performance and safety of structures under seismic events.

This study employs OpenSees 3.5 to construct a finite element model of the interested structure. IDA is performed using natural seismic waves motivating the structure, allowing for the identification of collapse points and their corresponding MIDR based on the defined collapse states. The probability distribution of the MIDRs is then fitted to serve as a criterion for determining the collapse limit states. Subsequently, stochastic seismic excitations of artificial samples are generated in accordance with Chinese regulations, followed by time history analyses of the structure. Using the defined collapse state, the distribution of the EDPs under various IMs is fitted using the fractional exponential moments-based maximum entropy method (FEM-MEM). Next, the failure probability of the structure under different IMs is calculated using the combined performance index (CPI), which is derived from the difference between the EDPs values and the distribution of MIDRs at collapse points. Finally, the collapse fragility curve of the structure is fitted using a shifted generalized lognormal distribution. The process of the method proposed in this article is shown in Figure 1.

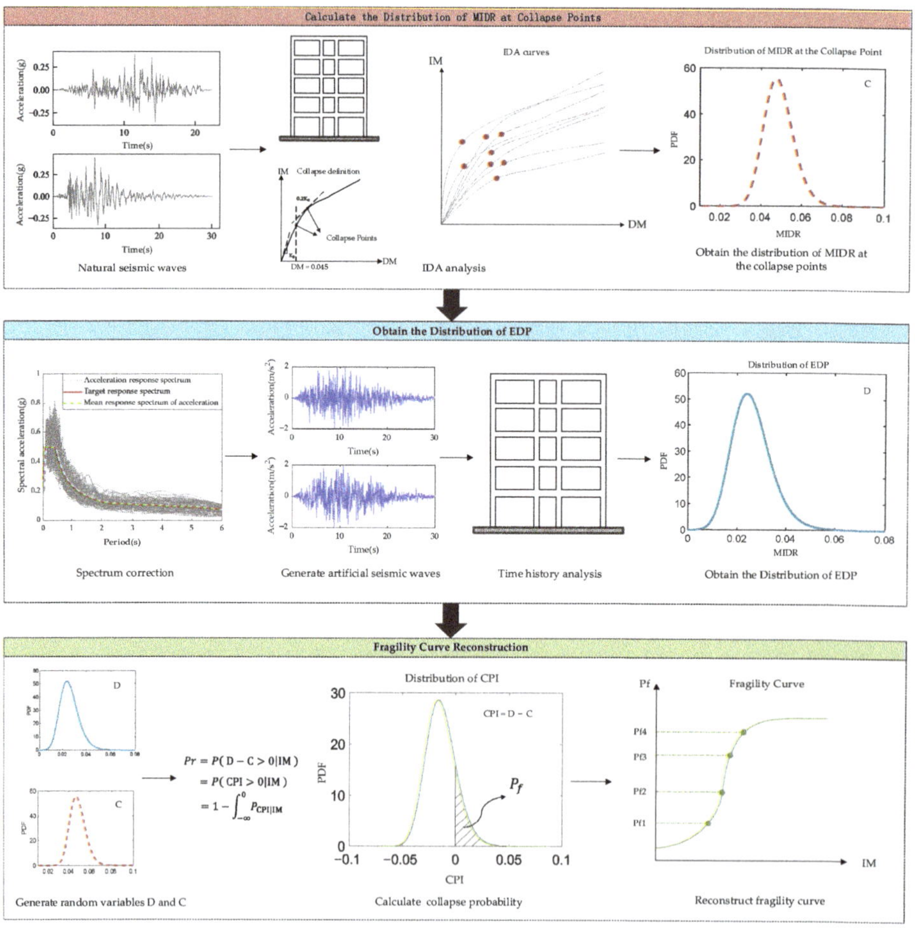

Figure 1. The flowchart of the proposed method.

2. Seismic Collapse Fragility Analysis Considering Uncertainty in the Bearing Capacity

2.1. Problem Description

IDA can provide valuable insights into the response and damage accumulation of structures under various IMs. According to the suggestion of FEMA350 [22], there are two main criteria for determining structural collapse. The first is that the tangent stiffness of the IDA curve reduces to 20% of the initial stiffness. The second is that the MIDR of the structure exceeds 0.1. These two criteria are derived from extensive testing and on-site observations, offering significant practical values for assessing the structural collapse state. In current reliability-based fragility analyses, it is essential to obtain the distribution of the EDPs and to obtain the failure probability on the basis of a predetermined threshold.

However, the structural responses under stochastic earthquake excitation (such as inter-story drift ratio, etc.) inherently exhibit randomness and uncertainty. Therefore, when performing structural seismic collapse fragility analysis, it is vital to comprehensively consider these uncertain factors to achieve more accurate and reliable results. This method provides a stronger scientific foundation for the seismic design and risk assessment of structures.

In existing fragility studies [27], many scholars have conducted in-depth discussions on the uncertainty related to structures and seismic movements. Cao [28] proposed an analytical framework for seismic hazard and fragility that comprehensively considers uncertainties in both structural capacity and demand. Similarly, Jia et al. [29] investigated the uncertainties in the mean and standard deviation of the EDPs and thresholds, successfully extending traditional probabilistic seismic risk analysis formulas to scenarios involving probability distributions.

It is worth noting that the demand uncertainty is generally more significant than the capacity uncertainty [28]. However, in numerous seismic studies, the uncertainty in the capacity is often overlooked or simplified, which may lead to bias in the fragility analysis results, potentially misguiding seismic design and decision-making.

2.2. Seismic Fragility Analysis of Structures Based on Seismic Reliability

2.2.1. Seismic Reliability Analysis Based on the Fractional Exponential Moments-Based Maximum Entropy Method

Accurately and efficiently determining the seismic demand distribution of a structure is crucial for calculating the failure probability and conducting a fragility analysis. This study examines structural fragility from the perspective of seismic reliability, using the FEM-MEM to derive the distribution of the MIDRs under stochastic seismic motions.

For structures under seismic events, the response of interest in this paper is represented by Equation (1). Subsequently, the structural failure probability is derived, followed by the fragility analysis.

$$Z(t) = W\left(\mathbf{X}(t), \dot{\mathbf{X}}(t)\right) = \mathcal{W}(\boldsymbol{\Phi}, t) \tag{1}$$

where W and $\mathcal{W}$ are the corresponding physical operators.

The maximum absolute value Z of the structural response is given by

$$Z = \max_{t \in [0,T]} [|Z(t)|] \tag{2}$$

Through this method, the maximum response of the structure under earthquake action can be more accurately captured, providing a solid foundation for the fragility analysis.

For a continuous random variable Z, its probability density function is expressed as $f_Z(z)$. The FEM-MEM shown in Equation (3) can be used to solve the probability density function of Z [30]:

$$\begin{cases} \text{Find}: & \overline{\alpha} \text{ and } \boldsymbol{\lambda} = [\lambda_1, \lambda_2, \ldots, \lambda_m] \\ \text{Minimize}: & \Gamma(\boldsymbol{\alpha}, \boldsymbol{\lambda}) = \log\left\{ \int_{\Omega_z} \exp\left(-\sum_{k=1}^{m} \lambda_k \exp(-\alpha_k z) \right) dz \right\} + \sum_{k=1}^{m} \lambda_k \mathcal{M}_Z^{\alpha_k} \\ \text{Subject to}: & -2 \le \overline{\alpha} \le 2 \end{cases} \tag{3}$$

where $\mathcal{M}_Z^{\alpha_k}$ is the k-th fractional exponential moments based on the random variable Z, and m is the number of constraints. As long as the values of $\boldsymbol{\alpha}$ and $\boldsymbol{\lambda}$ are determined, the probability density function of the random variable Z can be obtained. An effective univariate strategy is adopted to solve $\boldsymbol{\alpha}$ [31], and the formula is as follows:

$$\alpha_k = k\frac{\alpha^*}{m} = k\overline{\alpha}, k = 1, \ldots, m \tag{4}$$

where $\overline{\alpha}$ is a low fractional order, usually taken as [−2,2], which simplifies the solution of $\boldsymbol{\alpha}$. After the initial value of $\boldsymbol{\alpha}$ is determined, the Lagrange multiplier $\boldsymbol{\lambda}$ is determined using an estimator–corrector scheme [32]. After determining the values of $\boldsymbol{\lambda}$ and $\boldsymbol{\alpha}$, the probability density function of the random variable Z can be obtained through Equation (3).

Based on the above methods, this study utilizes the FEM-MEM to fit the structural response obtained from the time history analyses. The probability distributions of the EDPs (D) and the MIDRs at the collapse points (C) are derived.

The probability distribution of the CPI, defined as the difference value between the MIDR at collapse point C and the EDPs (D), is obtained. Finally, the structural failure probability can be calculated by

$$Pr(\text{DM} \mid \text{IM}) = 1 - \text{CDF}_{\text{CPI}|\text{IM}}(0) = 1 - \int_{-\infty}^{0} p_{\text{CPI}|\text{IM}} \tag{5}$$

In the fragility analysis, the focus is typically placed on large failure probability [16]. The method adopted in this paper is capable of capturing the main body of the extreme response and efficiently reconstructing the PDF of the CPI to calculate the failure probability. Consequently, the application of the FEM-MEM can enhance the fragility analysis process, enabling a more accurate and efficient quantification of seismic fragility.

2.2.2. Fragility Curve Fitting Based on Shift Generalized Lognormal Distribution

The lognormal cumulative distribution function is usually used to fit the fragility curves [33]:

$$P(\text{Collapse}|\text{IM} = x) = \Phi\left(\frac{\ln(x/\theta)}{\beta}\right) \tag{6}$$

where $P(\text{Collapse}|\text{IM} = x)$ represents the probability of structural collapse when $\text{IM} = x$; $\Phi(\cdot)$ is the cumulative distribution function of the standard normal distribution; θ is the median value of fragility function, representing the IM value corresponding to a 50% probability of structural collapse; and β represents the deviation of the IM.

Although fitting a fragility curve with a lognormal distribution can achieve certain results in the fragility analysis, the assumption of a lognormal distribution is subjective and may not accurately reconstruct the fragility curve. Moreover, current seismic collapse fragility analysis methods still rely on IDA or multiple strip analysis (MSA), which require extensive time history analyses. To address this issue, various non-lognormal methods have been proposed in recent studies to reconstruct fragility curves. For example, Jeon et al. [34] used logistic regression to obtain fragility curves for bridge components and systems. Mangalathhu et al. [35] applied machine learning to derive the fragility curve of concrete bridges. Xu et al. [16] reconstructed the fragility curve using a four-parameter shifted generalized lognormal distribution (SGLD), which does not rely on the lognormal assumption and requires only the failure probabilities at four IM values to obtain the fragility curve. The probability density function of the SGLD [36] is expressed as follows:

$$f_{\mathrm{SGLD}}(x) = \frac{\beta_s}{x-c}\exp\left(-\frac{1}{r\eta^r}\left|\ln\left(\frac{x-c}{\xi}\right)\right|^r\right) \quad (7)$$

where c and ξ are the position and scale parameters; η and r are the shape parameters. $\beta_s = \frac{1}{2\eta\Gamma(1+1/r)r^{1/r}}$, $\Gamma(\cdot)$ represents the Gamma function. The value range of the parameters in Equation (7) [37] are $r, \eta, \xi > 0, c \in R$.

$$F_{\mathrm{SGLD}}(x) = \frac{1}{2} + \frac{1}{2}\mathrm{sgn}\left(\frac{x-c}{\xi} - 1\right)h\left(\frac{1}{r}, \frac{\left|\frac{1}{\eta}\ln\left(\frac{x-c}{\xi}\right)\right|^r}{r}\right) \quad (8)$$

where sgn(·) represents the sign function, and $h(m,n) = \int_0^n t^{m-1}e^{-t}dt/\Gamma(m)$.

In this paper, the cumulative distribution function of the SGLD is used to reconstruct the fragility function. By calculating the failure probability of the structure under four IMs using seismic reliability methods and substituting these values into Equation (8), the fragility curve can be reconstructed with ease.

Compared to the traditional lognormal distribution model, the SGLD offers a more flexible form, allowing for more accurate fitting results and providing advantages in both generality and precision.

3. Proposed Method

3.1. Definition of the Collapse Point

According to previous studies and relevant standards [26], a structure is considered to collapse when the inter-story drift ratio exceeds 0.045. Therefore, in this study, the collapse point is defined as the point where the stiffness in the IDA curve decreases to 20% of the initial stiffness or where the inter-story drift ratio exceeds 0.045. The collapse point criteria used in this article are shown in Figure 2.

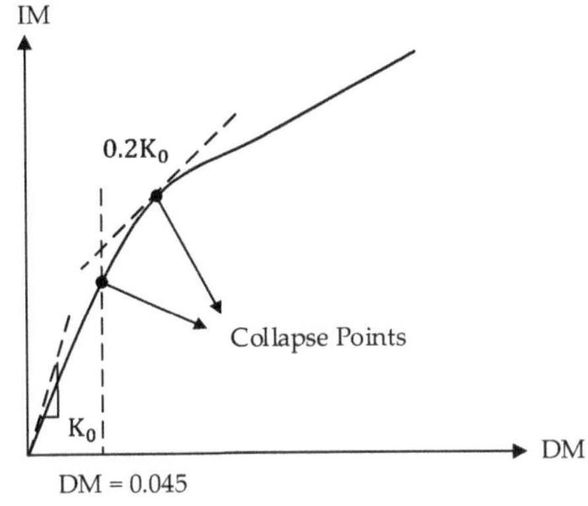

Figure 2. Definition of the collapse point.

This article investigates the collapse risk of building structures under seismic loads. Structures under earthquakes mainly bear dynamic loads, which are the main factor causing collapses. Hence, dynamic analysis is used in this article to investigate the collapse threshold of structures under earthquake loads. Compared to static analysis, using dynamic methods such as time history analysis can better reflect the dynamic characteristics of structures and explore the real response under earthquake loads. Hence, employing the time history analysis method can yield more precise collapse thresholds, a critical aspect for structural collapse fragility analysis, thereby enhancing the precision of the subsequent fragility analysis.

In this article, we choose a representative set of natural earthquake waves to conduct IDA analysis on a RC structure. Based on the collapse definition proposed in this article, the collapse points of the structure under dynamic loads are determined by continuously increasing the PGA of natural earthquake waves through IDA. In order to efficiently obtain the distribution of the MIDRs of the collapse points of the structure, the FEM-MEM is used in this paper to explore the distribution.

3.2. Engineering Demand Parameters Analyzing

Without a loss of generality, the kinematic equation of a multi-degree-of-freedom system under earthquake excitation can be expressed as

$$\mathbf{M}\ddot{\mathbf{X}}(t) + \mathbf{C}\dot{\mathbf{X}}(t) + \mathbf{F}(\mathbf{X}, t) = -\mathbf{MI}\ddot{\mathbf{U}}_g(t) \tag{9}$$

where $\mathbf{X}$, $\dot{\mathbf{X}}$, and $\ddot{\mathbf{X}}$ are the displacement, velocity, and acceleration vectors, respectively. $\mathbf{M}$ and $\mathbf{C}$ are the mass matrix and damping matrix, respectively. $\mathbf{F}(\mathbf{X}, t)$ is restoring the force vectors, and $\ddot{\mathbf{U}}_g(t)$ is stochastic ground motions. To obtain the distribution of the EDPs, this study performs time history analysis on the structure based on a group of artificial stochastic seismic waves. Fully non-stationary stochastic seismic waves, modeled through the spectral representation method [38–40], are expressed as

$$\ddot{\mathbf{U}}_g(t) = \sum_{k=0}^{\infty} \sqrt{2S_{\ddot{\mathbf{U}}_g}(\omega_k, t)\Delta\omega} cos(\omega_k t + \varphi_k) \tag{10}$$

where $S_{\ddot{\mathbf{U}}_g}(\omega_k, t)$ represents the evolutionary power spectral density function, which can be expressed as

$$S_{\ddot{\mathbf{U}}_g}(\omega_k, t) = A^2(\omega, t) \cdot S(\omega) \tag{11}$$

where $A(\omega, t)$ is the time–frequency modulation function, which can be expressed as [41]

$$A(\omega, t) = \exp\left(-\eta_0 \frac{\omega t}{\omega_g T}\right) \left[\frac{t}{c} \exp\left(1 - \frac{t}{c}\right)\right]^k \tag{12}$$

In this article, the target double-sided power spectral density $S(\omega)$ is selected as the Clough Penzien spectrum [42], which can be expressed as

$$S(\omega) = \frac{\omega_g^4 + 4\zeta_g^2\omega_g^2\omega^2}{\left(\omega_g^2 - \omega^2\right)^2 + 4\zeta_g^2\omega_g^2\omega^2} \times \frac{\omega^4}{\left(\omega_f^2 - \omega^2\right)^2 + 4\zeta_f^2\omega_f^2\omega^2} S_0 \tag{13}$$

where S_0 is the intensity factor of ground motions [43], expressed as

$$S_0 = \frac{\overline{a}_{\max}^2}{\gamma_0^2\left[\pi\omega_g\left(2\zeta_g + \frac{1}{2\zeta_g}\right)\right]} \tag{14}$$

The meanings of the parameters related to artificial seismic waves are provided in Table 1. Moreover, to enhance the correlation between the artificial seismic waves and real engineering scenarios, this study employs an iterative correction method [44] to refine the seismic waves. This correction process involves iterative calculations to ensure that the mean response spectrum of the artificial seismic accelerations aligns with the standard response spectrum.

Table 1. Parameters in the simulation of non-stationary seismic ground motions.

Parameters	Description	Parameters	Description
ω_g	Dominant frequency	$\Delta\omega$	Frequency interval
ω_f	Parameter of the second filter binding the low-frequency	c	Average arrival time of PGA
ζ_g	Damping ratio of the site soil	η_0	Frequency modulation factor
ζ_f	Parameter of the second filter binding the low-frequency component	T	Duration time
$\bar{a}_{max}$	PGA value	k	Shape control coefficient
γ_0	Peak value factor	ω_k	Discrete frequencies

3.3. *Fragility Analysis Based on the CPI*

The uncertainty in the bearing capacity significantly impacts the calculation results [27,28]. Relying solely on the deterministic thresholds recommended in the codes may lead to an underestimation of the structural performance levels, resulting in inaccurate seismic risk assessments [28]. In light of this, this study comprehensively considers both demand and capacity uncertainties in the seismic collapse fragility analysis.

This study derives the probability distributions of the structural demand parameters and bearing capacity through time history analysis. Based on these distributions, the CPI is introduced as a criterion for evaluating the seismic performance of the structure, defined by the following formula [28]:

$$Pr(\mathrm{DM} \mid \mathrm{IM}) = Pr(\mathrm{D} - \mathrm{C} > 0 \mid \mathrm{IM}) = Pr(\mathrm{CPI} > 0 \mid \mathrm{IM}) = 1 - Pr(\mathrm{CPI} \leq 0 \mid \mathrm{IM}) \quad (15)$$

where DM is the damage measure of the structure, IM is the intensity measure, represented by the peak ground acceleration (PGA), D is the EDPs, and C is the structural bearing capacity.

According to Equation (15) the failure probability of the structure under IM is related to the distribution of the CPI. $Pr(\mathrm{CPI} \leq 0 \mid \mathrm{IM})$ represents the probability of the CPI < 0 under the IM, which can be written as the value of the cumulative distribution function (CDF) of the random variable CPI at 0. Therefore, Equation (15) can be expressed in the form of Equation (5).

$P_{\mathrm{CPI}|\mathrm{IM}}$ represents the probability density function of the combined performance index (CPI) under a specific IM. Once the distribution of the CPI is determined, calculating the failure probability becomes simple and direct.

4. Case Analysis

4.1. *OpenSees Model of a RC Frame*

This study conducts a fragility analysis of a five-story three-span reinforced concrete frame structure [45]. According to the Chinese code [46], the fortification intensity of the structure is set at 8 degrees, with a design basic acceleration of 0.2 g. The design earthquake group is classified as the first group, and the design site category is categorized as Class II. The floor dead load is 4.5 $\mathrm{KN/m^2}$, while the live load is 2.0 $\mathrm{KN/m^2}$. The framework layout and reinforcement details of the beam and column sections are shown in Figure 3. OpenSees is used for the modeling and time history analysis. The beam and column members are modeled with the nonlinear force-based element with fiber sections. The concrete01 constitutive model is used for concrete, and the steel02 constitutive model is used for steel reinforcement. The constraint effect of stirrups on the core concrete is considered with the confined concrete model. An amplification factor is incorporated to account for the enhancing effects of stirrups on the concrete ductility, as well as strength, according to the modified Kent–Park model [47,48]. The compressive strength of the concrete is 26.8 MPa, and the elastic modulus of the concrete is 32,500 MPa. The yield strength of the longitudinal reinforcement and hoop reinforcement are 400 MPa and 235 MPa, respectively, and the elastic modulus of the steel reinforcement is 200 GPa.

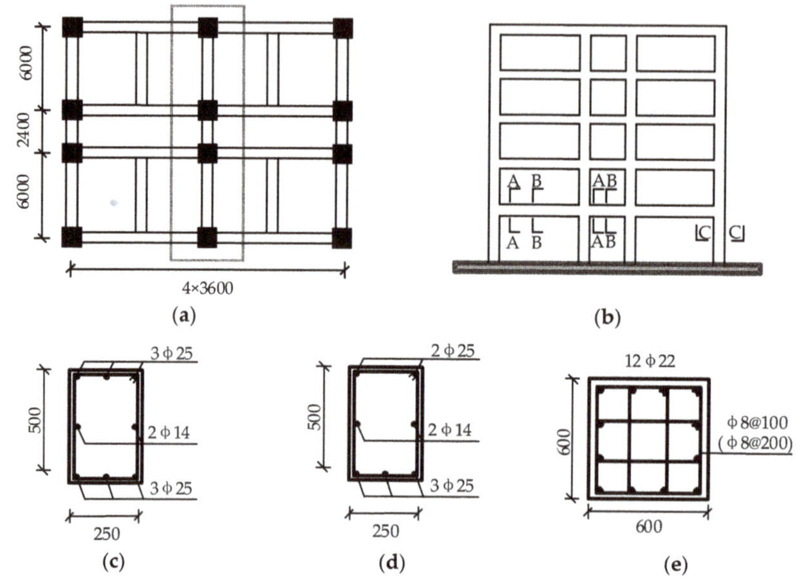

Figure 3. Layout of the frame and sectional reinforcements: (**a**) Plane arrangement of the structure. (**b**) Elevation arrangement of the structure. (**c**) Size and reinforcement of (A). (**d**) Size and reinforcement of (B). (**e**) Size and reinforcement of (C).

4.2. Stochastic Ground Motions Generation

In order to generate seismic motions that comply with the seismic design parameters of the structure, this article generates a total of 89 seismic motions based on the recommendations given in Reference [49]. The parameter values used to generate completely non-stationary stochastic seismic motion in this paper are as follows: $\zeta_g = \zeta_f = 0.6$, $\omega_g = 15.71$ rad/s, $\omega_f = 1.57$ rad/s, $\gamma_0 = 2.9$, $\bar{a}_{\max} = 200$ cm/s^2, $\Delta\omega = 0.1$, $T = 30$ s, $k = 2.0$, $\eta_0 = 0.15$, $c = 9.0$ s, and $\omega_k = k\Delta\omega(k = 1, 2, \ldots, 1000)$. The generated stochastic seismic samples are shown in Figure 4.

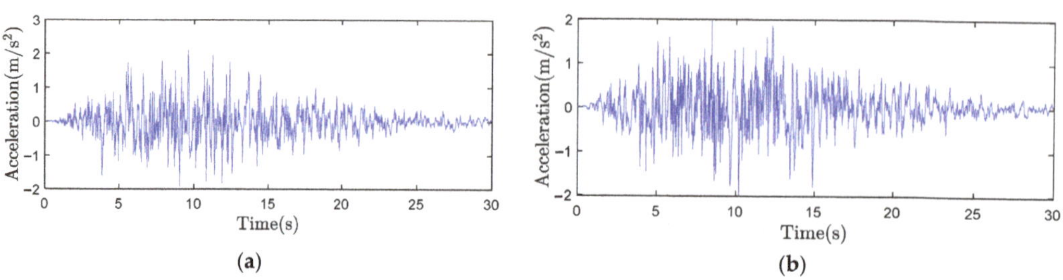

Figure 4. Artificial ground motion samples: (**a**) seismic wave 1; (**b**) seismic wave 2.

This study employs an iterative method to calibrate the seismic motion based on the acceleration response spectrum. The mean of the corrected acceleration response spectrum closely aligns with the standard response spectrum. Meanwhile, artificial seismic waves effectively capture the randomness of earthquakes, reflecting the characteristics of real ground motions. Figure 5 illustrates the acceleration response spectra before and after correction. This approach ensures that the stochastic seismic motions exhibit strong randomness while maintaining a high degree of correlation between their acceleration response spectra and the standard spectrum, thereby enhancing the accuracy of the subsequent fragility analyses.

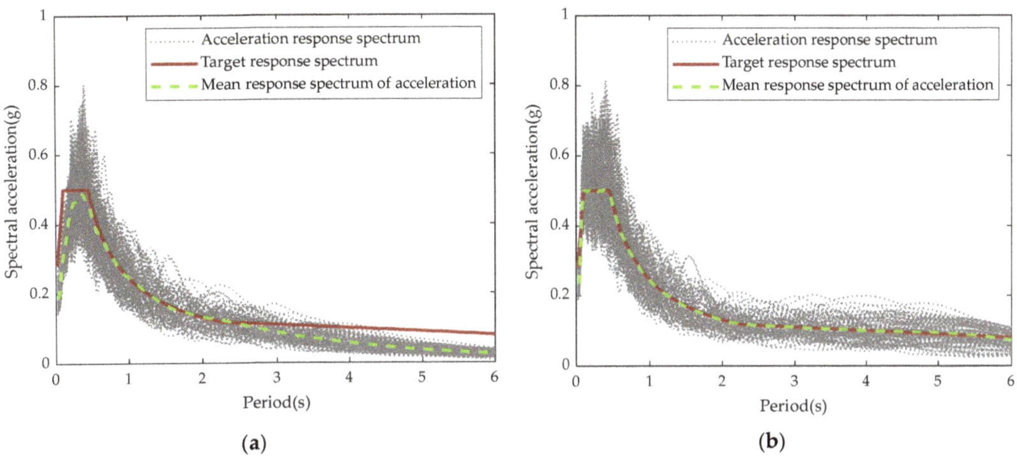

Figure 5. Acceleration response spectra comparisons: (**a**) original response spectra and mean spectrum; (**b**) corrected response spectra and mean spectrum.

4.3. Distribution of the Maximum Inter-Story Drift Ratios at the Collapse Point

In order to obtain the distribution of MIDRs at the collapse points of the structure under earthquakes, this study selects 100 natural earthquake waves based on the recommendations of FEMA P695 (ATC-63) [50]. The PGA is adjusted from 0 to 2.0 g in increments of 0.1 g for the initial incremental dynamic analyses. The IDA curves are shown in Figure 6a. With the increase in the IM, the stiffness of the IDA curves decreases, and then, the collapse points can be obtained based on the collapse determination proposed in this paper. Samples of the MIDRs at the collapse points, derived using the FEM-MEM, are fitted to yield the distribution shown in Figure 6b.

Next, 89 samples of fully non-stationary stochastic ground motions are generated using the spectral representation method, with their PGA adjusted to exceed 1.0 g for the IDA. The FEM-MEM is then employed to fit the samples of the EDPs, resulting in the distribution of these parameters under different IMs. Figure 7 shows the distribution of the EDPs(D) and MIDRs(C) at the collapse points for two typical IMs. In this study, FEM-MEM is utilized to derive unknown distributions with a limited number of samples [30,51], which can efficiently solve the failure probability in the fragility analysis.

4.4. Seismic Collapse Fragility Curve

Based on the obtained distribution of the structural EDPs and the MIDRs at collapse points under different IMs, a Monte Carlo simulation is performed to generate random variables D and C that match these distributions.

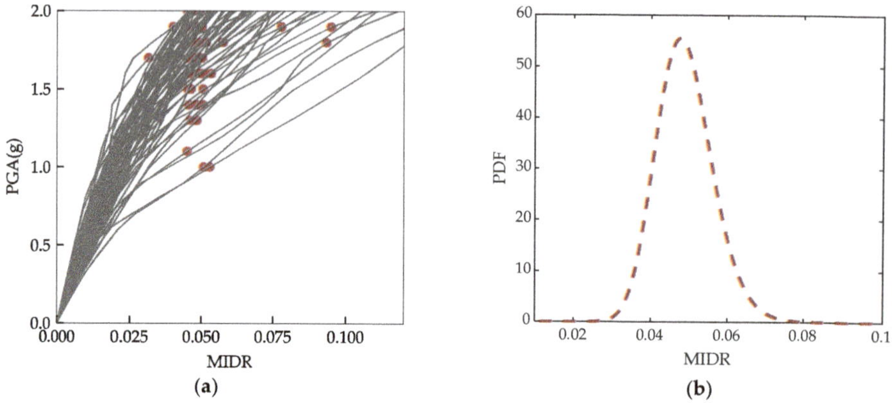

Figure 6. IDA curves and distribution of the collapse points: (**a**) IDA curves and collapse points; (**b**) the distribution of the collapse points.

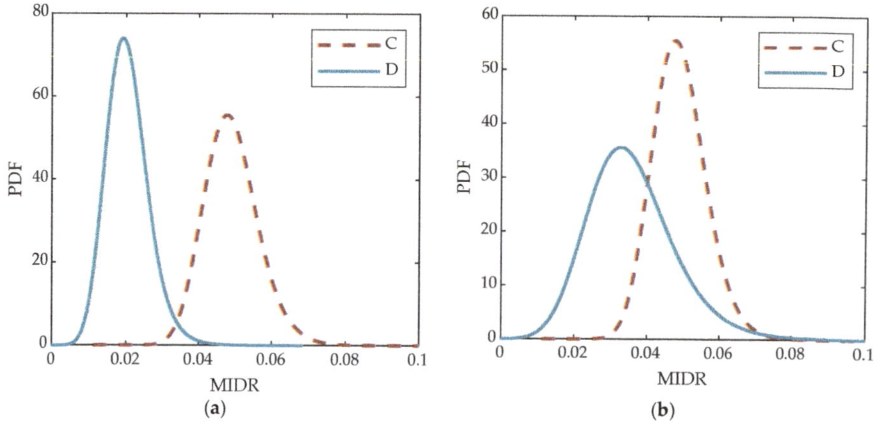

Figure 7. Distribution of the EDPs under different IMs: (**a**) PGA = 1.0 g; (**b**) PGA = 1.5 g.

The structural failure probability, accounting for bearing capacity uncertainty, is then calculated using Equation (5). Finally, the structural collapse fragility curve is reconstructed using the SGLD. For comparison, fragility curves are also derived under the assumption of a lognormal distribution, with the collapse limits defined at MIDRs of 1/50 and 1/25, as illustrated in Figure 8.

The results indicate that the MIDRs of the collapse points of the structure under seismic excitation show various values, which should be considered in the collapse fragility analysis. Table 2 presents the failure probabilities of the structure under different thresholds and IMs using the seismic reliability method. The failure probability accounting for bearing capacity uncertainty is lower than that of the two deterministic thresholds in this study. On the other hand, the lognormal fragility model does not accurately fit the failure probability obtained through the reliability-based method. Compared to the lognormal-based fragility model, the four-parameters SGLD fragility model provides a significantly better fit for the reference points. The fitting capacity of the lognormal distribution is relatively constrained, often failing to accurately match the target reference points. The method proposed in this paper is capable of comprehensively taking into account the bearing capacity of the structure and can offer a more precise estimation of the fragility curve.

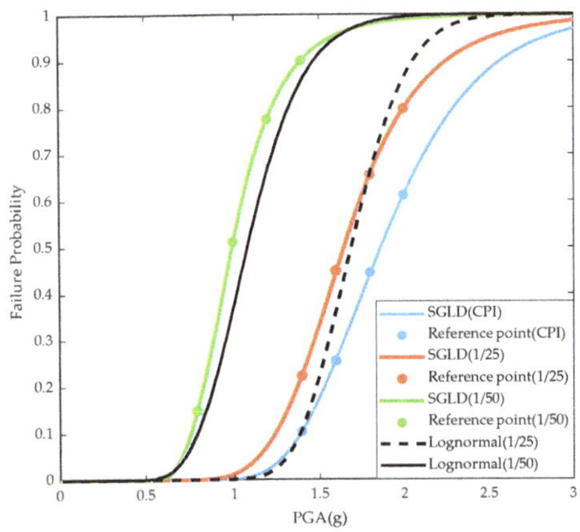

Figure 8. Comparison of fragility curves under different limits.

Table 2. Comparison of failure probability.

IM(g)	0.1	0.5	1.0	1.2	1.4	1.6	1.8	2.0
Failure Prob. (CPI)	0	0	0.0007	0.0240	0.1036	0.2551	0.4455	0.6113
Failure Prob. (1/25)	0	0	0.0037	0.0580	0.2232	0.4496	0.6556	0.7980
Failure Prob. (1/50)	0	0.0006	0.5121	0.7761	0.9019	0.9515	0.9759	0.9877

The analysis framework proposed in this paper can also be adopted for the fragility assessment of other types of structures. It is worth noting that the definition of the limit states for different types of structures should be carefully considered, which is crucial for the fragility analysis.

5. Results and Discussion

This article concurrently addresses the uncertainty of the EDPs and bearing capacity, introduces a fragility analysis framework based on a seismic reliability method, and employs the SGLD to model the failure probability, resulting in the seismic collapse fragility curve of the structure. The conclusions of the article are as follows:

(1) According to the time history analysis results, the structural bearing capacity is not a definite value, which shows a certain distribution pattern. The uncertainty of the bearing capacity has a significant impact on the fragility of the structure. Compared to the inter-story drift ratio limits of 1/25 and 1/50, the median collapse IMs considering the uncertainty of the bearing capacity has increased by 13.2% and 87.3%, respectively.

(2) The lognormal fragility model exhibits a poor fit for the failure probability obtained through the seismic reliability method. Compared to the fragility function assumed by the lognormal, using the SGLD as the fragility function model can more accurately reconstruct the fragility curve. The SGLD is more flexible and more universal in different fragility analysis scenarios.

(3) The failure probability obtained by considering the uncertainty of the bearing capacity is lower than the failure probability obtained based on the deterministic threshold, and structural seismic risk assessment should consider the uncertainty of the bearing capacity.

However, there are still some limitations in this study that need to be further investigated. (1) The artificial seismic waves should consider more characteristics of real earthquakes, and it is necessary to explore the influence of these aspects on the seismic risk assessment. (2) The IDA method adopted in this paper is time-consuming. A method of efficiently and accurately obtaining the collapse point of the structure is worth further research. (3) This study takes reinforced concrete structures as an example for seismic collapse analysis. However, further research is required for different types of structures.

Author Contributions: Methodology, T.Z. and Y.L.; investigation, T.Z.; data curation, T.Z. and Y.L.; writing—original draft preparation, T.Z. and Y.L.; writing—review and editing, T.Z. All authors have read and agreed to the published version of the manuscript.

Funding: This research received no external funding.

Data Availability Statement: The original contributions presented in this study are included in the article. Further inquiries can be directed to the corresponding author.

Conflicts of Interest: The authors declare no conflicts of interest.

References

1. Barbat, A.H.; Moya, F.Y.; Canas, J.A. Damage Scenarios Simulation for Seismic Risk Assessment in Urban Zones. *Earthq. Spectra* **1996**, *12*, 371–394. [CrossRef]
2. Fragility Analysis Methods: Review of Existing Approaches and Application. *Nucl. Eng. Des.* **2017**, *323*, 245–258. [CrossRef]
3. Ruggieri, S.; Calò, M.; Cardellicchio, A.; Uva, G. Analytical-Mechanical Based Framework for Seismic Overall Fragility Analysis of Existing RC Buildings in Town Compartments. *Bull. Earthq. Eng.* **2022**, *20*, 8179–8216. [CrossRef]
4. Whitman, R.V.; Reed, J.W.; Hong, S.T. *Earthquake Damage Probability Matrices*; IIT Kanpur: Kanpur, India, 1974.
5. Rossetto, T.; Elnashai, A. Derivation of Vulnerability Functions for European-Type RC Structures Based on Observational Data. *Eng. Struct.* **2003**, *25*, 1241–1263. [CrossRef]
6. Vamvatsikos, D.; Cornell, C.A. Incremental Dynamic Analysis. *Earthq. Eng. Struct. Dyn.* **2002**, *31*, 491–514. [CrossRef]
7. Vamvatsikos, D.; Cornell, C.A. The Incremental Dynamic Analysis and Its Application to Performance-Based Earthquake Engineering. In Proceedings of the 12th European Conference on Earthquake Engineering, London, UK, 9–13 September 2002.
8. Scozzese, F.; Tubaldi, E.; Dall'Asta, A. Assessment of the Effectiveness of Multiple-Stripe Analysis by Using a Stochastic Earthquake Input Model. *Bull. Earthq. Eng.* **2020**, *18*, 3167–3203. [CrossRef]
9. Ni, P.P.; Wang, S.H.; Jiang, L.; Huang, R.Q. Seismic Risk Assessment of Structures Using Multiple Stripe Analysis. *Appl. Mech. Mater.* **2012**, *226–228*, 897–900. [CrossRef]
10. Jalayer, F.; De Risi, R.; Manfredi, G. Bayesian Cloud Analysis: Efficient Structural Fragility Assessment Using Linear Regression. *Bull. Earthq. Eng.* **2015**, *13*, 1183–1203. [CrossRef]
11. Nettis, A.; Gentile, R.; Raffaele, D.; Uva, G.; Galasso, C. Cloud Capacity Spectrum Method: Accounting for Record-to-Record Variability in Fragility Analysis Using Nonlinear Static Procedures. *Soil Dyn. Earthq. Eng.* **2021**, *150*, 106829. [CrossRef]
12. Sainct, R.; Feau, C.; Martinez, J.-M.; Garnier, J. Efficient Methodology for Seismic Fragility Curves Estimation by Active Learning on Support Vector Machines. *Struct. Saf.* **2020**, *86*, 101972. [CrossRef]
13. Kazemi, F.; Asgarkhani, N.; Jankowski, R. Machine Learning-Based Seismic Fragility and Seismic Vulnerability Assessment of Reinforced Concrete Structures. *Soil Dyn. Earthq. Eng.* **2023**, *166*, 107761. [CrossRef]
14. Feng, D.-C.; Cao, X.-Y.; Wang, D.; Wu, G. A PDEM-Based Non-Parametric Seismic Fragility Assessment Method for RC Structures under Non-Stationary Ground Motions. *J. Build. Eng.* **2023**, *63*, 105465. [CrossRef]
15. Liu, Z.; Zhang, Z. Artificial Neural Network Based Method for Seismic Fragility Analysis of Steel Frames. *KSCE J. Civ. Eng.* **2018**, *22*, 708–717. [CrossRef]
16. Xu, J.; Liao, L.; Zhang, Y.; Wang, D. Efficient Seismic Fragility Analysis of Structures from Dynamic Reliability Perspective. *Eng. Struct.* **2022**, *271*, 114889. [CrossRef]
17. Mai, C.; Konakli, K.; Sudret, B. Seismic Fragility Curves for Structures Using Non-Parametric Representations. *Front. Struct. Civ. Eng.* **2017**, *11*, 169–186. [CrossRef]
18. Zheng, X.-W.; Li, H.-N.; Shi, Z.-Q. Hybrid AI-Bayesian-Based Demand Models and Fragility Estimates for Tall Buildings against Multi-Hazard of Earthquakes and Winds. *Thin-Walled Struct.* **2023**, *187*, 110749. [CrossRef]
19. Wang, Q.; Geng, P.; Wang, L.; He, D.; Shen, H. Machine Learning-Driven Feature Importance Appraisal of Seismic Parameters on Tunnel Damage and Seismic Fragility Prediction. *Eng. Appl. Artif. Intell.* **2024**, *137*, 109101. [CrossRef]

20. Wei, B.; Zheng, X.; Jiang, L.; Lai, Z.; Zhang, R.; Chen, J.; Yang, Z. Seismic Response Prediction and Fragility Assessment of High-Speed Railway Bridges Using Machine Learning Technology. *Structures* **2024**, *66*, 106845. [CrossRef]
21. Bertero, V.V. Strength and Deformation Capacities of Buildings under Extreme Environments. *Struct. Eng. Mech.* **1977**, *53*, 211–215.
22. FEMA. *Recommended Seismic Design Criteria for New Steel Moment-Frame Buildings*; Federal Emergency Management Agency: Washington, DC, USA, 2000.
23. Fattahi, F.; Gholizadeh, S. Seismic Fragility Assessment of Optimally Designed Steel Moment Frames. *Eng. Struct.* **2019**, *179*, 37–51. [CrossRef]
24. Cardone, D.; Perrone, G.; Piesco, V. Developing Collapse Fragility Curves for Base-Isolated Buildings. *Earthq. Eng. Struct. Dyn.* **2019**, *48*, 78–102. [CrossRef]
25. Dabaghi, M.; Saad, G.; Allhassania, N. Seismic Collapse Fragility Analysis of Reinforced Concrete Shear Wall Buildings. *Earthq. Spectra* **2019**, *35*, 383–404. [CrossRef]
26. Lu, X.; Gu, D.; Zhao, P.; Zhang, C.; Cheav, P.C. Performance-based design of building structures for collapse under earthquake. *Build. Struct.* **2022**, *52*, 50–57+133. (In Chinese) [CrossRef]
27. Pourreza, F.; Mousazadeh, M.; Basim, M.C. An Efficient Method for Incorporating Modeling Uncertainties into Collapse Fragility of Steel Structures. *Struct. Saf.* **2021**, *88*, 102009. [CrossRef]
28. Cao, X.-Y.; Feng, D.-C.; Beer, M. Consistent Seismic Hazard and Fragility Analysis Considering Combined Capacity-Demand Uncertainties via Probability Density Evolution Method. *Struct. Saf.* **2023**, *103*, 102330. [CrossRef]
29. Jia, D.-W.; Wu, Z.-Y. Seismic Risk Analysis Based on Imprecise Distribution and Failure Probability Function under Multidimensional Limit State. *Structures* **2023**, *50*, 963–977. [CrossRef]
30. He, S.; Xu, J.; Zhang, Y. Reliability Computation via a Transformed Mixed-Degree Cubature Rule and Maximum Entropy. *Appl. Math. Model.* **2022**, *104*, 122–139. [CrossRef]
31. Li, G.; He, W.; Zeng, Y. An Improved Maximum Entropy Method via Fractional Moments with Laplace Transform for Reliability Analysis. *Struct. Multidiscip. Optim.* **2018**, *59*, 1301–1320. [CrossRef]
32. Xu, J. A New Method for Reliability Assessment of Structural Dynamic Systems with Random Parameters. *Struct. Saf.* **2016**, *60*, 130–143. [CrossRef]
33. Baker, J.W. Efficient Analytical Fragility Function Fitting Using Dynamic Structural Analysis. *Earthq. Spectra* **2015**, *31*, 579–599. [CrossRef]
34. Jeon, J.-S.; Mangalathu, S.; Song, J.; Desroches, R. Parameterized Seismic Fragility Curves for Curved Multi-Frame Concrete Box-Girder Bridges Using Bayesian Parameter Estimation. *J. Earthq. Eng.* **2019**, *23*, 954–979. [CrossRef]
35. Mangalathu, S.; Jeon, J.-S. Stripe-Based Fragility Analysis of Multispan Concrete Bridge Classes Using Machine Learning Techniques. *Earthq. Eng. Struct. Dyn.* **2019**, *48*, 1238–1255. [CrossRef]
36. Low, Y.M. A New Distribution for Fitting Four Moments and Its Applications to Reliability Analysis. *Struct. Saf.* **2013**, *42*, 12–25. [CrossRef]
37. Yuan, P.-T. On the Logarithmic Frequency Distribution and the Semi-Logarithmic Correlation Surface. *Ann. Math. Statist.* **1933**, *4*, 30–74. [CrossRef]
38. Amin, M.; Ang Alfredo, H.S. Nonstationary Stochastic Models of Earthquake Motions. *J. Eng. Mech. Div.* **1968**, *94*, 559–584. [CrossRef]
39. Yeh, C.-H.; Wen, Y.K. Modeling of Nonstationary Ground Motion and Analysis of Inelastic Structural Response. *Struct. Saf.* **1990**, *8*, 281–298. [CrossRef]
40. Shinozuka, M.; Deodatis, G. Simulation of Stochastic Processes by Spectral Representation. *Appl. Mech. Rev.* **1991**, *44*, 191–204. [CrossRef]
41. Liu, Z.; Liu, W.; Peng, Y. Random Function Based Spectral Representation of Stationary and Non-Stationary Stochastic Processes. *Probabilistic Eng. Mech.* **2016**, *45*, 115–126. [CrossRef]
42. Clough, R.W.; Penzien, J.; Griffin, D.S. *Dynamics of Structures*; Prentice Hall: Hoboken, NJ, USA, 1990.
43. Xu, J.; Feng, D.-C. Stochastic Dynamic Response Analysis and Reliability Assessment of Non-Linear Structures under Fully Non-Stationary Ground Motions. *Struct. Saf.* **2019**, *79*, 94–106. [CrossRef]
44. Yang, Q.; Tian, Y. *Earthquake Ground Motions & Artificial Generation*; Science Press: Beijing, China, 2014.
45. Yu, X.; Lu, D.; Li, B. Relating Seismic Design Level and Seismic Performance: Fragility-Based Investigation of RC Moment-Resisting Frame Buildings in China. *J. Perform. Constr. Facil.* **2017**, *31*, 04017075. [CrossRef]
46. *GB50011*; Code for Seismic Design of Buildings. Standardization Administration of China: Beijing, China, 2010.
47. Scott, B.; Park, R.; Priestley, M.J.N. Stress-Strain Behavior of Concrete Confined by Overlapping Hoops at Low and High Strain Rates. *ACI J. Proc.* **1982**, *79*, 13–27. [CrossRef]
48. Cao, X.-Y.; Wu, G.; Ju, J.-W.W. Seismic Performance Improvement of Existing RCFs Using External PT-PBSPC Frame Sub-Structures: Experimental Verification and Numerical Investigation. *J. Build. Eng.* **2022**, *46*, 103649. [CrossRef]

49. Liu, Z.; Zeng, B.; Wu, L. Simulation of non stationary ground motion by spectral representation and random functions. *J. Vib. Eng.* **2015**, *28*, 411–417. (In Chinese) [CrossRef]
50. FEMA. *Quantification of Building Seismic Performance Factors*; Federal Emergency Management Agency: Washington, DC, USA, 2009.
51. Xu, J.; Dang, C. A Novel Fractional Moments-Based Maximum Entropy Method for High-Dimensional Reliability Analysis. *Appl. Math. Model.* **2019**, *75*, 749–768. [CrossRef]

buildings

MDPI

Article

Seismic Performance Research on a Graded-Yielding Metal Brace with Self-Centering Functions

Zhonghai An [1], Wenming Wang [2,3], Hui Wang [4], Zhe Li [1], Debin Wang [3,4,*] and Guangcai Xie [1]

1 Shandong Electric Power Engineering Consulting Institute Co., Ltd., Jinan 250013, China; anzhonghai@sdepci.com (Z.A.); lizhe@sdepci.com (Z.L.); xieguangcai@sdepci.com (G.X.)
2 School of Civil Engineering, Shandong Jianzhu University, Jinan 250101, China; wangwenmingmr@sdjzu.edu.cn
3 Key Laboratory of the Ministry of Education for Structural Strengthening and Retrofitting of Buildings and Underground Space Engineering, Shandong Jianzhu University, Jinan 250101, China
4 School of Traffic Engineering, Dalian Jiaotong University, Dalian 116028, China; iceberg2@djtu.edu.cn
* Correspondence: wdb1215@djtu.edu.cn

Abstract: With the aim of achieving a graded-protection braced frame structure and minimizing the excessive residual deformation of traditional metal dampers under intense seismic action, a graded-yield-type metal self-centering brace (SC-GYMB) is proposed. The brace is composed of X-shaped and U-shaped steel plates with different yield point displacements, which jointly dissipate energy. Additionally, it employs a composite disc spring as a self-centering element to provide restoring force for the brace. The brace's basic structure and working mechanism are described, and the theoretical model for its restoring force is derived. The ABAQUS finite element software (ABAQUS 2021) is utilized to investigate the hysteretic performance of the SC-GYMB under low-cycle reciprocating load, while thoroughly discussing the influence of various model parameters on its key mechanical behavior. The results demonstrate a strong agreement between the theoretical restoring force model and the numerical simulation results. The hysteretic curves of the braces exhibit a distinct "flag" characteristic, indicating excellent energy dissipation capacity and self-centering performance. Moreover, these curves display a hierarchical yield behavior that satisfies the seismic performance requirements for different intensity earthquakes. The deformation mechanism of X-shaped steel sheets transitions from bending deformation during the initial loading stage to tensile deformation in the subsequent loading stage. Increasing the initial pre-compression force of the combined disc spring enhances the restoration performance of the brace. Augmenting the thickness of X-shaped or U-shaped steel sheets modifies the displacement and load at both the first and second yield points, thereby enhancing energy dissipation capacity and bearing capacity of the brace; however, it also leads to increased residual deformation.

Keywords: graded yield; self-centering performance; energy dissipation; restoring model; hysteresis curve

1. Introduction

Metal dampers have gained widespread application in practical engineering due to their advantages, such as having a simple mechanical model, a clear energy dissipation mechanism, excellent durability, and low cost [1–5]. Under seismic action, dampers usually undergo plastic deformation prior to that of structural members, consuming a considerable amount of seismic input energy. They can not only furnish additional damping to the structure but also supply additional lateral stiffness, thereby being widely employed in various braced frame structures [6,7]. To achieve the goal of adequately utilizing the energy dissipation performance of metal dampers under various seismic intensities, in recent years, some scholars have embarked on the research and development of energy dissipation and seismic mitigation devices with multiple yield points, in order to meet the energy dissipation requirements of structures under seismic actions of different intensities [8–11]. Wu et al. [12] proposed a metal sleeve graded damper by using steel bars with different yield displacements and aspect ratios as energy dissipation components. The experimental results showed that the damper possesses significant multi-stage yielding and failure characteristics and has excellent energy dissipation performance and deformation capacity. Chen et al. [13] proposed a graded-yielding metal damper composed of two different sizes of U-shaped steel. This damper has prominent graded-yielding characteristics and stable hysteretic performance, and its hysteresis curve closely matches the theoretical three-fold-line restoring force model. Yang et al. [14] proposed a double-yield-point buckling-restrained brace based on traditional buckling-restrained braces. Experimental results showed that the two yielding steel cores can be activated for energy dissipation at different yield displacements. When applied to a ten-story steel frame, the braced structure achieves dual-objective control effectiveness at different seismic intensities.

To align with the overall demand for structural resilience development, scholars both domestically and internationally have developed numerous passive energy dissipation and seismic-mitigation devices that possess both self-centering functionality and energy-dissipation capabilities [15,16]. Zhang et al. [17] constructed an assembled self-centering buckling-restrained brace (ASC-BRB). Relevant research has shown that, after the disc spring of the brace is compressed, its stiffness significantly increases, and the ASC-BRB can effectively reduce the deformation concentration effect of the structure under strong seismic loads. Ding et al. [18] proposed a guide-rod-type self-centering buckling-restrained brace by pre-compressing a combined disc spring and buckling-restrained brace in parallel. The self-centering system of this brace not only provides a self-centering force but also supplements energy dissipation, and its hysteresis curve exhibits typical flag-shaped characteristics and excellent energy dissipation performance. Jin et al. [19] arranged the energy dissipation core plate and combined disc spring in parallel along the axial direction, proposing a fully assembled self-centering buckling-restrained brace, which significantly reduces the residual deformation of the brace.

Based on the aforementioned analysis, the existing seismic energy dissipation devices still lack the dual energy dissipation characteristics of self-centering and graded yielding, making it difficult to simultaneously meet the graded seismic energy dissipation requirements of traditional braced frame structures under both frequent and rare earthquakes. Therefore, this paper constructs a self-centering graded-yield metal brace (SC-GYMB), utilizing X-shaped and U-shaped steel plates as its energy dissipation components. Specifically, the X-shaped steel plate has a smaller yield displacement, allowing it to be activated and undergo yielding energy dissipation during minor earthquakes. The U-shaped steel plate, on the other hand, exhibits a larger yield displacement and excellent fatigue resistance, enabling it to undergo yielding energy dissipation during major earthquakes, thereby meeting the energy dissipation requirements of structures under seismic actions of varying intensities. Simultaneously, the composite disc springs are selected as the self-centering components, imparting a self-centering force to the brace and reducing the residual deformation of the brace.

2. Basic Structure and Working Mechanism of a SC-GYMB

2.1. Basic Structure of a SC-GYMB

The basic structure of a SC-GYMB is shown in Figure 1, primarily consisting of a brace loading shaft, peripheral restraint side plates, restraint end plates, as well as an energy dissipation system and a self-centering system.

The loading shaft of the SC-GYMB is shaped as a variable-cross-section stepped rod. The energy dissipation system at its midpoint is composed of X-shaped steel plates and U-shaped steel plates, both of which are symmetrically arranged around the outer periphery of the loading shaft. The self-centering system consists of the combined disc springs, limit nuts, pressure-bearing plates, and disc spring baffle affixed to the peripheral restraint side plates. Two sets of self-centering systems are symmetrically distributed on both sides of the energy dissipation system. In this paper, a disc spring is chosen as the self-centering element, whose main role is to supply a restoring force for the brace after the initial pre-compression force is applied. Consequently, it can also be substituted with a ring spring, a spring, or other elastic materials possessing self-centering functionality. The restraint side plates and the restraint end plates, as well as the connections between the restraint side plates, are all connected using high-strength bolts. Similarly, the X-shaped steel plates and the U-shaped steel plates are also connected to the restraint side plates using high-strength bolts.

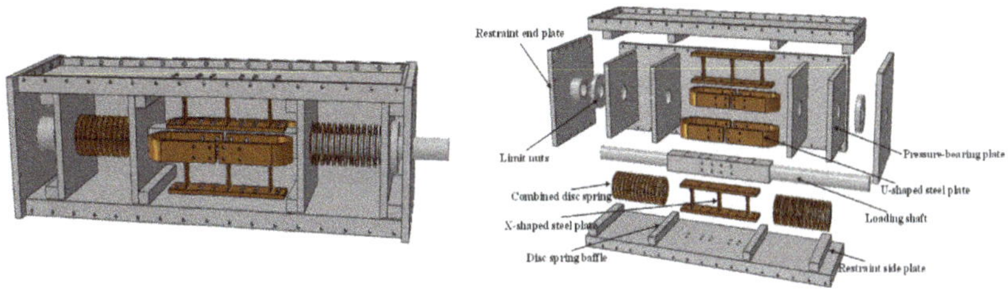

(a) (b)

Figure 1. Structure of the SC-GYMB: (**a**) Basic structure of the SC-GYMB; (**b**) Schematic diagram of components.

2.2. Working Mechanism of the SC-GYMB

The loading deformation process of the SC-GYMB is illustrated in Figure 2. Prior to loading the brace, the disc spring is positioned between the limit nuts and is compressed, thereby forming a self-centering system capable of exerting a restoring force. The pressure-bearing plate is tightly affixed to the limit nut due to the elastic force of the disc spring. At this moment, the loading shaft and the external restraint sleeve are in a self-balanced state, and they are connected in series.

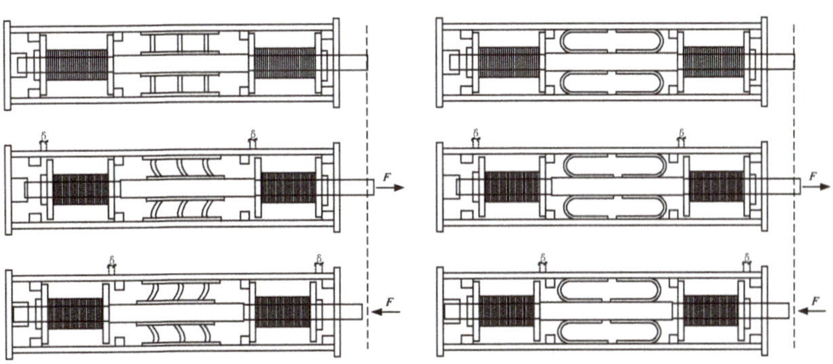

(a) (b)

Figure 2. Schematic diagram of the working principle of the SC-GYMB: (**a**) The deformation diagram of the X-shaped steel plate; (**b**) The deformation diagram of the U-shaped steel plate.

Taking the tension case of the SC-GYMB as an example, when the brace load is relatively small, the limit nut fixed on the loading shaft begins to drive the pressure-bearing plates to compress the combined disc spring, causing it to deform. Both the X-shaped and U-shaped steel plates fixed on the loading shaft undergo bending deformation. The loading shaft, the external restraint sleeve, the combined disc spring, the X-shaped steel plate, and the U-shaped steel plate only undergo linear elastic deformation, collectively providing stiffness for the brace. When the load exceeds the initial pre-compression force of the disc spring, relative movement occurs between the loading shaft and the external restraint sleeve, causing the combined disc spring to be compressed and the restoring force to increase. The X-shaped steel plate undergoes plastic deformation, activating the first yield point of the brace. At this point, the U-shaped steel plate has not yet yielded and is still in the linear elastic deformation stage. As the external load further increases, the restoring force of the brace increases, the plastic deformation area of the X-shaped steel plate expands further, and the U-shaped steel plate begins to enter the yield state, reaching the second yield point. At this point, both the X-shaped and U-shaped steel plates have entered the yield state, jointly dissipating the seismic input energy. When the brace is unloaded, the X- and U-shaped plates return to their initial equilibrium state under the restoring force. The deformation of the brace when compressed is exactly the same as when tensioned.

3. Restoring Force Model

3.1. Restoring Force Model of the X-Shaped Plate

The yield energy dissipation is achieved by an out-of-plane bending deformation of the X-shaped steel plate, whose dimensional parameters are shown in Figure 3. Here, $\boldsymbol{h}$ represents the height of the X-shaped steel plate, $\boldsymbol{t}$ denotes its thickness, and $\boldsymbol{b}$ signifies its width.

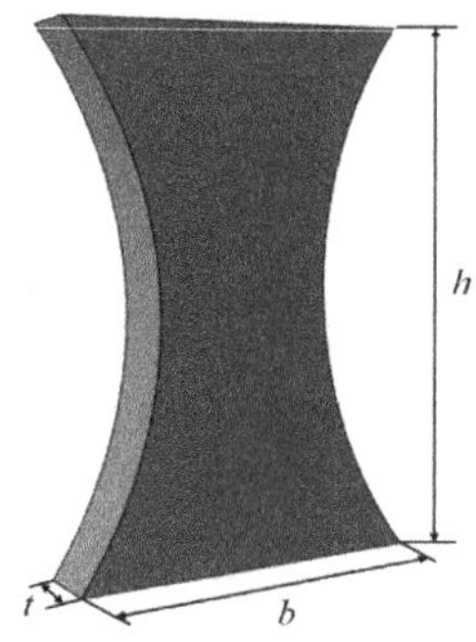

Figure 3. Schematic diagram of the X-shaped steel plate.

Compared to the bilinear material constitutive model featuring hardening characteristics, this paper opts for the ideal elastic-plastic model, which is more conducive to establishing the restoring force model. The derived restoring force model does not affect the overall variation rules of the mechanical properties of the brace. Additionally, the constitutive model of actual materials often exhibits hardening and strengthening properties, which are beneficial to the bearing capacity, energy dissipation capacity, and post-yield stiffness of the brace, ensuring a more adequate safety margin for the brace and not underestimating the damage incurred during the loading process of the brace. Therefore, the ideal elastic-plastic model is adopted for the constitutive model of the X-shaped steel plate in this paper. According to reference [20], the main mechanical performance parameters of the X-shaped steel plate, namely the yield load $\boldsymbol{F_{yx}}$, yield displacement $\boldsymbol{u_{yx}}$, and initial stiffness $\boldsymbol{k_x}$, are calculated using expressions (1), (2), and (3), respectively. The restoring force model is illustrated in Figure 4.

$$F_{yx} = \frac{n_x f_{yx} b t^2}{3h} \tag{1}$$

$$u_{yx} = \frac{f_{yx} h^2}{2E_x t} \tag{2}$$

$$k_x = \frac{F_x}{u_{yx}} = \frac{2n_x E_x b t^3}{3h^3} \tag{3}$$

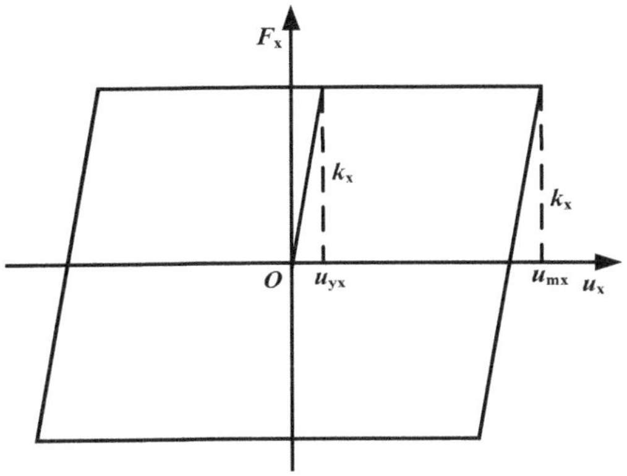

Figure 4. Restoring force model of the X-shaped steel plate.

In the formula, n_x denotes the number of X-shaped plates, f_{yx} denotes the yield strength of the X-shaped steel plate, and E_x signifies the elastic modulus of the X-shaped steel plate.

3.2. Restoring Force Model of the U-Shaped Plate

The U-shaped steel plate primarily consists of three parts: the upper straight segment, the semi-circular segment, and the lower straight segment, as illustrated in Figure 5. Specifically, the length of the straight segment is L, the effective length of the straight segment is l, the height is H, the width is B, the thickness is T, and the radius of the centerline of the semi-circular segment is R.

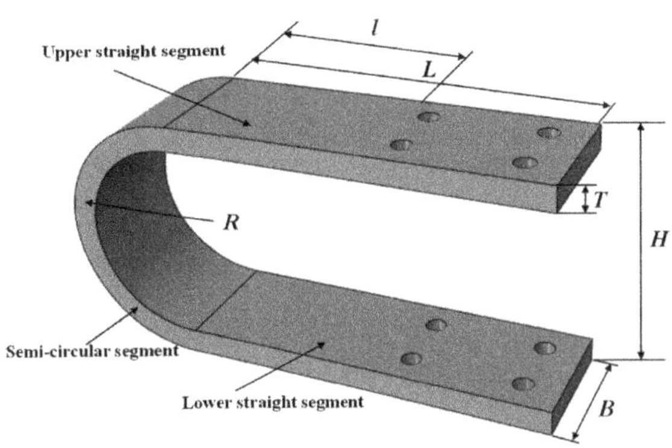

Figure 5. Schematic diagram of the U-shaped steel plate.

Existing experimental research findings indicate that U-shaped steel plates display excellent fatigue resistance [21,22]. Their performance remains relatively stable under limited displacement amplitudes, and the damage is also relatively minor. The ideal elastic-plastic model is also employed for the U-shaped steel plate. Its restoring force model is illustrated in Figure 6. The calculation formulas for its yield load F_{yU}, yield displacement u_{yU}, and initial stiffness k_U [23] are as follows.

$$F_{yU} = \frac{n_U f_{yU} B T^2}{4R + T} \tag{4}$$

$$u_{yU} = \frac{6 f_{yU} R^2 (\pi R + 4l)}{E_U T (T + 4R)} \tag{5}$$

$$k_U = \frac{F_{yU}}{u_{yU}} \quad (6)$$

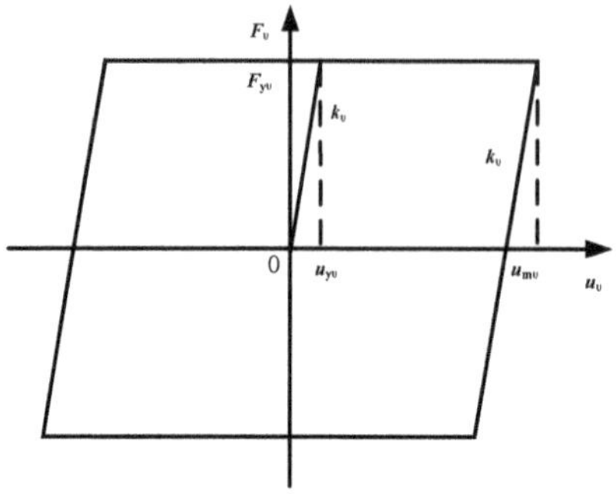

Figure 6. Restoring force model of the U-shaped steel plate.

In the formula: n_U represents the number of U-shaped steel plates; f_{yU} denotes the yield strength of the U-shaped steel plates; and E_U signifies the elastic modulus of the U-shaped steel plates.

3.3. Restoring Force Model of the Self-Centering System

The self-centering system comprises two sets of parallel-connected combined disc springs. Based on the disc spring specifications [24], A-series disc springs are chosen, and their constitutive model utilizes a bilinear elastic model, as illustrated in Figure 7.

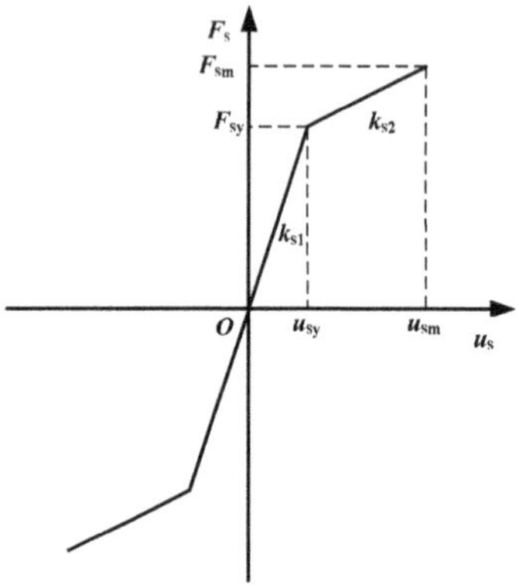

Figure 7. Restoring force model of self-centering system.

The total stiffness of the disc spring assembly, connected in parallel on both sides, is as follows:

$$k_S = \frac{2k_{dS}}{n_S} \quad (7)$$

The calculation formula for the stiffness of the loading shaft is as follows:

$$k_{in} = \frac{EA_1A_2A_3}{A_1A_2l_3 + A_1A_3l_2 + A_2A_3l_1} \quad (8)$$

The expression for calculating the stiffness of the restraint sleeve is as follows:

$$k_{out} = \frac{EA_0}{l_0} \quad (9)$$

The stiffness of the loading shaft and the restraint sleeve in series can be calculated using the following formula:

$$k_0 = \frac{k_{\text{in}} k_{\text{out}}}{k_{\text{in}} + k_{\text{out}}} \tag{10}$$

The calculation formula for the first and second stiffness of the self-centering system is as follows, respectively:

$$k_{S1} = k_0 + k_S \tag{11}$$

$$k_{S2} = k_S \tag{12}$$

In the equations above, k_{dS} denotes the stiffness of a single disc spring; n_S denotes the number of disc spring leaves on one side; E signifies the elastic modulus of the disc spring; A_1, A_2, and A_3 correspond to the cross-sectional areas of the loading shaft sections I, II, and III, respectively; l_1, l_2, and l_3 correspond to the lengths of the loading shaft sections I, II, and III, respectively; A_0 is the cross-sectional area of the restraint sleeve; and l_0 is the length of the restraint sleeve.

3.4. Restoring Force Model of the SC-GYMB

The restoring force model of the SC-GYMB can be formed by the linear superposition of the X-shaped steel plate restoring force model, the U-shaped steel plate restoring force model, and the self-centering system restoring force model. The overall brace restoring force model is shown in Figure 8.

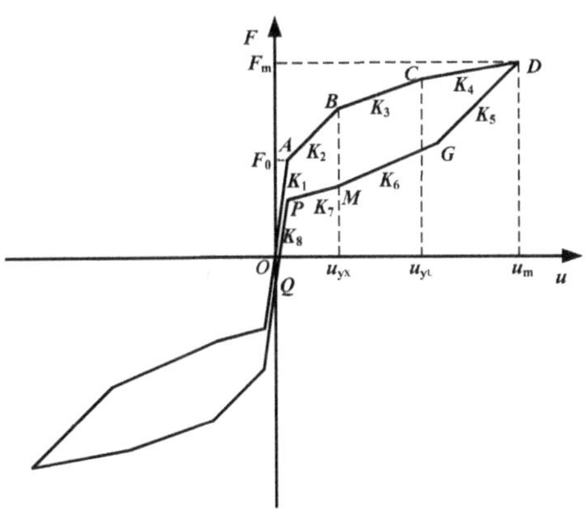

Figure 8. The restoring force model of the self-centering graded yield metal brace.

Taking the first cycle of tensile loading as an example, the initial pre-compression force of the self-centering system's combined disc spring is set as F_0. Its restoring force model can be divided into the following eight stages.

In the first stage (OA segment), when the tensile force applied to the brace precisely equals the initial pre-compression force of the combined disc spring, there is no relative displacement between the loading shaft and the restraint sleeve. The relationship between the two is in a series state, and the deformation of the brace is primarily due to the elastic deformation of both components. The X-shaped and U-shaped plates are inactive. At this stage, the brace stiffness K_1 is the sum of the pre-yield stiffness of the X-shaped steel plate, the pre-yield stiffness of the U-shaped steel plate, and the first stiffness of the self-centering system. The load and displacement at point A are expressed as follows:

$$K_1 = k_X + k_U + k_{S1} \tag{13}$$

$$F_{d0} = F_0 \tag{14}$$

$$u_0 = \frac{F_{d0}}{K_1} \tag{15}$$

In the second stage (AB segment), after overcoming the initial pre-compression force F_0 of the combined disc spring, continue to apply load. The brace loading shaft and the restraint sleeve undergo relative motion. At this point, the X- and U-shaped plates are in a state of linear elastic deformation until the X-plate enters the yield phase. Due to the extreme stiffness of the loading shaft and the restraint sleeve, their elastic deformation is relatively tiny compared to the total deformation of the brace, so their contribution to the overall deformation of the brace can be neglected. The stiffness of the brace is the parallel stiffness of the X-shaped plate, the U-shaped plate, and the combined disc spring. Thus, the load at point B is expressed as follows:

$$K_2 = k_X + k_U + k_{S2} \tag{16}$$

$$F_{dyX} = F_{d0} + K_2 \times (u_{yX} - u_0) \tag{17}$$

In the third stage (BC segment), as the brace continues to be loaded, the X-shaped steel plate begins to yield, while the U-shaped steel plate remains unyielding at this point. The compression deformation of the disc spring increases until the U-shaped steel plate yields. The stiffness of the brace during this stage is equal to the sum of the stiffness of the X-shaped steel plate after yielding, the stiffness of the U-shaped steel plate before yielding, and the stiffness of the combined disc spring. Since an ideal elastic-plastic model is used for the X-shaped steel plate, the stiffness after yielding is calculated to be zero. Thus, the load at point C is expressed as follows:

$$K_3 = k_U + k_{S2} \tag{18}$$

$$F_{dyU} = F_{dyX} + K_3 \times (u_{yU} - u_{yX}) \tag{19}$$

In the fourth stage (CD segment), as the brace continues to be loaded, the U-shaped steel plate begins to yield, and the combined disc spring is further compressed until it reaches the maximum design displacement of the first cycle. The stiffness at this stage is the sum of the stiffness of the X-shaped plate and the U-shaped plate after yielding, and the stiffness of the combined disc spring. Since the U-shaped steel plate also adopts an ideal elastic-plastic model, the stiffness after yielding is calculated to be zero. Therefore, the load expression at point D is as follows:

$$K_4 = k_{S2} \tag{20}$$

$$F_{dm} = F_{dyU} + K_4 \times (u_m - u_{yU}) \tag{21}$$

In the fifth stage (DG segment), the brace begins to unload, with both the X-shaped steel plate and the U-shaped steel plate undergoing simultaneous unloading until the X-shaped steel plate reversely yields. The displacement in this stage consists of two parts: the elastic unloading of the X-shaped steel plate and the reverse elastic loading. Both parts exhibit the same stiffness as the loading stiffness in the second stage. Hence, the displacement and load at point G, respectively, are as follows:

$$K_5 = K_2 \tag{22}$$

$$u_G = u_m - 2u_{yX} \tag{23}$$

$$F_G = F_{dm} - K_5 \times (u_m - u_G) \tag{24}$$

In the sixth stage (GM segment), unloading continues until the U-shaped steel plate yields. The total displacement of this stage and the fifth stage (DG segment) is the combined displacement of the elastic unloading and reverse elastic loading of the U-shaped steel plate.

The stiffness is the same as the loading stiffness in the third stage. Thus, the displacement and load at point M, respectively, are as follows:

$$K_6 = K_3 \tag{25}$$

$$u_M = u_m - 2u_{yU} \tag{26}$$

$$F_M = F_G - K_6 \times (u_G - u_M) \tag{27}$$

In the seventh stage (MP segment), the brace continues to unload until the pressure-bearing plate just contacts the disc spring baffle. At this point, the deformation of the X- and U-shaped plates essentially returns to zero. At this time, the stiffness of the brace is equivalent to the loading stiffness in the fourth stage. Then, the displacement and load at point P, respectively, are as follows:

$$K_7 = K_4 \tag{28}$$

$$u_P = u_0 \tag{29}$$

$$F_P = F_M - K_7 \times (u_M - u_P) \tag{30}$$

In the eighth stage (PQ segment), upon contact between the pressure-bearing plate and the disc spring baffle, the loading shaft and the restraint sleeve cease to undergo relative motion. Both the X-shaped and U-shaped plates are in the yield regime. The external load is gradually unloaded to zero, and the stiffness is slightly less than that in the first stage. It is the sum of the stiffness of the loading shaft and the restraint sleeve in series, the stiffness of the combined disc spring, and the stiffness after the X-shaped and U-shaped energy dissipation segments yield. Since both of these energy dissipation units adopt an ideal elastic-plastic model, the stiffness after yielding is calculated to be zero. Therefore, the displacement and load at point Q are, respectively, as follows:

$$K_8 = k_0 + k_S \tag{31}$$

$$u_Q = 0 \tag{32}$$

$$F_Q = F_P - K_8 \times (u_P - u_Q) \tag{33}$$

From Equation (33), it can be seen that when the initial pre-pressure of the self-centering system exceeds the yield load of the energy dissipation system, $F_P \geq 0$, indicating that there is no residual deformation in the brace. Otherwise, $F_P \leq 0$, indicating that there is residual deformation in the brace.

4. Establishment and Analysis of SC-GYMB Finite Element Model

4.1. Establishment of the SC-GYMB Finite Element Model

4.1.1. Design Parameters of SC-GYMB

To verify the seismic performance of SC-GYMB, the mechanical properties of the brace under low-cycle reciprocating loads were studied using the ABAQUS finite element software. The total length of the brace component is 1.16 m. The self-centering system consists of two sets of paired disc springs, with each set containing 28 discs, achieving a total pre-pressure of 47.68 kN. The energy dissipation system consists of 6 X-shaped plates and 4 U-shaped plates. The design parameters and dimensions of the disc springs, X-shaped steel plates, and U-shaped steel plates are shown in Table 1, where D, d, and t represent the outer diameter, inner diameter, and thickness of the disc springs, respectively.

Table 1. Main design parameters of the SC-GYMB.

Disc Spring/mm			X-Shaped Steel Plate/mm			U-Shaped Steel Plate/mm			
D	*d*	*t*	*h*	*b*	*t*	*B*	*T*	*R*	*l*
100	51	6	80	50	5	50	10	40	40

4.1.2. Establishment of a Finite Element Model

The SC-GYMB model is established based on the ABAQUS/Standard module. The disc spring is simulated using three-dimensional shell elements S4R, with the actual thickness of the shell elements taken into account. During assembly, a gap is reserved based on the central axis of the disc spring. The combined disc spring is stacked in a single-piece manner, with constraints imposed on the disc spring itself to limit its rotational freedom, ensuring that the combined disc spring can only undergo axial movement. The pre-pressure of the disc spring is simulated by applying displacement to the pressure-bearing plate. The remaining components of the brace were simulated using the 8-node hexahedral reduced integral solid element C3D8R. The mutual fixation relationships between the X-shaped steel plate, U-shaped steel plate, loading shaft, and restraint sleeve are all simulated using the "Tie" command. Normal contact is applied between the remaining components, with no friction set in the tangential direction. To enhance computational efficiency, an integrated modeling approach is adopted for all components of the restraint sleeve, simulating the consolidation relationships among them. Although this modeling approach disregards the effect of inter-component friction, its effect on the overall performance of the brace is extremely minimal and thus it is neglected in this paper. The external restraint sleeve is fixedly constrained, with a reference point RF established on the end face of the loading shaft and controlled using motion coupling. The finite element model of the SC-GYMB is shown in Figure 9.

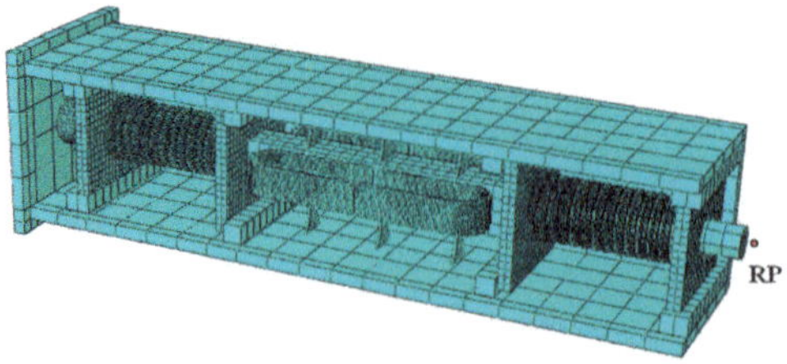

Figure 9. Finite element model of the SC-GYMB.

Both the X- and U-shaped plates were constructed from Q235 steel using an ideal elastic-plastic constitutive model. The elastic modulus is set at $E = 2.06 \times 10^5$ MPa, and the Poisson's ratio is $v = 0.3$. The constitutive model and parameters of the disc spring material align with the elastic modulus and Poisson's ratio of the energy-dissipation steel plate. The yield strength of both X-shaped and U-shaped steel plates is designated at 235 MPa, while the ultimate strength is set at 375 MPa. The loading shaft, restraint sleeve, and various connecting components are crafted from Q345 grade steel, boasting yield strength of 345 MPa and an ultimate strength of 554 MPa. Low-cycle reciprocating loads are exerted on the RF located at the end of the loading shaft, utilizing a displacement-controlled loading mode. The initial loading displacement is 5 mm, and the loading amplitude increment is set at 5 mm until loading reaches 20 mm, with each amplitude level subjected to one cycle of loading.

4.2. Verification of the Restoring Force Model for the SC-GYMB Reset System

In light of the relatively straightforward mechanical model of metal materials, which eases numerical simulations, and based on the experimental outcomes from references [25,26], it can be deduced that the finite element modeling approach adopted in this paper can effectively simulate the hysteretic behavior of U-shaped steel plates and X-shaped steel plates under low-cycle reciprocating loads. Therefore, this article does not further separately verify the hysteretic behavior of X-shaped and U-shaped steel plates. Employing this modeling method, we derived the hysteresis curve for the SC-GYMB self-centering system and juxtaposed it against theoretical calculations, as illustrated in Figure 10. Obviously, the theoretical outcomes derived from disc spring specifications [24] align closely with the numerical simulations, underscoring the precision and efficacy of the finite element simulations for the self-centering system.

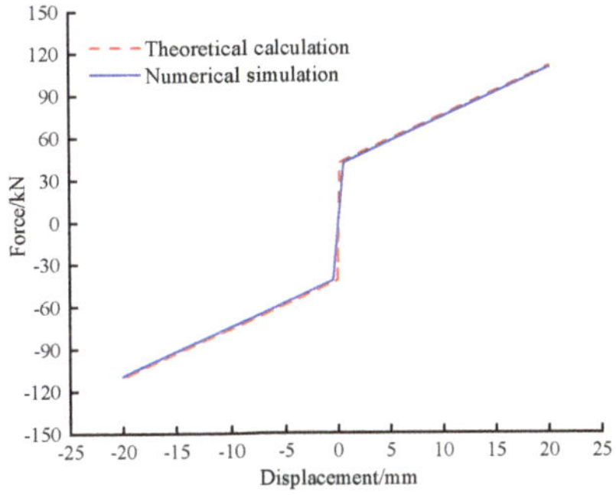

Figure 10. Theoretical calculation and simulation results of the self-centering system.

4.3. Verification of the Restoring Force Model for the SC-GYMB

Figure 11 presents a comparison chart of the hysteresis curves from the SC-GYMB finite element simulation and the theoretical calculation results. It can be obviously observed that the hysteresis curves from the numerical simulation and the theoretical calculation are generally consistent, effectively verifying the accuracy of the theoretical restoring force model. The initial stiffness obtained from the theoretical calculation is slightly higher than the simulation results, and the numerical simulation results are higher than the theoretical calculation values at various displacement amplitudes. From Figure 11b, which depicts the tensile deformation process during the first loading cycle, it can be observed that both hysteresis curves have two distinct yield points, corresponding to the yield displacements of the X-shaped steel plate and the U-shaped steel plate, respectively. This reflects the design concept of graded energy dissipation for braces, which can meet the seismic design requirements for graded energy dissipation under different intensities of seismic excitation. The comparison results of yield displacement and yield load are shown in Table 2. The main reasons for the errors between numerical simulation and theoretical calculation are as follows: ① In the theoretical calculation, it is assumed that each component is a rigid body, without considering the elastic deformation of each part, resulting in a higher initial stiffness in the theoretical calculation; and ② in the theoretical calculation process, the transition of the deformation mechanism for the X-shaped steel plate is not considered, while in the numerical simulation process, it is found that the deformation of the X-shaped steel plate transitions from bending deformation to tensile deformation in the later loading stages, thus increasing the brace load.

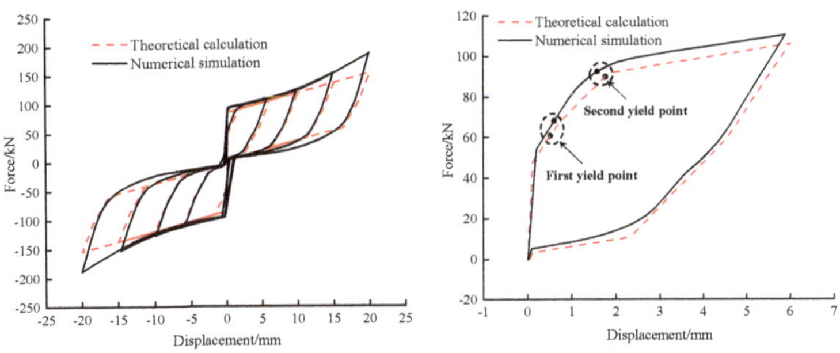

Figure 11. Comparison of theory and numerical simulation results: (**a**) Comparison of hysteresis curves; (**b**) Comparison of the first half-cycle.

Table 2. Comparison of theoretical calculation and numerical simulation results.

Performance Parameters	K_1/(kN/mm)	u_{yx}/mm	F_{dyx}/kN	u_{yu}/mm	F_{dyu}/kN	F_{dm}/kN
Theoretical calculation	453.45	0.73	67.49	1.84	91.79	154.08
Numerical simulation	427.23	0.90	77.73	1.75	93.93	188.19

The simulated brace hysteresis curve depicted in Figure 11 is smooth and plump, exhibiting stable hysteretic characteristics and displaying typical "flag" features. During the initial loading phase, the deformation mechanism of the X-shaped plate manifests itself as bending deformation. As the displacement amplitude continues to increase, its deformation mechanism transitions from bending deformation to tensile deformation. The stiffness and bearing capacity after yielding significantly increase, which helps enhance the energy dissipation performance of the brace after yielding. The increased stiffness after yielding ensures that the brace can still maintain a high stiffness at this time, which is extremely beneficial for reducing the risk of collapse of the structure during a major earthquake. At the same time, the brace exhibits excellent self-centering performance with a maximum residual displacement of only 0.99 mm.

Figures 12a and 12b, respectively, present the stress and equivalent plastic strain cloud maps of the X-shaped steel and U-shaped steel plates when the displacement is loaded to 0.9 mm and 1.8 mm. As can be seen from the figures, when the displacement is loaded to 0.9 mm, the X-shaped steel plate yields first, with its yield point located at the fixed end of the steel plate. At this time, the U-shaped plate is in the linear elastic phase and has not yet yielded. When the displacement is loaded to 1.8 mm, the U-shaped steel plate begins to yield, with the yield point located at the junction of the straight segment and the circular arc segment. By this time, the X-shaped plate had yielded extensively, thus validating the graded yield properties of the brace.

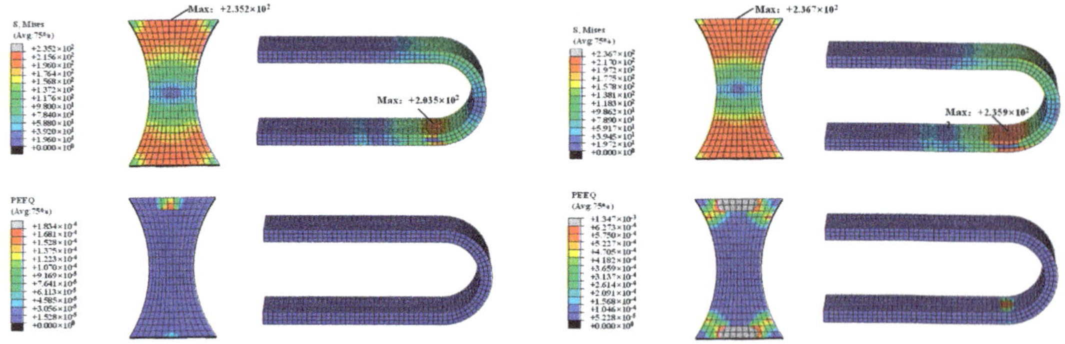

(a) (b)

Figure 12. Stress and stress nephogram of the energy dissipation parts in different loading stages: (**a**) The X-shaped steel plate enters the yield state; (**b**) The U-shaped steel plate enters the yield state.

5. Analysis of the Influence of Design Parameters on SC-GYMB

When designing SC-GYMB, it is imperative to comprehensively consider the effects of various structural parameters on the hysteretic performance of the brace. Based on the mutual verification between numerical simulation and theoretical calculation results, we have taken into account the effects of four key parameters: the initial pre-compression force of the combined disc spring, the thickness of the X-shaped steel plate, the thickness of the U-shaped steel plate, and the width of the U-shaped steel plate on the hysteretic performance of the brace. The specific structural parameters of the brace are detailed in Table 3.

Table 3. The main parameters of the SC-GYMB.

Parameter Types	Model Number	F_0 (kN/mm)	t/mm	B/mm	T/mm
Initial pre-compression force	D-$F_0$32-t5-B50-T10	32	5	50	10
	D-$F_0$40-t5-B50-T10	40	5	50	10
	D-$F_0$48-t5-B50-T10	48	5	50	10
	D-$F_0$56-t5-B50-T10	56	5	50	10
Thickness of X-shaped steel plate	D-$F_0$48-t2-B50-T10	48	2	50	10
	D-$F_0$48-t5-B50-T10	48	5	50	10
	D-$F_0$48-t8-B50-T10	48	8	50	10
	D-$F_0$48-t11-B50-T10	48	11	50	10
Thickness of U-shaped steel plate	D-$F_0$48-t5-B50-T7	48	5	50	7
	D-$F_0$48-t5-B50-T10	48	5	50	10
	D-$F_0$48-t5-B50-T13	48	5	50	13
	D-$F_0$48-t5-B50-T16	48	5	50	16
Thickness of U-shaped steel plate	D-$F_0$40-t5-B50-T10	48	5	10	10
	D-$F_0$48-t5-B30-T10	48	5	30	10
	D-$F_0$48-t5-B50-T10	48	5	50	10
	D-$F_0$48-t5-B60-T10	48	5	60	10

5.1. The Influence of the Combined Disc Spring Pre-Compression Force on the Hysteretic Behavior of SC-GYMB

The hysteresis curves of SC-GYMB under various initial pre-pressures are illustrated in Figure 13, while the simulation results for the main performance parameters are presented in Table 4. An increase in initial pre-compression force did not alter the initial stiffness of the brace. This is because, prior to the activation of the self-centering system, the stiffness of the brace primarily relies on the combined stiffness of the loading shaft and restraint sleeve in series, as well as the stiffness of the X-shaped and U-shaped steel plates prior to yielding. The stiffness contribution from the combined disc spring is negligible. Both the first and second yield point loads of the brace increase with the rise in initial pre-compression force, and the rate of increase closely aligns with the increase in initial pre-compression force. Similarly, the maximum load capacity of the brace exhibits a comparable trend.

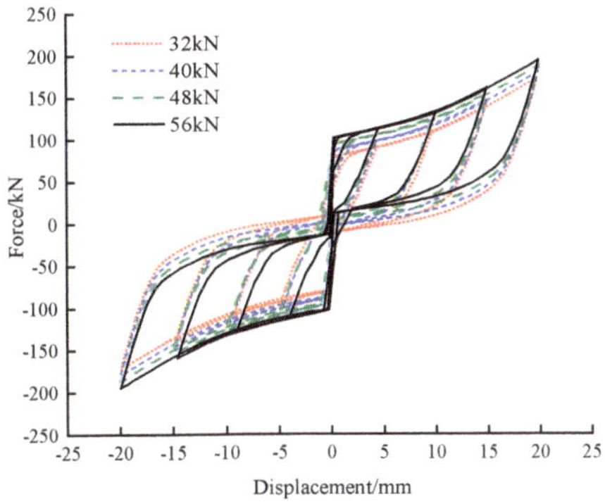

Figure 13. The effect of the self-centering system's initial pre-compression force.

Table 4. Parameter analysis results of the SC-GYMB.

Parameter Types	Model Number	K_1 (kN/mm)	u_{yx}/mm	F_{dyx}/kN	u_{yu}/mm	F_{dyu}/kN	F_{dm}/kN	u_{dr}/mm
Initial pre-compression force	D-$F_0$32-t5-B50-T10	427.23	0.90	61.97	1.75	78.67	172.62	6.12
	D-$F_0$40-t5-B50-T10	427.23	0.90	69.82	1.75	86.23	180.63	1.60
	D-$F_0$48-t5-B50-T10	427.23	0.90	77.73	1.75	93.93	188.03	0.76
	D-$F_0$56-t5-B50-T10	427.23	0.90	85.69	1.75	101.99	195.93	0.57
Thickness of X-shaped steel plate	D-$F_0$48-t2-B50-T10	417.85	1.54	79.55	1.75	82.06	147.82	0.18
	D-$F_0$48-t5-B50-T10	427.23	0.90	77.73	1.75	93.93	188.03	0.76
	D-$F_0$48-t8-B50-T10	458.37	0.55	79.38	1.75	129.82	217.27	6.14
	D-$F_0$48-t11-B50-T10	524.27	0.43	90.25	1.75	157.08	256.91	11.02
Thickness of U-shaped steel plate	D-$F_0$48-t5-B50-T7	417.35	0.90	62.38	2.28	76.33	165.88	0.40
	D-$F_0$48-t5-B50-T10	427.23	0.90	77.73	1.75	93.93	188.03	0.76
	D-$F_0$48-t5-B50-T13	445.21	0.90	91.50	1.31	122.12	223.75	9.09
	D-$F_0$48-t5-B50-T16	473.75	0.90	122.62	1.13	162.66	277.76	13.56
Width of U-shaped steel plate	D-$F_0$48-t5-B10-T10	415.20	0.90	63.82	1.75	70.12	154.62	0.09
	D-$F_0$48-t5-B30-T10	421.22	0.90	72.54	1.75	82.03	169.37	0.10
	D-$F_0$48-t5-B50-T10	427.23	0.90	77.73	1.75	93.93	188.03	0.76
	D-$F_0$48-t5-B60-T10	430.23	0.90	79.65	1.75	101.02	194.74	1.28

Note: u_{dr} represents the residual displacement of brace.

5.2. The Influence of the Thickness of the X-Shaped Steel Plate on the Hysteretic Behavior of the SC-GYMB

The SC-GYMB hysteresis curves for various X-shaped steel plate thicknesses are depicted in Figure 14. Upon consulting Table 4, it becomes evident that as the thickness of the brace's X-shaped plate increases from 2 mm to 11 mm, the enclosed region of the brace's hysteresis curve is significantly enlarged, indicating a substantial enhancement of the energy dissipation capacity. Both the first and second yield point loads of the brace have increased, with the increase being more pronounced at the second yield point. The maximum load increase occurs when the displacement reaches its maximum value. The primary reason for this is that, at higher displacements, the deformation mechanism of the X-shaped steel plate transitions from bending to tensile deformation, making the steel plate thickness a particularly influential factor. As the thickness of the X-shaped plate increases, the initial stiffness of the brace also increases, and the unloading stiffness exhibits a similar trend to the initial stiffness of the brace.

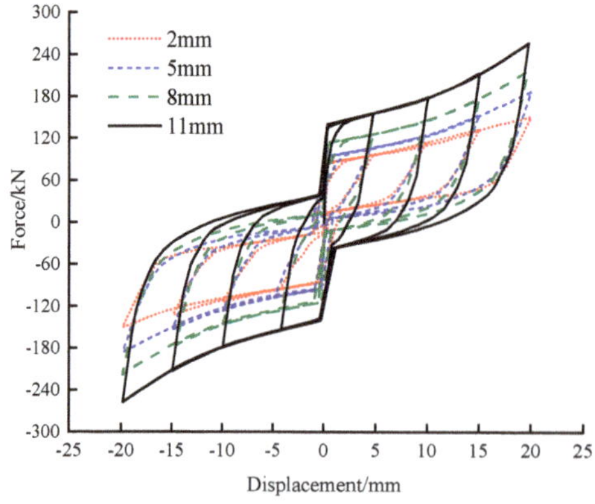

Figure 14. The effect of the X-shaped steel plate's thickness.

Increasing the thickness of the X-shaped steel plate reduces the displacement of the first yield point of the brace, which is consistent with the conclusion derived from the theoretical Formula (2) mentioned earlier. The displacement of the second yield point remains unchanged, as it primarily depends on the U-shaped steel plate and is independent of the thickness of the X-shaped steel plate. With the initial pre-compression force remaining unchanged, the residual displacement of the brace increases significantly as the thickness of the X-shaped steel plate increases. As the thickness of the X-shaped steel plate increases from 2 mm to 11 mm, the residual displacement of the brace increases from 0.18 mm to 11.02 mm, representing a maximum increase of over 100%. The main reason for this is that the increase in the thickness of the X-shaped steel plate leads to an increase in the yield load of the energy dissipation system, which in turn results in a decrease in the brace reset ratio.

5.3. The Influence of the Thickness and Width of the U-Shaped Steel Plate on the Hysteretic Behavior of the SC-GYMB

The SC-GYMB hysteresis curves under different U-shaped steel plate thicknesses are shown in Figure 15a. Combined with Table 4, it can be seen that as the U-shaped steel plate thickness increases from 7 mm to 16 mm, the shape of the brace hysteresis curve remains unchanged, but the area of the hysteresis loop significantly increases, significantly enhancing the energy dissipation capacity. Both the first and second yield point loads of the brace have increased, with the increase in the second yield point load being greater than that of the first yield point load. The initial stiffness of the brace increases with the increase in the thickness of the U-shaped steel plate, but its effect is less significant than that of the thickness of the X-shaped steel plate on the brace stiffness. The unloading stiffness exhibits a similar variation pattern to the initial stiffness of the brace. The increase in U-shaped steel plate thickness has a similar effect on the displacement at different yield points of the brace. As the thickness increases, the displacement at the first yield point remains unchanged, while the displacement at the second yield point decreases. The reason for this is detailed in Formula (5) in the previous theoretical section. The residual displacement of the test specimen D-$F_0$48-t5-B50-T7, which features a U-shaped steel plate with a thickness of 7 mm, is 0.40 mm. However, when the thickness is augmented to 16 mm, the residual displacement of the test specimen D-$F_0$48-t5-B50-T16 escalates to 13.56 mm. This substantial increase is mainly attributed to the fact that, under the identical initial pre-compression force of the brace, the first yield load of the brace energy dissipation system rises from 14.38 kN to 74.62 kN, and the brace reset ratio drops significantly. At this juncture, the initial pre-compression is no longer adequate to meet the requirements for the restoring force of the brace.

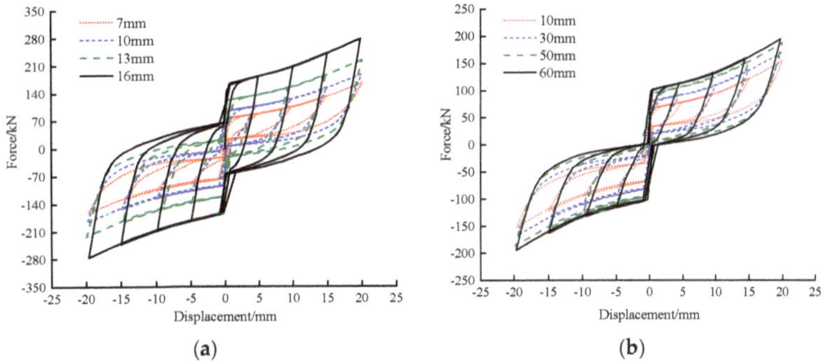

Figure 15. The effects of the U-shaped steel plate's thickness and width: (**a**) The effects of the U-shaped steel plate's thickness; (**b**) The effects of the U-shaped steel plate's width.

The hysteresis curves of the SC-GYMB with varying widths of U-shaped steel plates are depicted in Figure 15b. By referring to Table 4, it becomes evident that, in comparison to the impact of U-shaped steel sheet thickness on its hysteresis performance, the influence of its width is notably diminished. Altering the width of the U-shaped steel plate did not impact the yield point displacement of the brace, with both yield point displacements remaining unaltered. As the width of the U-shaped steel plate increases, the brace load rises during various loading stages. However, the increase in load at different displacement amplitudes is relatively minor. The initial stiffness is also minimally influenced by the width of the U-shaped steel plate. In comparison with a U-shaped steel plate with a width of 10 mm, when the width is augmented to 60 mm, its initial stiffness merely increases by 3.6%. Although the energy dissipation capacity has slightly improved, the residual displacement has not undergone significant changes, merely increasing from 0.09 mm to 1.28 mm, which is still negligible.

6. Conclusions

This paper introduces a metal self-centering brace, the SC-GYMB, featuring graded yielding characteristics. Leveraging the distinct yield point displacement properties of X-shaped and U-shaped steel plates, it achieves the graded yielding function of the brace. Additionally, it employs pre-compression combined disc springs to realize its self-centering capability. Through theoretical derivation, a restoring force model for the brace was derived. Based on ABAQUS software, a finite element model of the brace was constructed. The hysteretic performance of the brace was investigated by applying low-cycle reciprocal loading to the brace. The simulation results were compared with the theoretical restoring force model. Further parametric studies of the key model parameters of the brace have been carried out. The main conclusions are as follows:

(1) A novel type of self-centering brace, the SC-GYMB, featuring hierarchical yield behavior, is proposed. In this system, a composite disc spring serves as the self-centering element, while X-shaped and U-shaped steel plates function as energy dissipation components. This brace is capable of meeting the seismic demands of structures under varying levels of seismic excitation. The restoring force model derived from the theoretical derivation closely aligns with the hysteresis curves obtained from numerical simulations, thereby validating the rationality and accuracy of the proposed restoring force model.

(2) The hysteresis curve of the SC-GYMB is smooth and plump, exhibiting typical flag-shaped characteristics. X-shaped steel plates and U-shaped steel plates, featuring different yield point displacements, are capable of fulfilling the requirements of graded yielding. During the later stages of loading, the deformation mechanism of the X-shaped steel plate shifts from bending deformation to tensile deformation, which effectively enhances the stiffness and load-bearing capacity after the brace yields.

(3) Increasing the initial pre-compression force of the self-centering system can enhance the self-centering performance of the SC-GYMB. By increasing the thickness of the X-shaped or U-shaped plate, one can alter the displacement of the first and second yield points of the brace and elevate the brace load during various loading stages. However, this also results in an increase in the residual deformation of the brace.

Currently, only preliminary research has been conducted on the seismic performance of the SC-GYMB proposed in this paper, focusing solely on numerical simulation and theoretical analysis. There are still discrepancies between the simulation results and theoretical predictions, and the selection of material constitutive models is also on the conservative side. Therefore, in subsequent research, model tests will be conducted on the aforementioned SC-GYMB. Additionally, a nonlinear dynamic time-history analysis will be performed on the frame structure with the additional SC-GYMB, based on the grade design concept, to study the seismic control effect of the braced frame structure.

Author Contributions: Conceptualization, Z.A. and D.W.; methodology, W.W. and Z.L.; software, H.W. and D.W.; validation, H.W. and G.X.; formal analysis, W.W. and D.W.; investigation, Z.A. and D.W.; resources, Z.A.; data curation, H.W.; writing—original draft preparation, Z.A. and H.W.;

writing—review and editing, D.W.; visualization, W.W.; supervision, Z.A.; funding acquisition, W.W. and D.W. All authors have read and agreed to the published version of the manuscript.

Funding: This study was supported by the Key Laboratory of Building Structural Retrofitting & Underground Space Engineering (Grant No. MEKL202309), the National Natural Science Foundation of China (Grant No. 51908340) and the financial support from Major Scientific & Technological Innovation Projects of Shandong Provence (Grant No. 2021CXGC011204).

Data Availability Statement: Data are contained within the article.

Conflicts of Interest: Authors Zhonghai An, Zhe Li, and Guangcai Xie are employed by the Shandong Electric Power Engineering Consulting Institute Co., Ltd. The remaining authors declare that the research was conducted in the absence of any commercial or financial relationships that could be construed as a potential conflict of interest.

References

1. Wang, D.B.; Fu, X.; Liu, H.J. Study on the performance of a novel rotation-amplified brace and its vibration control for a frame structure. *Structures* **2023**, *57*, 105142. [CrossRef]
2. Oh, S.H.; Park, H.Y. Experimental study on seismic performance of steel slit damper under additional tensile load. *J. Build. Eng.* **2022**, *50*, 104110. [CrossRef]
3. Titirla, M.D. A state-of-the-art review of passive energy dissipation systems in steel braces. *Buildings* **2023**, *13*, 851. [CrossRef]
4. Lie, W.; Wu, C.; Luo, W.; Wu, C.; Li, C.; Li, D.; Wu, C. Cyclic behavior of a novel torsional steel-tube damper. *J. Constr. Steel Res.* **2022**, *188*, 107010. [CrossRef]
5. Nuzzo, I.; Losanno, D.; Caterino, N.; Serino, G.; Bozzo Rotondo, L. Experimental and analytical characterization of steel shear links for seismic energy dissipation. *Eng. Struct.* **2018**, *172*, 405–418. [CrossRef]
6. Mazza, F. Dissipative steel exoskeletons for the seismic control of reinforced concrete framed buildings. *Struct. Control Health Monit.* **2021**, *28*, e2683. [CrossRef]
7. Tahmasebi, A.; Rahimi, M. Evaluation of nonlinear static and dynamic analysis of steel braced frame buildings subjected to near-field earthquakes using FBD and DBD. *Structures* **2021**, *34*, 1364–1372. [CrossRef]
8. Yang, J.; Liang, S.T.; Zhu, X.J. Development and performance research of an Xadas damper with a double-phased yield mechanism. *Structures* **2023**, *54*, 1420–1439. [CrossRef]
9. Lu, Y.J.; Lv, Q.F.; Liu, Y. A friction-strip hybrid damper with multi-phase energy dissipation mechanism: Cyclic test and numerical verification. *Thin-Wall Struct.* **2021**, *200*, 111913. [CrossRef]
10. Li, G.Q.; Sun, Y.Z.; Jiang, J.; Sun, F.F.; Ji, C. Experimental study on two-level yielding buckling-restrained braces. *J. Constr. Steel Res.* **2019**, *159*, 260–269. [CrossRef]
11. Hu, B.; Min, Y.; Wang, C.; Xu, Q.; Keleta, Y. Design, analysis and application of the double-stage yield buckling restrained brace. *J. Build. Eng.* **2022**, *48*, 103980. [CrossRef]
12. Wu, S.; He, H.X.; Lan, B.J. Experimental study and analysis on performance of metal tube damper with multi-stage yield and failure. *China Civ. Eng. J.* **2022**, *55*, 36–46. (In Chinese)
13. Chen, Y.; Jiang, H.J.; Liu, T. Study on the seismic behavior of graded yielding metal dampers. *Eng. Mech.* **2019**, *36*, 53–62. (In Chinese)
14. Yang, C.T.; Xie, L.L.; Liu, Q. Dual-objective control of braced steel frame using asynchronized parallel two-stage yielding BRB. *Soil Dyn. Earthq. Eng.* **2024**, *179*, 108522. [CrossRef]
15. Xie, Q.; Zhou, Z.; Meng, S.P. Experimental investigation of the hysteretic performance of self-centering buckling-restrained braces with friction fuses. *Eng. Struct.* **2020**, *203*, 109865. [CrossRef]
16. Chou, C.C.; Huang, L.Y. Mechanics and validation tests of a post-tensioned self-centering brace with adjusted stiffness and deformation capacities using disc springs. *Thin-Walled Struct.* **2024**, *195*, 111430. [CrossRef]
17. Zhang, C.Z.; Guo, X.N.; Zhu, S.J. Hysteretic behavior of assembled self-centering buckling-restrained brace. *J. Tongji Univ. (Nat. Sci.)* **2021**, *49*, 8–19.
18. Ding, Y.K.; Liu, Y.T. Cyclic tests of assembled self-centering buckling-restrained braces with pre-compressed disc springs. *J. Constr. Steel Res.* **2020**, *172*, 106229. [CrossRef]
19. Jin, S.S.; Li, Y.K.; Zhou, J.T. Hysteresis model and experimental investigation of assembled self-centering buckling-restrained braces. *Eng. Mech.* **2022**, *39*, 49–57.
20. Lu, L.; Liu, X.; Dai, G.X. An experimental study on the mechanical properties of an axial compression-tension metallic damper and its application. *J. Vib. Shock* **2017**, *36*, 141–147.
21. Qu, B.; Dai, C.X.; Qiu, J.; Hou, H.T.; Qiu, C.X. Testing of seismic dampers with replaceable U-shaped steel plates. *Eng. Struct.* **2019**, *179*, 625–639. [CrossRef]
22. Ebadi Jamkhaneh, M.; Ebrahimi, A.H.; Shokri Amiri, M. Experimental and numerical investigation of steel moment resisting frame with U-shaped metallic yielding damper. *Int. J. Steel Struct.* **2019**, *19*, 806–818. [CrossRef]

23. Jia, J.F.; Zhao, L.Y.; Bai, Y.L. Design of self-centering energy dissipation brace with U-Shape steel plates and analysis on hysteretic behavior. *J. Southeast Univ. (Nat. Sci. Ed.)* **2021**, *51*, 754–761.
24. *GB/T 1972-2005*; Disc Spring. China Standards Press: Beijing, China, 2005.
25. Chen, Y.; Chen, C.; Jiang, H.J. Study of an innovative graded yield metal damper. *J. Constr. Steel Res.* **2019**, *160*, 240–254. [CrossRef]
26. Qiu, C.X.; Huang, T.Y.; Liu, J.W. Cyclic loading tests and numerical simulations of an assembled X-shaped flexural steel damper. *Adv. Struct. Eng.* **2024**, *27*, 301–313. [CrossRef]

buildings

MDPI

Article

Hysteretic Performance of Composite Damper with Yielding Reserve Stiffness

Xiaorui Zhang [1], Lihua Zhu [1,2,*], Libo Liu [1] and Jialong Li [3]

[1] School of Civil Engineering, Hebei University of Engineering, Handan 056038, China; xiaoruizhangheu@163.com (X.Z.); liulibo@hebeu.edu.cn (L.L.)

[2] State Key Laboratory of Coastal and Offshore Engineering, Dalian University of Technology, Dalian 116024, China

[3] Chengdu Surveying Geotechnical Research Institute Co., Ltd. of MCC, Chengdu 610023, China; wekalee@163.com

* Correspondence: zhulihua@hebeu.edu.cn

Abstract: Previous research on composite dampers has rarely addressed the issue of large deformations of structures under limit state. However, the proposed damper in this paper takes this issue into account and could provide yielding reserve stiffness for structures, ensuring structural resilience. A composite damper with yielding reserve stiffness (YRSD), consisting of a friction unit and a metal yield unit, was proposed. Low cyclic loading tests with different energy-dissipating steel plate thicknesses and bolt preloads were carried out and experimental results were compared with that of numerical simulation. This paper focuses on the synergistic energy dissipation mechanism of the proposed damper and the effects of various factors on its hysteretic performance, including the bolt preload and thickness of X-shaped steel plates. The results show that the synergistic energy dissipation mechanism of the proposed damper is well, exhibiting the behavior of hardening post-yielding stiffness and multi-stage energy dissipation characteristics, which could provide yielding reserve stiffness for the structure. The experimental hysteresis curve of YRSD is full, indicating its strong energy dissipation capacity, and the skeleton curve of experiment is consistent with that of the theoretical model. The envelope area of the rectangular hysteresis curve of YRSD increases by 107.3% with the preload increased by 100%. When the thickness of the X-shaped steel plates is increased by 2 mm, the resistance of YRSD increases by 26.2% and the post-yield stiffness increases by 37.9%. The stiffness degradation trend of all specimens initially decreases and then increases. The energy dissipation capacity of the friction unit increases by 53.8% as the preload is doubled. The capacity of the metal yield unit increases by 31.7% as the thickness of the X-shaped steel plates is increased by 2 mm. When the energy dissipation capacities of the friction unit and the metal yield unit are close to equal, the optimal energy dissipation capacity of the proposed damper is achieved. The error of results between the numerical analysis and experimentation is less than 10%, providing a basis for the parametric analysis of similar composite damper with yielding reserve stiffness.

Keywords: yielding reserve stiffness; multi-stage energy dissipation; hysteretic performance; low cyclic loading; numerical analysis

1. Introduction

By absorbing the seismic input energy, energy dissipation devices could minimize the seismic damage of the main components of structures and improve the overall seismic performance of entire structures. Such devices have gradually been applied to various practical engineering structures. Currently, global researchers [1–13] have conducted extensive experimental and theoretical research on traditional energy dissipation devices, such as viscous dampers, viscoelastic dampers, friction dampers, and metal dampers. Among these devices, the metal dampers [14,15] maintain good ductility even after fully developing plasticity, while the friction dampers [16–18] exhibit a full hysteresis curve with resistance that is relatively stable and easy to control. Therefore, both types of devices are currently widely used.

Displacement-dependent dampers have a single energy dissipation mechanism, and under major earthquakes, the difference between their pre-yield stiffness and post-yield stiffness is relatively large when the damped structures enter into limit state, resulting in significant interstory deformation of the structure after yielding. Therefore, researchers have combined dampers with different yield strengths to achieve multi-stage yielding. For instance, Chen et al. [19] proposed an annular metal yield damper, which combines two metal rings of different sizes in parallel to achieve inner ring yield under small displacement and outer ring yield under large displacement. The results showed that the proposed damper exhibits excellent deformation performance and multi-stage yielding behavior. Jarrah et al. [20] introduced a piston metallic damper, and experimental and numerical simulation results indicated that this damper reveals multi-stage energy dissipation characteristics and has stable hysteresis performance to dissipate seismic input energy, meeting specification requirements. Wu et al. [21] designed a metal sleeve damper consisting of two steel strips with different depth/width ratios. The study indicated that the proposed damper has multiple yield points and multiple failure stages. However, under major earthquakes, among metal dampers composed of the same energy dissipation mechanism, the energy dissipation unit with a lower yield point first fails, leading to a decreased trend in post-yield stiffness of the damper. Some scholars have combined different energy dissipation mechanisms to fully utilize their advantages. For example, Zhu et al. [22] developed a lattice-shaped friction device and conducted its experimental and theoretical research. The quasi-static test indicated that both friction mechanisms could work together effectively and the proposed device exhibits its behavior of hardening post-yielding stiffness. Golmoghany et al. [23] connected friction dampers with vertical shear panels, using stopper bolts to alter the energy dissipation cases to realize two-level seismic energy dissipation. Cai et al. [24] combined shear-shaped metal dampers with friction dampers in series and achieved multi-stage energy dissipation by setting stop sliding displacement to trigger different dampers. Parham et al. [25] proposed a piston hybrid frictional metallic damper, consisting of two independent energy-dissipating units controlled by a piston. The results showed that all specimens have endured accumulated displacement of over 2000 mm, dissipating a significant amount of input energy in the structure. Chukka et al. [26] combined X-shaped metal plates with friction dampers in parallel and conducted cyclic loading tests, improving structural seismic performance under different earthquake levels. Xu et al. [27] developed rotational friction dampers with buckling restrained brace and confirmed its excellent comprehensive performance of friction and plastic energy-dissipating ability through low cycle tests, providing a new seismic measure for prefabricated structures.

In summary, previous research on composite dampers has rarely addressed the issue of large structural deformations under limit state. However, the proposed YRSD considers this issue and provides yielding reserve stiffness for structures to ensure their resilience. This paper combines the advantages of the metal yielding damper and the friction damper and develops the YRSD composed of a friction unit and a metal yield unit in series. Compared with other composite dampers, the YRSD is easy to disassemble and assemble and has a synergistic energy dissipation mechanism. Only the friction unit dissipates energy under minor earthquakes, while, under medium to major earthquakes, the metal yielding unit is activated by a stop block and works in conjunction with the friction damper to dissipate energy, providing yielding reserve stiffness for the structure. The effects of parameters such as the thickness of X-shaped steel plates and bolt preloads on the energy dissipation capacity and hysteretic performance of the YRSD will be investigated through low cyclic loading tests, comparing with numerical simulation results. The research flowchart is found in Figure 1.

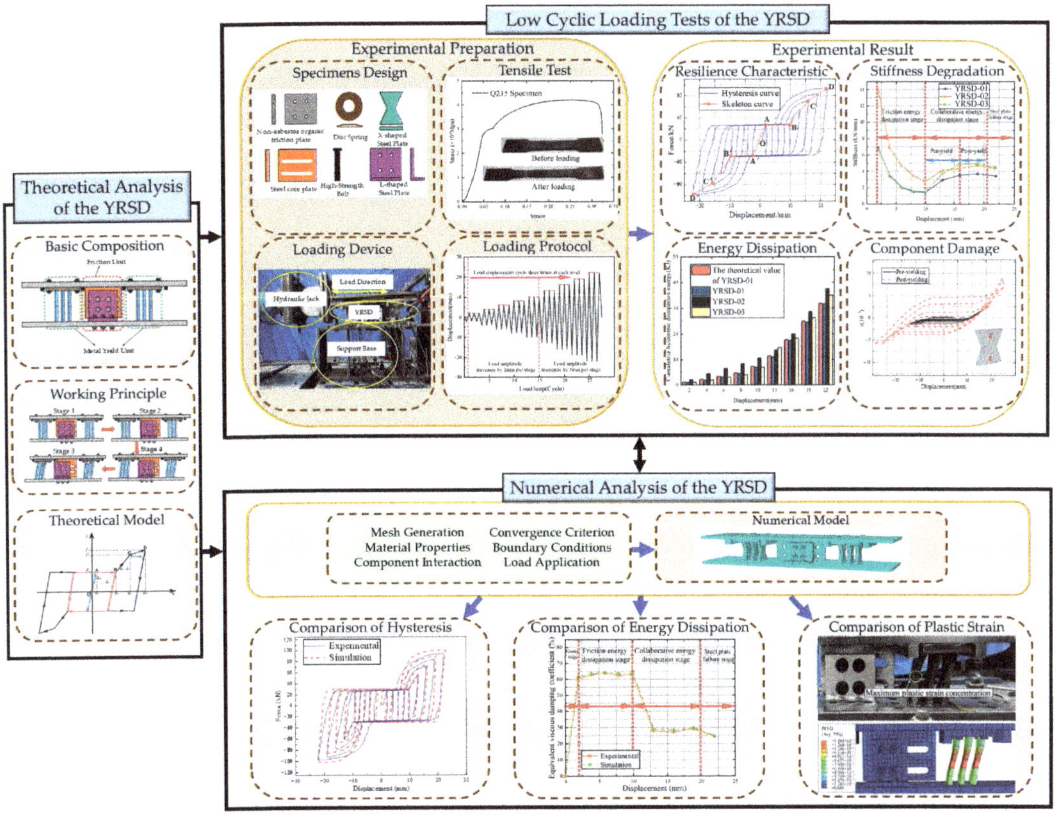

Figure 1. Research flowchart.

2. Property of YRSD

2.1. Basic Composition

Figure 2 shows the basic structure of the YRSD consisting of a friction unit and a metal yield unit connected in series. The friction unit includes an upper plate, four stop blocks, a steel core plate, two L-shaped steel plates, and two non-asbestos organic friction plates. These components are connected by high-strength bolts and disc springs. The metal yield unit includes six X-shaped steel plates, two positioning plates, and a lower plate. These are connected by welding. The metal yield unit is bolted to the upper plate for easy replacement after earthquakes.

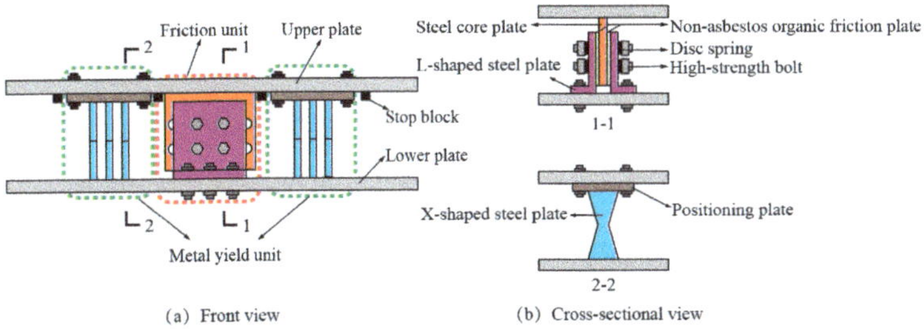

Figure 2. Basic composition of YRSD.

2.2. Working Principle

The YRSD undergoes four stages based on different loading displacements: elastic, friction energy dissipation, synergistic energy dissipation, and steel plate failure, as shown in Figure 3. (1) Elastic stage: At small displacements, the lateral stiffness of the YRSD is provided mainly by the friction force. Since the external force is smaller than the sliding force, the damper does not slide. At this stage of the YRSD, the structure equipped with the YRSD experiences small deformation and the main structure obtains some lateral stiffness provided by the YRSD. (2) Friction energy dissipation stage: As the loading displacement increases, the loading force reaches the sliding force of the friction unit, causing relative sliding between the steel core plate and the non-asbestos organic friction plate. The stiffness of the YRSD is determined by the magnitude of the friction force. In this stage, the displacement remains below the threshold displacement of the stop block, and the metal yield unit does not deform. The structure is in a moderate deformation state and seismic input energy of the structure is mainly dissipated by the friction unit. (3) Synergistic energy dissipation stage: The loading displacement increases until it reaches the threshold displacement of the stop block. The increase in displacement drives the metal yield unit to deform, dissipating energy with the friction unit. Friction dampers have good energy dissipation capacity, while metal yield dampers still have a certain post-yield stiffness after entering the post-yield stage. The energy dissipation capacity of damped structures under strong earthquakes needs to be given priority consideration. When seismic input energy is dissipated, the seismic response of structures will decrease. Therefore, the YRSD first dissipates energy through the friction unit. When stop blocks come into contact with positioning plates, the YRSD enters the synergistic energy dissipation stage, two energy dissipation units jointly dissipate energy, and the metal yield unit provides yield reserve stiffness for the structure. (4) Steel plate failure stage: When the displacement exceeds the ultimate displacement of the X-shaped steel plates, the steel plates will fracture. The metal yield unit stops working, but the friction unit continues to dissipate energy. In the third and fourth stages of the YRSD, the structure will experience significant deformation compared with the first two stages. Friction and metal yield units jointly dissipate energy and structural deformation will be controlled within the specified limit due to yielding reserve stiffness provided by the metal unit. The structural components will not be damaged or will be slightly damaged, ensuring overall structural resilience. In summary, the position of the stop block could be adjusted based on the requirements of the structural resilience to increase or decrease the threshold displacement. The YRSD exhibits controllable multi-stage energy dissipation characteristics and the structure equipped with the YRSDs could meet various seismic requirements from minor to major earthquakes.

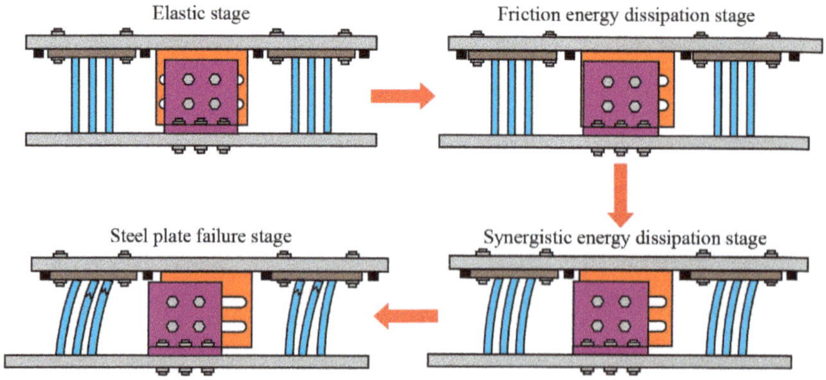

Figure 3. Working principle of YRSD.

2.3. Simplified Mechanical Model

Figure 4 shows the simplified mechanical model of the YRSD. Point A (u_i, F_i) represents the first yield point, where u_i is the first yield displacement and F_i is the first yield force. Point B (u_t, F_i) represents the first limit point, where u_t is the first limit displacement. Point C (u_y, F_y) represents the second yield point, where u_y is the second yield displacement and F_y is the second yield force. Point D (u_m, F_m) represents the second limit point, where u_m is the second limit displacement and F_m is the second limit force. α is the post-yield stiffness ratio of the YRSD. In the figure, O-A is the elastic stage as described in Section 2.2, with an initial stiffness of k_0, provided only by the friction unit. A-B is the friction energy dissipation stage, with a stiffness of k_1. B-C-D is the synergistic energy dissipation stage. Before the yielding of the X-shaped steel plates, the stiffness of the YRSD is $k_1 + k_2$. After the yielding of the X-shaped steel plates, the stiffness is $k_1 + \alpha k_2$. When the loading displacement exceeds Point D, the YRSD enters the steel plate failure stage and the X-shaped steel plates fracture.

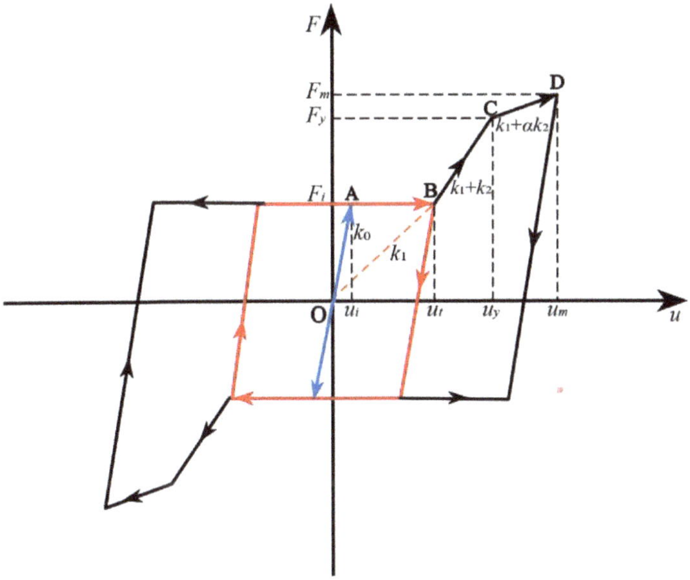

Figure 4. Theoretical hysteresis model of YRSD with yielding reserve stiffness.

3. Low Cyclic Loading Tests of YRSD

3.1. Specimens Design

To study the hysteretic performance of dampers at different stages, three specimens are designed for low cycling loading. The thickness of the X-shaped steel plates and the preload of the friction unit are varied. The detailed dimensions of each component are shown in Figure 5. The main parameters of the specimens are shown in Table 1. In Table 1, L is the distance from the stop blocks to the positioning plate. M_c is the size of the friction plate. H_c is the size of the upper and lower plates. X_c is the size of the steel core plate. Φ is the specification of the high-strength bolt. D is the size of the disc spring. F is the preload of a single bolt in the friction unit. t is the thickness of the X-shaped steel plates. Except for the X-shaped steel plates, which are made of Q235 steel, the other steel plates are made of Q345 steel. The disc spring is made of 60SiMn steel. The bolts are high-strength bolts of grade 8.8. The preload loading process of the specimens is shown in Figure 6. To obtain metallic material parameters such as elastic modulus, yield strength, ultimate strength, and elongation after fracture, tensile tests completed using an electronic universal testing machine are conducted according to the testing method specified in the standards [28].

Three tensile testing specimens named Q235-01, Q235-02, and Q235-03 are taken from the same batch of steel used for the X-shaped steel plates. The tensile testing results provide a basis for subsequent component damage analysis and numerical simulations. The average stress–strain curves of the specimens and the states before and after the tensile tests are shown in Figure 7. The results of the tensile tests are shown in Table 2. In Table 2, f_y is the yield strength of testing specimens; f_u is the ultimate strength of testing specimens; E is the elastic modulus of testing specimens; δ is the elongation after fracture of testing specimens; T is the thickness of the tensile specimens.

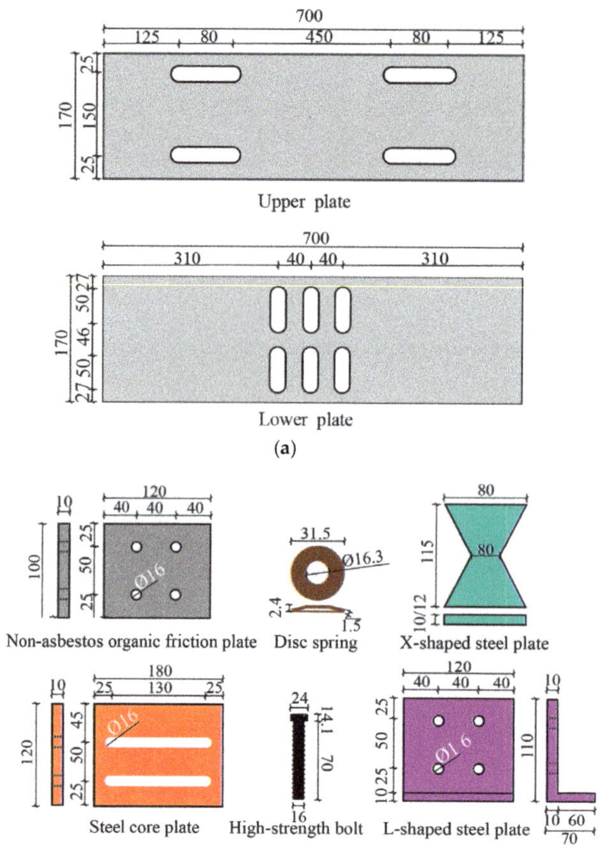

Figure 5. Details of YRSD main components: (**a**) upper plate and lower plate; (**b**) other components.

Table 1. Main parameters of specimens for low cyclic loading tests.

No.	L (mm)	M_c (mm)	H_c (mm)	X_c (mm)	Φ	D (mm)	F (kN)	t (mm)
YRSD-01	10	120 × 100 × 10	700 × 170 × 15	180 × 120 × 10	M16	31.5 × 16.3 × 2.4	5	10
YRSD-02	10	120 × 100 × 10	700 × 170 × 15	180 × 120 × 10	M16	31.5 × 16.3 × 2.4	10	10
YRSD-03	10	120 × 100 × 10	700 × 170 × 15	180 × 120 × 10	M16	31.5 × 16.3 × 2.4	5	12

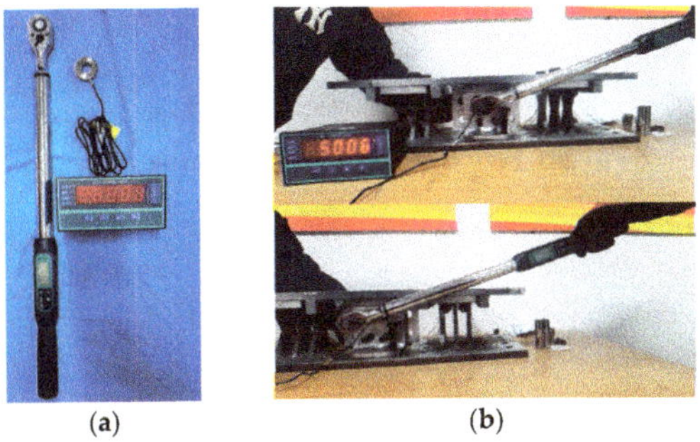

Figure 6. Preload loading process: (**a**) torque wrench and preload sensor; (**b**) apply preload.

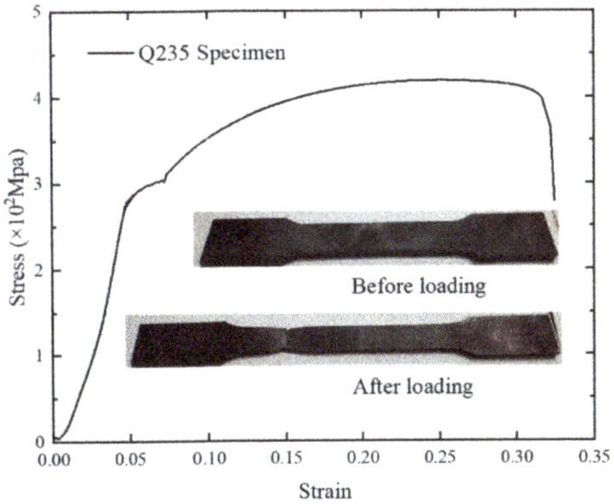

Figure 7. Average stress–strain curve of metallic materials' tensile tests.

Table 2. The results of tensile tests.

No.	f_y (MPa)	f_u (MPa)	E (1 × 10^5 N/mm^2)	δ (%)	T (mm)
Q235-01	274.9	418.9	2.06	32.7	6
Q235-02	296.4	422.7	2.02	33.1	6
Q235-03	292.8	420.6	2.10	29.5	6
Average	288	420.7	2.06	31.8	6

3.2. Loading Device and Protocol

The low cyclic loading tests are conducted on the ZT-FY50 loading device at Hebei Province Key Laboratory of Low Carbon Construction and Resilience Enhancement in Civil Engineering, as shown in Figure 8. The specimens are attached to a double-acting hydraulic jack via a hinge joint. Horizontal loads are applied by using an electric oil pump. The lower part of the specimens are fixed to the support base. The testing method of mechanical performance for the YRSD is carried out according to the national standards [29]. After the specimens yield, displacement-controlled loading is applied. The estimated first yield displacement is approximately 1 mm. Subsequent loading displacements are set as multiples of the yield displacement. The loading increment is set at 2 mm for the first 10 mm. For exceeding 10 mm, the increment increases to 3 mm. Each level is cycled 3 times. The loading protocol is shown in Figure 9.

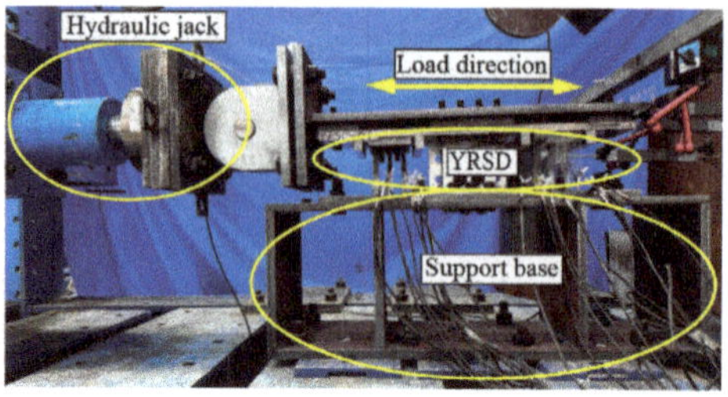

Figure 8. Experimental loading device.

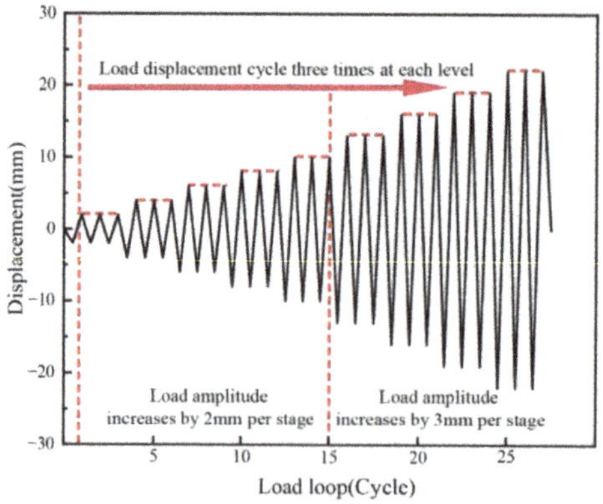

Figure 9. Experimental loading protocol.

3.3. Arrangement of Measuring Points

Three types of data are collected during the experiment: load, displacement, and strain. The loads are measured using a force sensor. Displacements are measured with YWD-100 m, and two displacement measuring points a set up: one on the right side of the upper plate to measure the total loading displacement, and the other on the right side of the positioning plate to measure the relative deformation of the X-shaped steel plates. Strains of the X-shaped steel plates are measured by using uniaxial and triaxial strain gauges attached to both sides of the X-shaped steel plates. Twelve strain measuring points are set up, with uniaxial gauges at the upper and lower roots, and triaxial gauges at the middle corner. The arrangements of measuring points are shown in Figure 10.

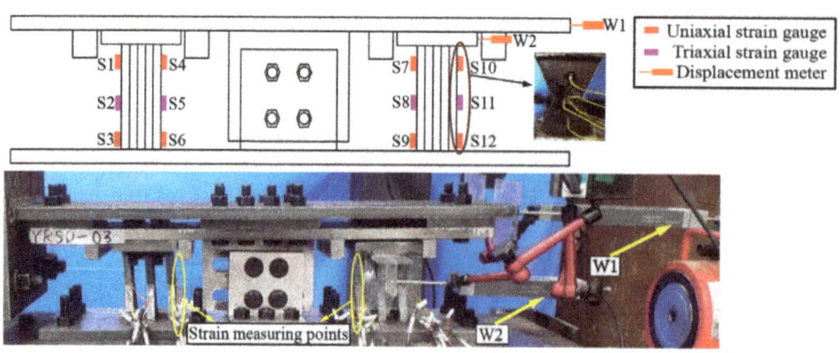

Figure 10. Measuring points of YRSD.

4. Analysis of Test Results of YRSD

4.1. Experimental Phenomenon

During the loading process, all specimens exhibit similar experimental phenomena. Figure 11 shows the specific loading process of YRSD-03. When the displacement loading is between 0 and 1 mm, no relative sliding of the friction unit occurs in any specimens. The YRSDs remain in the elastic stage (Figure 11a). When the loading displacement exceeds 1 mm, all specimens enter the friction energy dissipation stage. The friction units exhibit relative sliding accompanied by clear noises of friction. The resistances of YRSD-01, YRSD-02, and YRSD-03 fluctuate around 12.6 kN, 26.2 kN, and 12.8 kN, respectively (Figure 11b). When the load reaches 13 mm, all specimens enter the synergistic energy dissipation stage. In this stage, the stop blocks of the specimens contact the positioning plates, and drive the X-shaped steel plates to work together. At this point, the steel plates of all specimens show elastic deformation and maintain stability without out-of-plane instability during loading. The relative deformation of the X-shaped steel plates in YRSD-03 is 2.94 mm. When the displacement is increased to 19 mm, the X-shaped steel plates of all specimens reach a fully yielding state. Plastic deformation occurs at the corner of the YRSD-03 steel plate, and noticeable scratches appear on the steel core plate (Figure 11c). When the displacement exceeds 22 mm, all specimens reach the failure stage, the X-shaped steel plates fracture, and the experiment is terminated. Cracks appear at the ends and corners of the X-shaped steel plates in YRSD-03 after loading (Figure 11d).

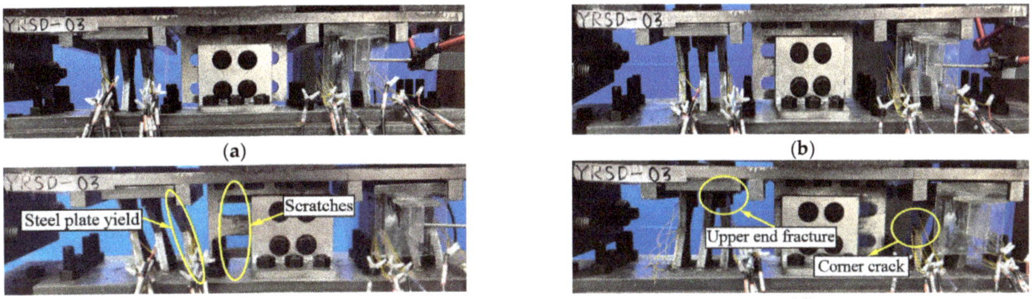

Figure 11. Loading state of the specimen: (**a**) elastic stage; (**b**) friction energy dissipation stage; (**c**) synergistic energy dissipation stage; (**d**) steel plate failure stage.

4.2. Resilience Characteristic

Figure 12 illustrates the hysteresis and skeleton curves of three specimens. The hysteresis curves of all specimens exhibit a full shape and excellent energy dissipation performance. The skeleton curves are consistent with those shown in Figure 4, exhibiting multi-stage energy dissipation characteristics. The results of the hysteresis curves reveal that the envelope area of the rectangular hysteresis curve for YRSD-02 increases by 56.1% compared with YRSD-01 due to the difference in initial bolt preload between YRSD-01 and YRSD-02. The resistance of the second limit point of YRSD-03 increases by 26.2%, and its post-yield stiffness increases by 37.9% compared with YRSD-01 because of varying thicknesses of X-shaped steel plates, indicating that YRSD-03 has a better seismic performance compared with YRSD-01 under major earthquakes. By comparing the skeleton curves, it is evident that the loading forces are less than the sliding force of friction units and the YRSDs remain in the elastic stage, as shown in the O-A of the figure. When the hysteresis curves show a clear rectangular shape, the YRSDs enter the friction energy dissipation stage, represented by the A-B in the figure. When the stop blocks contact the positioning plate, the dampers begin to synergistically dissipate energy and the resistance of the YRSDs increase, leading to noticeable hardening post-yielding stiffness. After loading displacement passes Point C, the X-shaped steel plates yield and the resistance of the metal yield units play a controlling role in the YRSD, resulting in a secondary stiffness increase, as illustrated in the B-C-D of Figure 12.

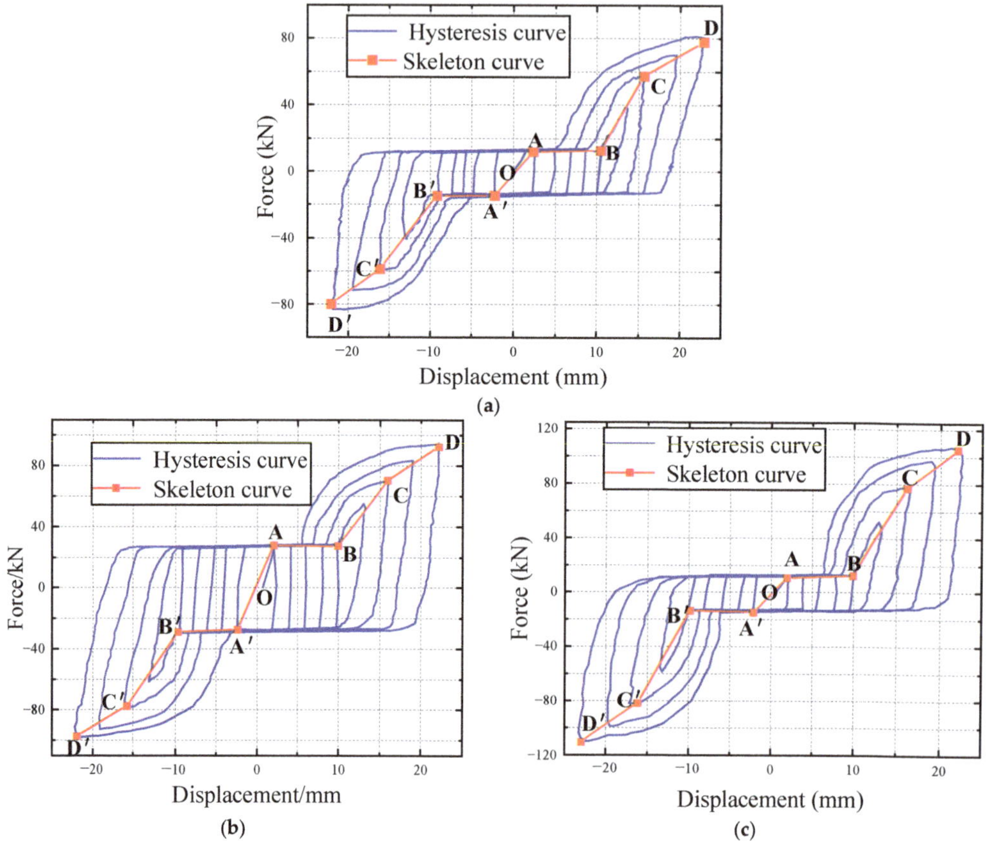

Figure 12. Hysteresis curve and skeleton curve of YRSD: (**a**) YRSD-01; (**b**) YRSD-02; (**c**) YRSD-03.

4.3. Stiffness Degradation

The secant stiffnesses [30] of each level of displacement amplitude are used to obtain the stiffness variation curves for the three sets of specimens. These curves reflect the ability of the YRSD to resist deformation under cyclic loading, as shown in Figure 13. The three curves in Figure 13 exhibit a generally consistent trend. Taking the 10 mm loading displacement as the turning point, the stiffness of the YRSDs initially decrease, then increase, and eventually stabilize, with a relatively smooth transition. When the loading displacement amplitude ranges from 2 to 10 mm, all specimens are in the friction energy dissipation stage. Since the initial bolt preload of the specimens remains unchanged, the stiffness begins to decrease with the increase in displacement. The initial bolt preload of YRSD-01 and YRSD-03 is the same, and the first 10 mm curves basically coincide. As the preload of the friction unit in YRSD-02 increases, its stiffness degradation increases by 4.49% compared with YRSD-01, indicating that the initial stiffness degradation rate of YRSD is proportional to the bolt preload of the friction unit. When loading displacement amplitude ranges from 10 to 16 mm, all specimens are in the synergistic energy dissipation stage, the metal yield unit begins to dissipate energy synergistically, and the resistance per unit displacement significantly increases, causing an increase in the stiffness after the initial degradation. Before 16 mm loading displacement, the X-shaped steel plates are still in an elastic state. Once loading displacement exceeds 16 mm, the X-shaped steel plates gradually yield, the increase in the resistance per unit displacement slows down, and the stiffness change becomes relatively moderate. After the X-shaped steel plates enter the failure stage, the stiffness of YRSD-01, YRSD-02, and YRSD-03 stabilizes at 3.43, 4.35, and 4.55 kN/mm.

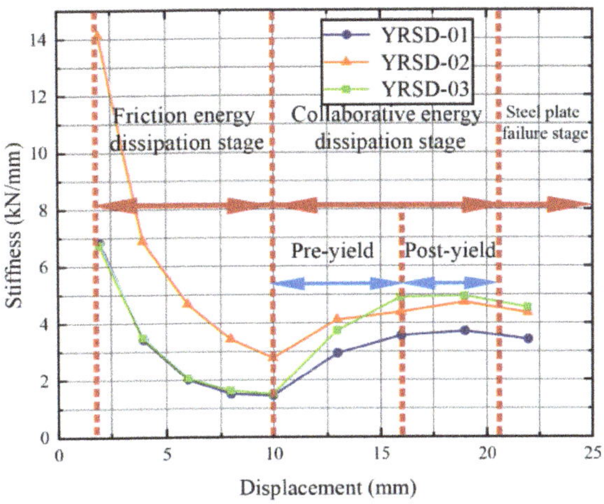

Figure 13. Degradation stiffness curve of YRSD.

4.4. Energy Dissipation Capacity

The cumulative hysteretic dissipation energy (E_q) and equivalent viscous damping coefficient (ζ_{eq}) [31] are used to evaluate the energy dissipation capacity of YRSD. The theoretical values of YRSD-01 are compared with the experimental values. The results indicate that the curve trends are consistent, with errors not exceeding 10%. In the elastic stage, the error of the energy dissipation capacity between theoretical and experimental values is 3.7%. In the friction energy dissipation stage, the error is 6.3%. In the synergistic energy dissipation stage, the error is 5.2%. In the steel plate failure stage, the error of the energy dissipation capacity is 3.3%. The errors may be caused by processing the accuracy of the specimens or gaps in the bolt connections. Figure 14a presents the cumulative hysteretic dissipation energy of three sets of specimens. As the loading displacement increases, the cumulative hysteretic dissipation energy of all specimens increases. Among these specimens, the energy dissipation capacity of YRSD-02 is optimal, indicating that when the energy dissipation of the friction unit and the metal yield unit is close to equal, the energy dissipation capacity of YRSD is optimal. The maximal cumulative hysteretic energy dissipation of YRSD-02 and YRSD-03 increases by 18.7% and 11.2%, respectively, compared with YRSD-01. A comparison of the friction energy dissipation stages of YRSD-02 and YRSD-01 shows that, when the bolt preload increases by 100%, the friction energy dissipation increases by 51.4%. This phenomenon indicates that friction energy dissipation is proportional to bolt preload. After entering the synergistic energy dissipation stage, the energy dissipation of YRSD-03 increases by 31.7% compared with YRSD-01. This increase in energy dissipation indicates that increasing the thickness of the X-shaped steel plates significantly enhances the yielding reserve stiffness of YRSD.

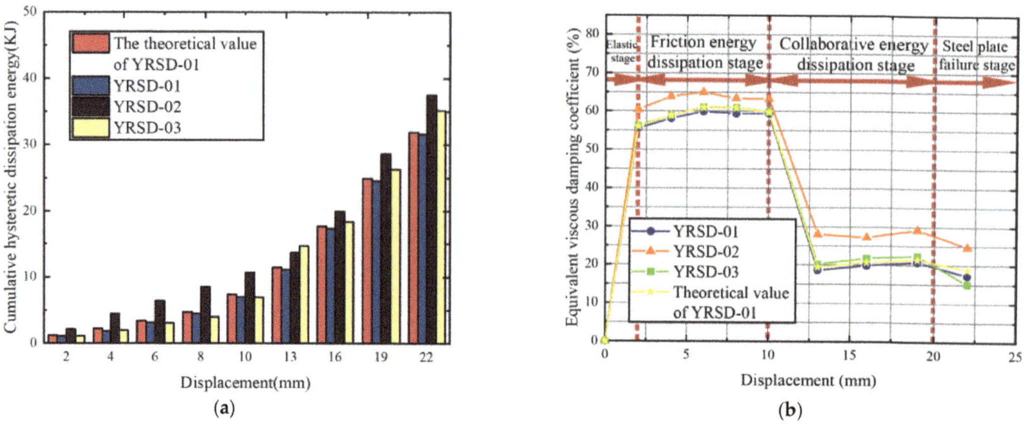

Figure 14. Energy dissipation capacity of YRSD: (**a**) cumulative hysteretic dissipation energy; (**b**) equivalent viscous damping coefficient.

Figure 14b shows equivalent viscous damping coefficient curves for three sets of specimens. When the loading displacement ranges from 0 to 2 mm, all specimens are in the elastic stage. The equivalent viscous damping coefficient increases rapidly. When the loading displacement exceeds 2 mm, all specimens enter the friction energy dissipation stage. The equivalent viscous damping coefficient changes smoothly. The coefficient of YRSD-02 is 8.7% and 7.2% higher than that of YRSD-01 and YRSD-03, respectively. This increase is due to the larger sliding force of the friction unit in YRSD-02. When the loading displacement exceeds 10 mm, the specimens enter the synergistic energy dissipation stage. The equivalent viscous damping coefficients of the specimens decrease, which is due to the relative sliding of the friction damper being obstructed when the stop blocks contact the X-shaped steel plates, resulting in a reduction in friction energy dissipation. When the loading displacement exceeds 13 mm, the X-shaped steel plates gradually yield. The equivalent viscous damping coefficient change flattens, and energy dissipation capacity remains stable. When the loading displacement exceeds 20 mm, the X-shaped steel plates enter the failure stage. The energy dissipation capacity of all specimens slightly decreases. Table 3 compares the initial and final test results.

Table 3. Comparison of the initial and final values of the experimental results.

No.	Initial Value			Final Value		
	F_i (kN)	k_0 (kN/mm)	ζ_{eq} (%)	F_y (kN)	$k_1+\alpha k_2$ (kN/mm)	ζ_{eq} (%)
YRSD-01	12.64	6.84	55.50	57.33	3.43	17.25
YRSD-02	28.65	14.13	60.34	70.52	4.35	24.68
YRSD-03	12.81	6.75	56.24	77.80	4.55	15.01

4.5. Component Damage

The strain magnitude of the X-shaped steel plates reflects their trends in damage changes. The strain development processes of all specimens are consistent. Therefore, taking YRSD-03 as an example, component damage analysis is carried out. Figure 15 shows the strain variation curve of measuring points of X-shaped steel plates with loading displacement. When the loading displacement is between 0 and 10 mm, the strain values at the root and corner of the X-shaped steel plates remain around 0, and no energy dissipation occurs. When the loading displacement exceeds 10 mm, the stop blocks contact the positioning plate. The specimen enters the synergistic energy dissipation stage and the strain alternates between positive and negative. The left and right sides of the X-shaped steel plates alternate between tension and compression. When the loading displacement exceeds 16 mm, the strain develops rapidly, exceeding the yield strain (1.44×10^{-3}) determined in the material properties test, and the X-shaped steel plates begin to yield. When the loading displacement reaches 22 mm, the maximum plastic strain (9.84×10^{-3}) occurs at the corner of the X-shaped steel plates, causing the steel plates to fracture. This fracture behavior is consistent with experimental phenomenon. Figure 16 shows the component conditions of all specimens after loading. As shown in the figure, except for the X-shaped steel plates that suffer varying degrees of damage, all other components do not exhibit any plastic deformation and could still be used. This result verifies the initial design concept.

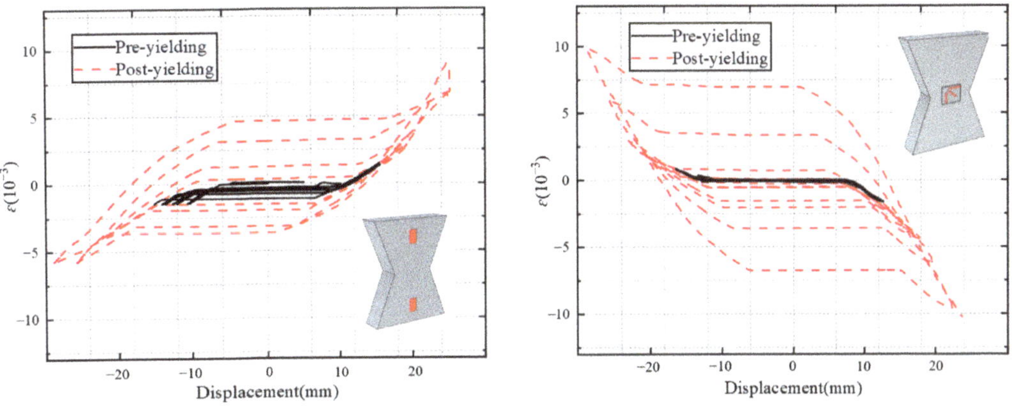

Figure 15. Curve of strain variation of X-shaped steel plates with loading displacement: (**a**) the root; (**b**) the corner.

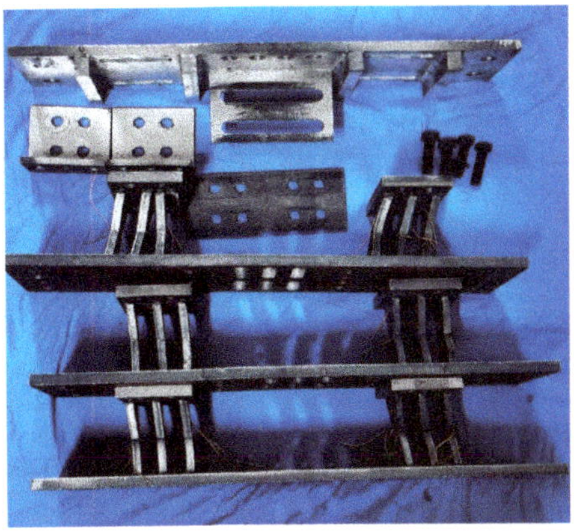

Figure 16. Component condition after loading.

5. Comparative Analysis with Numerical Simulation of YRSD

5.1. Numerical Model

The finite element software ABAQUS 2024 [32] is used to simulate the low cycling loading test of all specimens. The modeling process is divided into six steps. (1) In terms of part creation, C3D8R solid elements are used for all parts. (2) In terms of material properties, the elastic modulus is defined as 201 GPa and the Poisson's ratio as 0.3 for all components in the elastic stage. In the plastic stage, the X-shaped steel plates use the

Q235 standard specimen results from Section 3.1, while the other steel plates use the ideal elastoplastic model of Q345. The yield strength of the high-strength bolts is 640 MPa. (3) In terms of part interaction, only normal hard contact is applied between the positioning plate and the stop blocks. The tangential friction between the steel core plate and the friction plate is modeled as Coulomb friction with a coefficient of 0.5. Normal contact is also set for the steel core plate. The welding and bolt connections are replaced with tie constraints. (4) In terms of load, the bolt load is selected to apply the preload. The cross-section at the middle of the bolt is used as the preload section. The preload is applied in multiple analysis steps. A smaller value is applied in the initial step, and then it is fully applied and fixed at the current length in subsequent steps to simulate the actual force on bolts. (5) In terms of boundary conditions, the bottom of the lower plate is fully constrained. A reference point is established on the upper surface of the upper plate and coupled with both. The degrees of freedom in other directions are constrained to 0. Horizontal translational degrees of freedom are retained. Displacement amplitude is applied at the reference point. The loading protocol is consistent with that of the experiment. (6) In terms of the analysis steps, this analysis has two steps, with a minimum increment step size of 1 × 10-7. The first step lasts 1 s for applying the load, and the second step duration is based on the test loading protocol. The model consists of 2392 hexahedral meshes. After increasing the mesh scale by an order of magnitude, the error of results does not exceed 5%, which confirms the mesh independency of this model. The YRSD numerical model and meshing are shown in Figure 17.

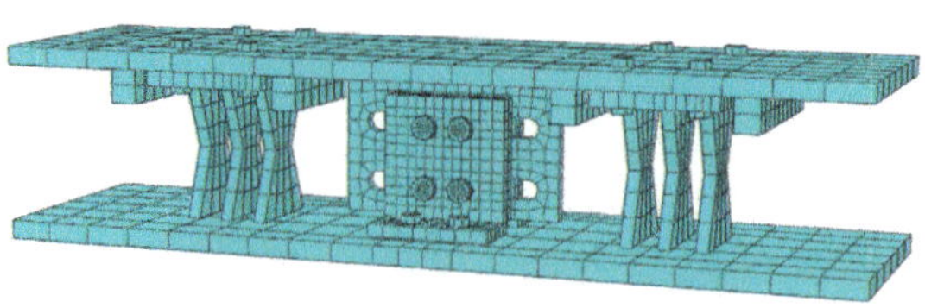

Figure 17. Numerical model and meshing of YRSD.

5.2. Hysteresis Curve

Figure 18 shows the comparison of hysteresis curves between the numerical simulations and experiments of three specimens. As evident from the figure, the overall shape of simulation curves of all specimens is in good agreement with experimental results, which could accurately simulate the behavior of hardening post-yielding stiffness of the YRSDs. The key points of the numerical simulation of the specimens are compared with those of the experiment in Table 4. As shown, YRSD-01 and YRSD-02 have smaller data errors than YRSD-03. YRSD-03 is analyzed as an example. The error of the first yield stiffness between the numerical simulation and the experimental results is 6.1% and 4.1% for the second yield stiffness. The errors of the first and second yield forces are 9.6% and 2.8%, respectively. The errors of maximum resistance in the tensile and compressive directions are 9.8% and 7.1%. These errors may be caused by force loss of the jack during the transform of loading direction or preload loss caused by loose bolts. The errors in all key points are within 10%. The result shows that the numerical simulation closely matches the test results and effectively simulates the mechanical properties of the YRSD.

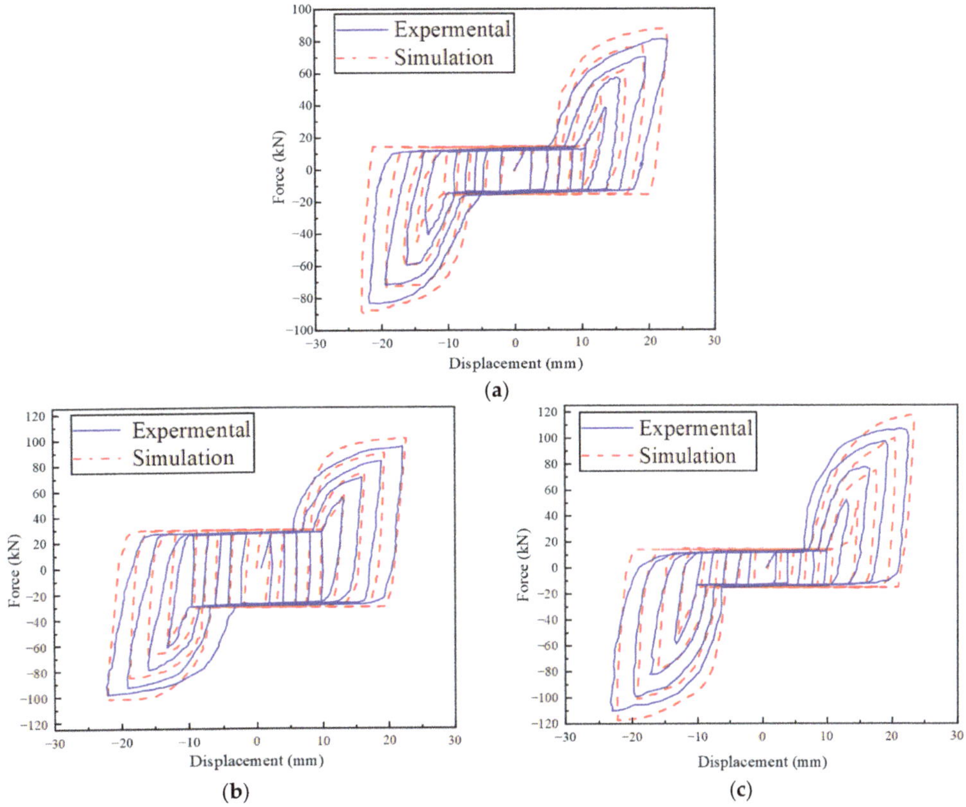

Figure 18. Comparison of hysteresis curves between numerical simulations and experiment: (**a**) YRSD-01; (**b**) YRSD-02; (**c**) YRSD-03.

Table 4. Comparison of key parameters between numerical simulations and experiment.

No.	Comparison	k_0 (kN/mm)	$k_1 + \alpha k_2$ (kN/mm)	F (kN)	F_y (kN)	F_t (kN)	F_c (kN)
YRSD-01	Experimental	6.84	3.56	12.64	57.33	80.51	82.90
	Simulation	7.13	3.96	13.71	60.19	86.76	87.58
YRSD-02	Experimental	14.13	4.40	28.65	70.52	98.08	102.42
	Simulation	15.39	4.74	30.31	75.57	103.59	104.96
YRSD-03	Experimental	6.75	4.92	12.81	77.80	106.10	107.68
	Simulation	7.16	5.12	14.05	75.67	117.60	116.41

5.3. Energy Dissipation Capacity

The maximum cumulative hysteretic dissipation energy (E_{qm}) and the equivalent viscous damping coefficient (ζ_{eq}) are compared between the numerical simulations and the experimental results (Figures 19 and 20). The simulated values for all specimens generally match the experimental results, although the simulated values are slightly higher. The maximum cumulative hysteretic energy dissipation in the numerical simulations for YRSD-01, YRSD-02, and YRSD-03 is 8.1%, 7.5%, and 9.4% higher than in the experimental results, respectively. During the elastic and friction energy dissipation stages, the numerical simulation results for the equivalent viscous damping coefficients of all specimens are slightly higher than the experimental values. After entering the synergistic energy dissipation stage, when all specimens have completely yielded, the simulation results align with the experimental results.

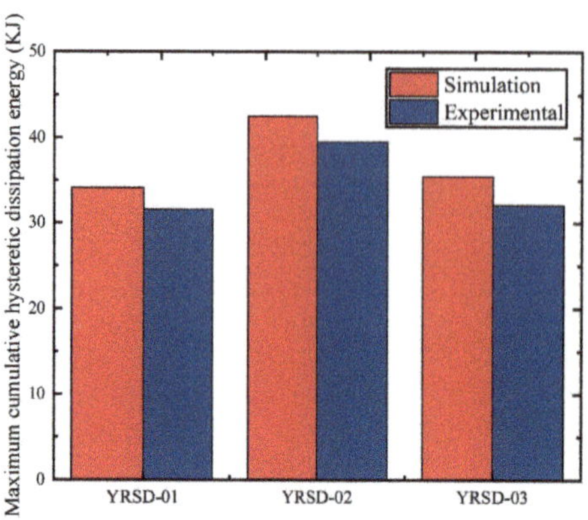

Figure 19. The numerical simulations and experiment of the maximum cumulative hysteretic energy dissipation.

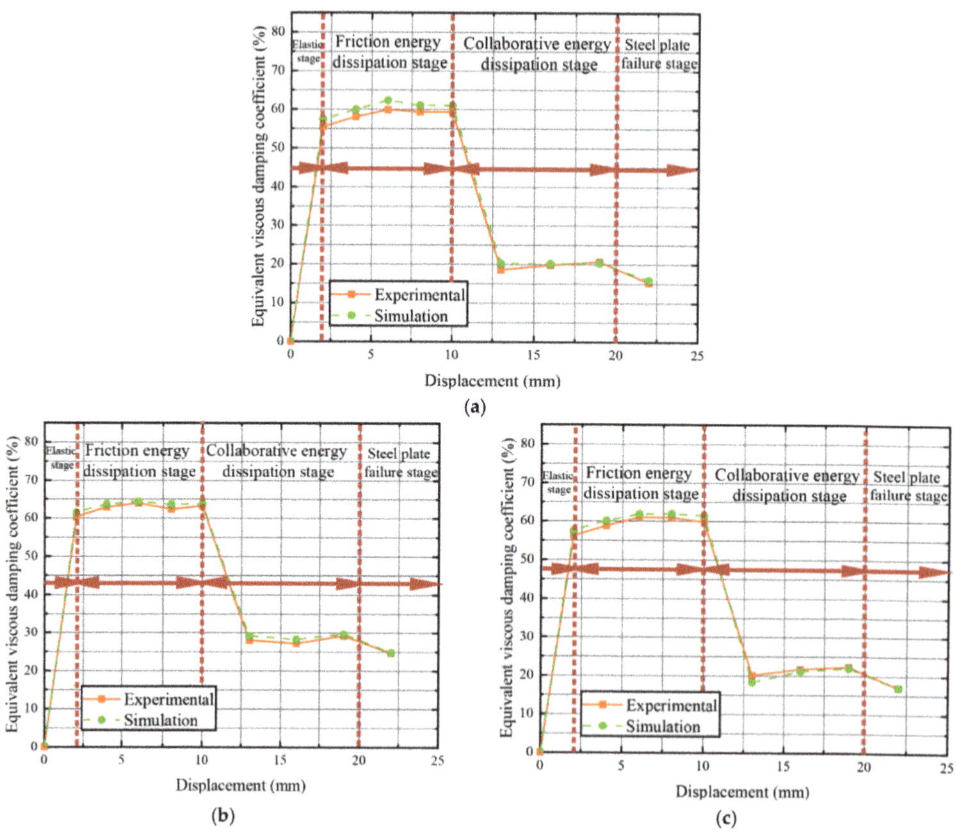

Figure 20. The numerical simulations and experiment of equivalent viscous damping coefficient: (**a**) YRSD-01; (**b**) YRSD-02; (**c**) YRSD-03.

5.4. Plastic Strain

The equivalent plastic strain (PEEQ) nephogram of all specimens is compared with the experimental results, as shown in Figure 21. The plastic deformation of all specimens is consistent. This numerical model could accurately simulate the experimental process. In the elastic and friction energy dissipation stages, the X-shaped steel plates do not participate in the work. No plastic strain is observed in any of the specimens (Figure 20a). After entering the synergistic energy dissipation stage, a comparison of the equivalent plastic strain from the experimental and numerical simulations shows that the maximum strain occurs in the middle of the X-shaped steel plates, the weakest part of the plates. This phenomenon demonstrates the energy-dissipating effect of the metal yield unit. It also provides yielding reserve stiffness, limiting the interstory drift of the structure (Figure 20b). After entering the steel plate failure stage, the plastic deformation of the X-shaped steel plates in all specimens gradually spreads from the corner to the end. The maximum plastic strain (9.15×10^{-3}) has an error of 7.5% compared with that of the experiment (9.84×10^{-3}). The failed X-shaped steel plates could be easily replaced because of bolt connections between the proposed dampers and structure. The YRSD could continue to function after replacement (Figure 20c).

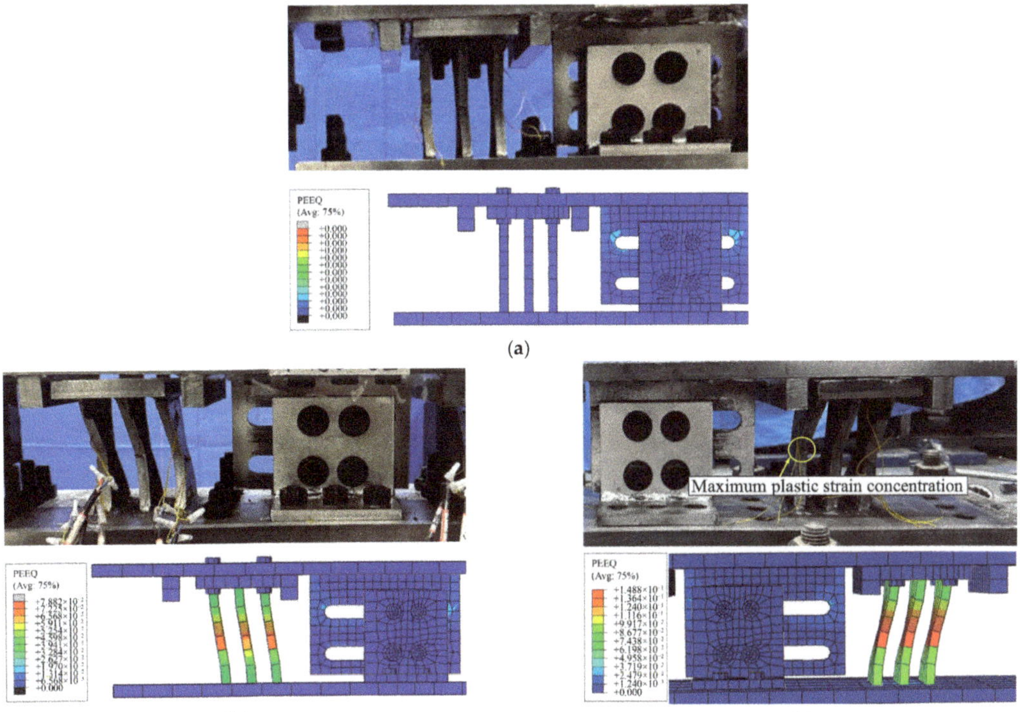

Figure 21. The numerical simulation and test of deformation: (**a**) elastic and friction energy dissipation stage; (**b**) synergistic energy dissipation stage; (**c**) steel plate failure stage.

6. Conclusions

A composite damper with yielding reserve stiffness was designed, and low cyclic loading test of three specimens were carried out and verified through numerical simulation. The following conclusions were obtained:

(1) The YRSD consists of a friction unit and a metal yield unit in series, which are easy to install and disassemble. The YRSD exhibits controllable multi-stage energy dissipation characteristics, which are achieved by adjusting the threshold displacement of the stop blocks to meet the entire process requirements of energy dissipation throughout minor to major earthquakes. During the loading process, all specimens exhibit stable performance. In the friction stage, the friction unit makes a clear noise of friction. The steel core plate shows obvious scratches. Eventually, all X-shaped steel plates in three specimens are completely yielded and damaged.

(2) The experimental hysteresis curve of YRSD is full, indicating a strong energy dissipation capacity. The experimental skeleton curve matches the theoretical model. The preload of the bolt significantly affects the energy dissipation performance of YRSD during the friction energy dissipation stage. The envelope area of the hysteresis curve increases by 56.1% when the preload increases by 100%. The thickness of the X-shaped steel plates directly affects the resistance of YRSD. When the thickness of the X-shaped steel plates increases by 2 mm, the resistance of YRSD increases by 26.2%, and the post-yield stiffness increases by 37.9%.

(3) The stiffness degradation curves for all specimens exhibit a consistent trend. The curves show a trend of decreasing, then increasing, and finally flattening out, with a turning point at a loading displacement of 10 mm. The transition of the curves is smooth, and the stiffness degradation rate is proportional to the preload during the first 10 mm.

(4) When the energy dissipation capacities of the friction unit and the metal yield unit are close to equal, the optimal energy dissipation capacity of the proposed damper is achieved. The energy dissipation capacity of the friction unit increases by 53.8% as the preload is doubled. The capacity of the metal yield unit increases by 31.7% as the thickness of the X-shaped steel plates is increased by 2 mm. Except for the X-shaped steel plates, all other components could continue to be used. The maximum plastic strain (9.84×10^{-3}) in the numerical model occurs at the corner of the X-shaped steel plates, consistent with the experimental results.

(5) The hysteresis curves and energy dissipation capacity from the numerical analysis are consistent with the experimental results. The data errors at all key points are within 10%, and the strain development trends match. These results confirm that the YRSD exhibits the expected behavior of hardening post-yielding stiffness, providing yielding reserve stiffness for the structure. The proposed YRSD could be applied to seismic isolation structures and self-centering structures in bridges or buildings with large deformations under limit state. The YRSD could compensate for post-yield stiffness of the structures to alleviate adverse effects of large deformation. Meanwhile, this type of composite damper could be designed according to different performance objectives, providing a new attempt for performance-based seismic design methods promoted by current regulations in various countries. This paper only compared the results of low cyclic tests and numerical simulations of the YRSD, lacking a shaking table test and dynamic time history analysis of the structure equipped with YRSDs. Moreover, this paper does not discuss the influence of the stiffness degradation behavior of the YRSD on the long-term durability of the structure, which is one of our future directions.

Author Contributions: Conceptualization, L.Z.; Methodology, L.Z.; Software, J.L.; Validation, X.Z.; Formal analysis, X.Z.; Investigation, X.Z.; Resources, L.L.; Data curation, X.Z.; Writing—original draft, X.Z.; Writing—review & editing, L.Z.; Visualization, X.Z.; Supervision, L.Z. and L.L.; Project administration, L.Z.; Funding acquisition, L.Z. All authors have read and agreed to the published version of the manuscript.

Funding: This research was funded by the Natural Science Foundation of Hebei Province of China (grant No. E2021402074); Science and Technology Project of Hebei Education Department (grant No. QN2023133); the Open Fund of State Key Laboratory of Coastal and Offshore Engineering, Dalian University of Technology (grant No. LP2123); Construction Technology Research Project of

Department of housing & urban-rural development, Hebei (2023-2075) and Sichuan Science and Technology Program, China (grant No. 2024NSFSC0156).

Data Availability Statement: The original contributions presented in this study are included in the article. Further inquiries can be directed to the corresponding author.

Conflicts of Interest: Author Jialong Li was employed by the company Chengdu Surveying Geotechnical Research Institute Co., Ltd. of MCC. The remaining authors declare that the research was conducted in the absence of any commercial or financial relationships that could be construed as a potential conflict of interest.

References

1. González-Sanz, G.; Escolano-Margarit, D.; Benavent-Climent, A. A New Stainless-Steel Tube-in-Tube Damper for Seismic Protection of Structures. *Appl. Sci.* **2020**, *10*, 1410. [CrossRef]
2. Mrad, C.; Titirla, M.D.; Larbi, W. Comparison of Strengthening Solutions with Optimized Passive Energy Dissipation Systems in Symmetric Buildings. *Appl. Sci.* **2021**, *11*, 10103. [CrossRef]
3. Arvind, R.; Santhi, M.H. A state of art review on hybrid passive energy dissipating devices. *J. Vib. Eng. Technol.* **2022**, *10*, 1931–1954. [CrossRef]
4. Almajhali, K.Y.M.; Alhaddad, W. A systematic review of hybrid passive energy dissipating devices. *Proc. Inst. Civ. Eng. Struct. Build.* **2024**, 1–43. [CrossRef]
5. Martinez-Rueda, J.E. On the evolution of energy dissipation devices for seismic design. *Earthq. Spectra.* **2002**, *18*, 309–346. [CrossRef]
6. Symans, M.D.; Charney, F.A.; Whittaker, A.S.; Constantinou, M.C.; Kircher, C.A.; Johnson, M.W.; McNamara, R.J. Energy dissipation systems for seismic applications: Current practice and recent developments. *J. Struct. Eng.* **2008**, *134*, 3–21. [CrossRef]
7. Li, G.; Li, H.N. Research on damping performance of new mild steel damper. *J. Vib. Shock* **2006**, *3*, 66–72+206–207. (In Chinese)
8. Priya, D.S.; Cinitha, A.; Umesha, P.K.; Iyer, N.R. A critical review on enhancing the seismic response of buildings with energy dissipation methods. *J. Struct. Struct.* **2015**, *42*, 78–88.
9. Di Cesare, A.; Ponzo, F.C.; Lamarucciola, N.; Nigro, D. Dynamic Seismic Response of Nonlinear Displacement Dependent Devices versus Testing Required by Codes: Experimental Case Studies. *Appl. Sci.* **2020**, *10*, 8857. [CrossRef]
10. Sujit, J.; Feng, Y.; Hao, Y.O. A state-of-the-art review on passive friction dampers and their applications. *Eng. Struct.* **2021**, *235*, 112022.
11. Shu, Z.; You, R.; Zhou, Y. Viscoelastic materials for structural dampers: A review. *Constr. Build. Mater.* **2022**, *342*, 127955. [CrossRef]
12. Men, J.; Wang, J.; Zhang, Q.; Fan, D.; Zhou, Q.; Huang, C.H. Experimental and numerical study on cyclic behavior of a shape-optimized composite metallic yield damper with two-phase energy dissipation. *Structures* **2023**, *53*, 1012–1029. [CrossRef]
13. Nyabongo, E.; Li, X. Development of energy-early-dissipated braces based on web buckling of low-yield I-section steel. *J. Constr. Steel Res.* **2024**, *217*, 108672. [CrossRef]
14. Zhai, Z.; Guo, W.; Yu, Z.; He, C.; Zeng, Z. Experimental and numerical study of S-shaped steel plate damper for seismic resilient application. *Eng. Struct.* **2020**, *221*, 111006. [CrossRef]
15. Li, Z.; Wang, J.; Li, C.; Cao, J. Optimum Arrangement of TADAS Dampers for Seismic Drift Control of Buildings Using Accelerated Iterative Methods. *Buildings* **2023**, *13*, 2720. [CrossRef]
16. Quintana, H.C.; Petkovski, M. Optimum performance of structural control with friction dampers. *Eng. Struct.* **2018**, *172*, 154–162. [CrossRef]
17. Aldea, S.; Bazáez, R.; Heresi, P.; Astroza, R. Effect of Bidirectional Hysteretic Dampers on the Seismic Performance of Skewed Multi-Span Highway Bridges. *Buildings* **2024**, *14*, 1778. [CrossRef]
18. Qu, J.; Liu, X.; Bai, Y.; Wang, W.; Li, Y.; Pu, J.; Zhou, C. A Novel Dual Self-Centering Friction Damper for Seismic Responses Control of Steel Frame. *Buildings* **2024**, *14*, 407. [CrossRef]
19. Chen, Y.; Jiang, H.J.; Liu, T.; Wan, Z.W.; Lu, Z. Study on the seismic behavior of graded yielding metal dampers. *J. Eng. Mech.* **2019**, *36*, 53–62.
20. Jarrah, M.; Khezrzadeh, H.; Mofid, M.; Jafari, K. Experimental and numerical evaluation of piston metallic damper (PMD). *J. Constr. Steel Res.* **2019**, *154*, 99–109. [CrossRef]
21. Wu, S.; He, H.; Chen, Y.; Cheng, S.; Wang, Z. Mechanical performance of multi-stage yield and failure metal sleeve damper: Experimental and numerical study. *Thin-Walled Struct.* **2022**, *176*, 109–315. [CrossRef]
22. Zhu, L.H.; Li, G.; Li, H.N. A lattice-shaped friction device and its performance in weak-story prevention. *Struct. Des. Tall Spec. Build.* **2018**, *27*, e1535. [CrossRef]
23. Golmoghany, M.Z.; Zahrai, S.M. Improving seismic behavior using a hybrid control system of friction damper and vertical shear panel in series. *Structures* **2021**, *31*, 369–379. [CrossRef]
24. Cai, Y.; Xu, W.; Ren, J. Experimental study on mechanical properties of friction-metal coupled damper with staged energy dissipation. *J. Build. Struct.* **2023**, *44*, 231–239. (In Chinese)

25. Parham, A.; Hamed, K. Experimental and numerical assessment of Piston Hybrid Frictional Metallic Damper (PHFMD). *Eng. Struct.* **2021**, *243*, 112669.
26. Chukka, N.D.K.R.; Krishnamurthy, M. Seismic performance assessment of structure with hybrid passive energy dissipation device. *Structures* **2020**, *27*, 1246–1259. [CrossRef]
27. Xu, G.; Ou, J. Seismic performance of combined rotational friction and flexural yielding metallic dampers. *J. Build. Struct.* **2022**, *49*, 104059. [CrossRef]
28. *GB/T228.1-2021*; Metallic Materials-Tensile Testing-Part 1: Method of Test at Room Temperature. Standards Press of China: Beijing, China, 2021. (In Chinese)
29. *JG/T 209-2012*; Dampers for Vibration Energy Dissipation of Buildings. Standards Press of China: Beijing, China, 2012. (In Chinese)
30. *JGJ/T 101-2015*; Specification for Seismic Test of Buildings. China Architecture & Building Press: Beijing, China, 2015. (In Chinese)
31. Priestley, M.J.N.; Grant, D.N. Viscous dampers in seismic design and analysis. *J. Earthq. Eng.* **2005**, *9*, 229–255. [CrossRef]
32. ABAQUS Inc. *ABAQUS Analysis User's Manual*; Dassault Systemes: Providence, RI, USA, 2013.

buildings

MDPI

Article

Multiscale Damage Identification Method of Beam-Type Structures Based on Node Curvature

Kai Ye [1,2], Shubi Zhang [1,*], Qiuzhao Zhang [1,*], Rumian Zhong [3] and Wenda Wang [1]

1 School of Environment Science and Spatial Infomatics, China University of Mining and Technology, Xuzhou 221116, China; lb23160003@cumt.edu.cn (K.Y.); wwd@cumt.edu.cn (W.W.)
2 Shenzhen Cheng'an Softcom Technology Group Co., Ltd., Shenzhen 518033, China
3 School of Transportation and Environment, Shenzhen Institute of Information Technology, Shenzhen 518172, China; 2023000106@sziit.edu.cn
* Correspondence: zhangsbi@vip.sina.com (S.Z.); qiuzhao.zhang@cumt.edu.cn (Q.Z.)

Abstract: This paper proposes a multiscale damage identification method for beam-type structures based on node curvature. Firstly, based on the assumption that micro-damage has little effect on stress redistribution and the basic relationship between structural bending moment and curvature, combined with the denoising function of wavelet analysis, the linear matrix equation before and after node curvature damage is solved using the singular value decomposition (SVD) method. Then, the theoretical feasibility of this method is verified with laboratory tests of a simply supported beam. Finally, the damage sensitivity and noise resistance of this method are verified using field measurements of a beam bridge. The results show that the nodal curvature serves as an indicator parameter for damage identification in beam-type structures, enabling the precise localization of damage within these structures. When utilizing a multiscale finite element model for analysis, the nodal curvature enhances the ability to identify both the location and severity of damage within small-scale elements. Furthermore, this method can provide a reference for the damage identification and health monitoring of other types of bridges.

Keywords: node curvature; denoising function; singular value decomposition; beam-type structures

1. Introduction

With the progress of society and the development of civil engineering technology, the health monitoring and safety assessment of large civil engineering structures, especially large bridge structures, have become hot topics in both academic and engineering circles at home and abroad. How to use collected data and signals to reconstruct the working status and health status of bridge structures, establish accurate finite element models, accurately identify the location and degree of structural damage, and serve for the damage and safety prognosis of bridge structures is one of the most core and critical issues in bridge structural health monitoring, damage diagnosis, and safety assessment [1–3].

The early damage identification of structures is essential. For example, in mechanical and steel structures, in order to ensure the normal operation of the entire building structural system, it is crucial to detect early cracks in the structure [4]. At present, there are two main purposes for damage identification at home and abroad, namely, to identify the location and degree of damage in order to support owners in taking active maintenance and reinforcement measures. Due to the fact that the core components of most civil engineering and mechanical structures can be simplified into beams or slabs for calculation and analysis, a large number of structural damage identification methods are first experimentally validated in beam or slab structures [1].

Currently, the research on damage identification methods mainly focuses on model-based and data-driven methods. Among them, the data-driven damage identification method is based on the massive time series data collected by structural health monitoring systems. By using the nonlinear characteristics of structural elements derived from the data or their variations, damage can be identified. The main methods include time series model methods [5,6], wavelet analysis methods, Hilbert–Huang transforms (HHT), Kalman filter methods [7,8], and so on. This method can quickly identify the time and location of damage occurrences and even achieve real-time damage identification, which has a positive effect on real-time early warning based on online health monitoring systems [9].

The most commonly used method in data-driven damage identification research is artificial neural networks. The main reason is that they have learning abilities, parallel processing capabilities for multiple inputs, nonlinear mapping and fault tolerance capabilities, and the ability to obtain adaptive learning networks through new samples, thus ensuring good recognition capabilities of artificial neural networks [10].

In the process of constructing different neural network structures for damage identification, Byington et al. [11] described an algorithm combining Bayesian network models and applied it to aircraft fault diagnosis. They also demonstrated its application to prediction. Przytula et al. [12] further utilized Bayesian network models for system diagnosis and identification frameworks, based on which the probability distribution of structural identification can be obtained. In terms of wavelet neural networks, Baskshi and Stephanopolous [13], Zhang et al. [14], and Zhang [15] were the first to integrate wavelet technology into neural networks. They discussed the theoretical framework of wavelet neural network methods based on gradient descent for identification, forming a relatively stable algorithm flow, which laid the foundation for the application of wavelet neural network technology in the engineering field. Subsequently, Wang and Vachtsevanos [16] proposed a dynamic wavelet neural network combined with reinforcement learning and genetic algorithms for identification and applied it to industrial chillers, demonstrating the superiority of dynamic wavelet neural network algorithms. Chinnam et al. [17,18] combined the signal processing advantages of wavelets with the recognition advantages of neural networks to analyze signals sampled from structures.

In other aspects combining neural network methods, Zhang and Lee et al. [19] applied neural network methods to damage identification and analysis processes in industrial vessels. Through damage experiments on three industrial vessels, they verified the effectiveness of neural network methods in damage identification; Wegerich et al. [20] applied similarity methods for data prediction, which uses a set of average training data and weights similarity. They asserted that this method can be applied to establish similarity relationships between individual data streams among any multivariate data streams and applied it to the processing of rotating machinery vibration signals. Bonissone et al. [21] established a damage identification analysis system for mechanical equipment through data analysis, principal component analysis, adaptive neural network fuzzy inference, trend analysis, and indicator establishment. Necati et al. [22] analyzed uncertainty factors in structural finite element analysis. By using random sampling techniques to obtain a set of corresponding parameter response values (such as strain, displacement, velocity, etc.) for different parameters (such as bearing conditions, elastic modulus, structural stiffness, etc.), they took such a set of corresponding parameter-response values as learning samples. Based on the Levenberg–Marquardt algorithm, they constructed artificial neural networks. Finally, they applied the verified neural network model to structural reliability prediction. Guan and Jha [23] applied maximum entropy methods to fatigue damage identification in structures, obtaining a corresponding relationship between aluminum alloy loading cycles and crack size. They also used experimental data to verify the effectiveness of the algorithm. Hu et al. [24] discussed the application of SVM and swarm intelligence algorithms in the damage identification of beam structures and focused on analyzing their uncertainty impact. However, the data-driven damage identification method highly relies on high-quality and large-scale data samples. If the data volume is insufficient or the data quality is poor, the performance and accuracy of the model may be seriously affected.

The model-based damage identification method mainly refers to the use of measured information of the structure to continuously correct the finite element model for structural damage identification. This method can understand the monitored structure from a holistic perspective and effectively identify damage. However, its computational and data storage requirements are relatively high, making real-time identification nearly impossible [25]. Luchinsky et al. [26] proposed a hybrid probabilistic method based on models for damage identification in solid rocket boosters. This method can widely consider the variability of structural parameters and validate the effectiveness of the damage identification method through experiments with sudden and slow openings in the boosters. Bjarni and Baldvin [27] pointed out that factors such as the dead load of bridge deck pavement layers, elastic modulus of cables, and main cable and deck line shapes can become uncertain factors in finite element analysis, leading to uncertainty in accurately simulating the response characteristics of bridges. Christian et al. [28] integrated an uncertainty analysis of structures into finite element codes, suggesting that although the Monte Carlo method can solve most random problems, its computational cost is too high. Therefore, the response surface method that combines improved Monte Carlo sampling technology is more practical. Luo et al. [29] proposed that when uncertain information is insufficient to accurately describe a probabilistic model, the probabilistic reliability theory based on convex models provides an assessment method for engineering safety. They used multi-ellipsoid models to describe uncertainties in factors such as materials, geometric dimensions, and load sizes and applied standardized transformations and standard space-generalized infinity norm metrics to define non-probabilistic reliability indicators. They derived explicit iterative formulas for optimization algorithms through linearization approximations of limit state equations and simplified solutions for reliability indexes. Finally, numerical examples demonstrated the correctness and effectiveness of the algorithm. Thomas [30] analyzed the limit states of structures by introducing a probability distribution function to parameterize uncertainties in structural analysis. They discussed the failure probabilities of structures and applied them to engineering examples. Zárate and Caicedo et al. [31] proposed a basic method for modifying and identifying crack lengths under fatigue loads. First, they determined the probability density function of fracture mechanics parameters through model modification. Then, they used Markov chain Monte Carlo (MCMC) to identify specimen crack lengths. Crack lengths were periodic functions of loading cycle counts and were validated using experimental data from tensile specimens under constant amplitude loads. Ling and Mahadevan [32] considered uncertainty quantification using Gaussian process models instead of finite element models to simulate fatigue crack growth. Real-time external variable amplitude load history monitoring data were used to build Bayesian combined auto-regressive integrated moving average (ARIMA) models. They proposed a Bayesian probabilistic approach to integrate model-based fatigue damage identification with online and offline structural health monitoring systems for fatigue damage identification. However, model-based damage identification methods typically rely on the accuracy of mathematical modeling of the structure. When the complexity of the actual bridge structure exceeds the range assumed by the model, the accuracy and reliability of the model may be affected [33–36].

With the establishment of multiscale mathematical theoretical models, there has been widespread discussion on multiscale physical and numerical models for structures. The effective simulation of scale coupling has solved the key problem in the process of multiscale finite element modeling, making it possible to establish numerical models that are efficient and precise. It also provides a way to couple model-based and data-driven damage identification methods. Based on the fundamental relationship between the structural bending moment and curvature and the assumption that structural micro-damage has minimal impact on stress redistribution, this article derives a damage index based on node curvature. By combining the denoising capabilities of wavelet analysis, a damage identification method is proposed by solving linear matrix equations of node curvature before and after damage. Finally, the effectiveness and noise resistance of the damage identification method based on node curvature are verified through a multiscale finite element model of a continuous beam bridge.

2. Damage Identification Method Based on Node Curvature

2.1. Basic Principle of Wavelet Denoising

For a given health monitoring system, the vector $X \in R^n$ represents the actual measured signals from all n sensors. It can be decomposed into the following form:

$$X = X^* + KF \tag{1}$$

where, X^* represents random noise signals caused by noise interference; F represents the magnitude of external environment and load effects; and K represents the correlation matrix between external action conditions and structural responses. By selecting the signal of the ith sensor represented by Equation (1) for wavelet transformation, one can obtain the following equation:

$$W_i(a,b) = \frac{1}{\sqrt{a}}\int_{-\infty}^{+\infty} x_i(t)\psi(\frac{t-b}{a})dt = [x_i(t), \psi_{a,b}(t)] \tag{2}$$

where a is the scale factor ($a > 0$), b is the translation factor, and the basic wavelet or mother wavelet $\psi_{a,b}(t)$ is translated and scaled. By further expanding Equation (2), one can obtain the following equation:

$$\begin{aligned} W_i(a,b) &= \frac{1}{\sqrt{a}}\int_{-\infty}^{+\infty} [x_i^*(t) + k_i f_i(t)]\psi(\frac{t-b}{a})dt \\ &= [x_i^*(t), \psi_{a,b}(t)] + [k_i f_i(t), \psi_{a,b}(t)] = W_{x_i^*}(a,b) + W_{k_i f_i}(a,b) \end{aligned} \tag{3}$$

According to the basic principles of wavelet analysis, the true signal that has a functional correspondence with the external action and the white noise signal $N(0, \sigma^2)$ that follows the normal distribution have a proportional relationship between the wavelet transform modulus maxima of the function and the scale factor a after wavelet transformation. On the other hand, the wavelet transform modulus maxima of the function are inversely proportional to the scale factor a. Based on this, wavelet transformation can be used to realize the denoising of measured signals. Considering the complexity and contingency in damage identification, this paper chooses the RBF kernel function.

2.2. Node Curvature

As shown in Figure 1, the undamaged and damaged beams are both composed of m units and n nodes. The node k is the intermediate node between the ith and $(i-1)$th units, and the distance from node k to the left end of the beam is l_k. Under external loads and environmental effects, the bending moment of unit i is $M(l_i)$, and the continuity of bending moments can be expressed as follows:

$$M(l_k) = EI_i y''_l(l_k) = EI_{i-1} y''_r(l_k) \tag{4}$$

where EI_i and EI_{i-1} represent the bending stiffness of the ith and $(i-1)$th units, respectively, and $y''_l(l_k)$ and $y''_r(l_k)$ are the left and right node curvatures at the end of the beam.

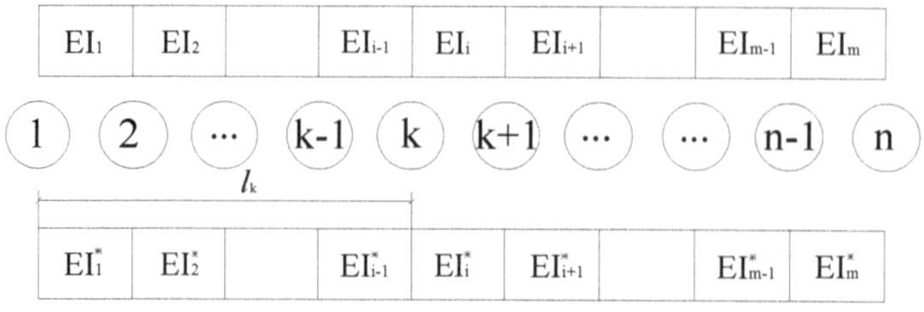

Figure 1. Schematic diagram of undamaged and damaged beams.

Similarly, the bending moment continuity equation for the damaged beam can be obtained as follows:

$$M^*(l_k) = EI_i^* y''_l(l_k)^* = EI_{i-1}^* y''_r(l_k)^* \tag{5}$$

where the bending stiffness and node curvature of node k are EI_k and $y''(l_k)$, respectively.

$$y''(l_k) = \frac{1}{2}(y''_l(l_k) + y''_r(l_k)) \tag{6}$$

By combining Equations (4)–(6), we obtain the following equation:

$$M(l_k) = EI_k y''(l_k) = \frac{1}{2}EI_k(\frac{M(l_k)}{EI_i} + \frac{M(l_k)}{EI_{i-1}}) \tag{7}$$

Therefore, it follows that:

$$EI_k = 2(\frac{EI_i EI_{i-1}}{EI_i + EI_{i-1}}) \tag{8}$$

$$EI_k^* = 2(\frac{EI_i^* EI_{i-1}^*}{EI_i^* + EI_{i-1}^*}) \tag{9}$$

Under minor damage conditions, it can be assumed that structural damage has a minimal impact on stress redistribution [6]. After wavelet denoising processing of the measured signals, the time domain integral equation of node k can be obtained as follows:

$$\int_{t=t_1}^{t=t_2} M(l_k)dt = \int_{t=t_1}^{t=t_2} M^*(l_k)dt \tag{10}$$

By combining Equations (8)–(10) and assuming that the stiffness of the undamaged structure is a constant *EI*, we obtain the following equation:

$$EI(\frac{EI_i^* + EI_{i-1}^*}{EI_i^* EI_{i-1}^*}) \int_{t=t_1}^{t=t_2} y''_k(t)dt = 2\int_{t=t_1}^{t=t_2} y''_k(t)^* dt \tag{11}$$

When $a_i = \frac{EI}{EI_i^*}$, $b_k = \int_{t=t_1}^{t=t_2} y''_k(t)dt$, $b_k^* = \int_{t=t_1}^{t=t_2} y''_k(t)^* dt$, we obtain the following equation:

$$(a_i + a_{i-1})b_k = 2b_k^* \tag{12}$$

where $k = 2, 3 \ldots n-1$. For a structure composed of n nodes and $n-1$ units, there are n−2 linear equations, which can be integrated into a matrix as follows:

$$B_{(n-2)\times(n-1)} A_{(n-1)\times 1} = 2B^*_{(n-2)\times 1} \tag{13}$$

where matrix A represents the stiffness changes before and after damage, and matrices B and B^* represent the structural curvatures before and after damage. Therefore, we obtain the following equation:

$$A = 2B^{-1}B^* \tag{14}$$

where matrix B is a non-symmetric matrix, and its inverse matrix is solved using the singular value decomposition (SVD) method. The left and right sides of Equation (14) are multiplied by the device matrix of B, as follows:

$$B'BA = 2B'B^* \tag{15}$$

Subsequently, we obtain the following equation:

$$(B'B)^{-1} = V(S)^{-1}U' \quad (16)$$

where

$$S_i^{-1} = \begin{cases} 1/S_i & S_i > 0 \\ 0 & S_i = 0 \end{cases} \quad (17)$$

Thus, matrix A can be obtained, as follows:

$$A = 2V(S)^{-1}U'B'B^* \quad (18)$$

By combining with the MATLAB 2023 program, matrix A can be calculated, and the damage index can be obtained.

$$a_i = \frac{EI}{EI_i^*} = \frac{EI}{EI - \Delta EI_i} \quad (19)$$

3. Simply Supported Beam Model Test

3.1. Introduction to the Numerical Model

As shown in Figure 2, the length of beam L = 0.7 m, the width b = 0.03 m, and the height h = 0.006 m. A transverse crack with a uniform depth $h_1 = \alpha h$ through the width of the beam is considered here. The width of the crack is d_1 and located at $x = L_1$ from the left end.

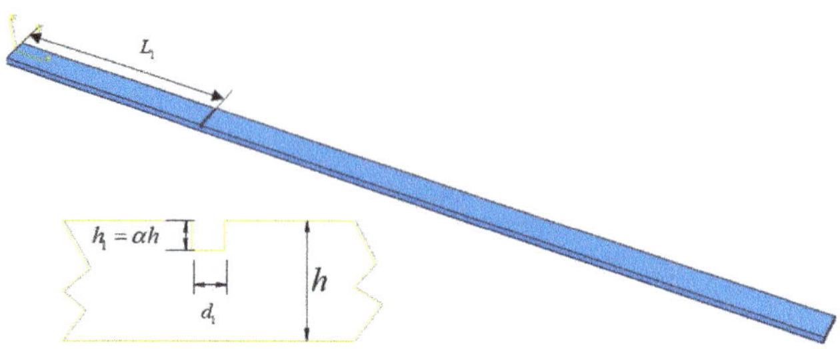

Figure 2. Beam structure with a transverse crack.

The numerical model test of simply supported steel beams is shown in Figure 3. The elastic modulus of the steel beam material is 210 GPa, the density is 7800 kg/m^3, and the external load P = 10 kN, and it passes through the steel beam at a speed of 1 m/s. The simulated conditions are divided into the following four types:

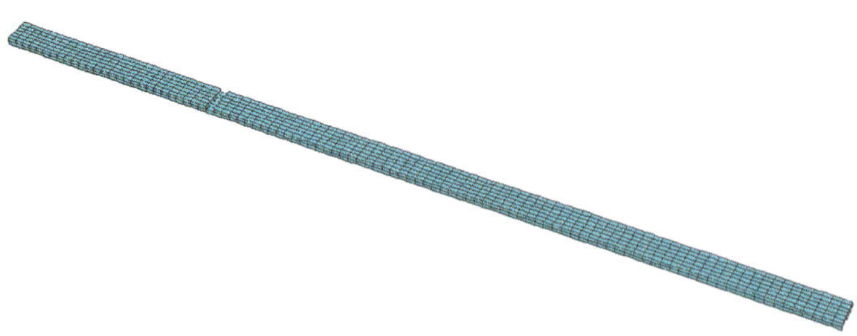

Figure 3. Numerical model test of simply supported steel beam.

Case 1: The stiffness of the 10th element of the steel beam is reduced by 1%;
Case 2: The stiffness of the 33rd element of the steel beam is reduced by 1%;
Case 3: The stiffness of the 10th and 33rd elements of the steel beam is reduced by 1%;
Case 4: There is no damage to the steel beam.

3.2. Damage Sensitivity Analysis

Based on the damage identification method discussed in this paper, which is based on nodal curvature, the stiffness matrix A and damage indicators under various damage conditions can be solved and obtained, as shown in Figure 4. It can be seen in Figure 4 that the damage indicator based on nodal curvature is particularly sensitive to steel beam damage and can effectively identify damage, whether it is singular or multiple.

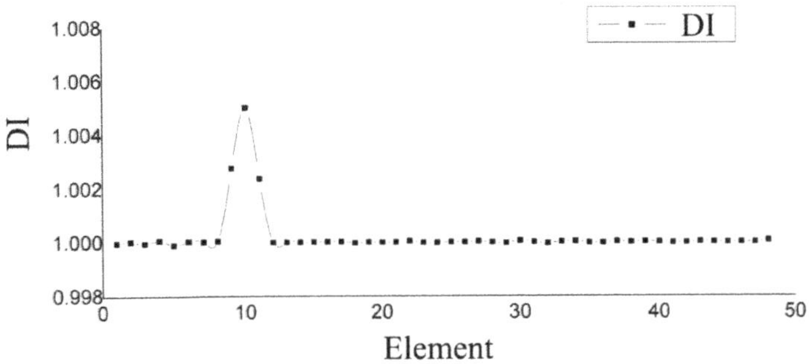

(**a**) Damage identification results of Case 1.

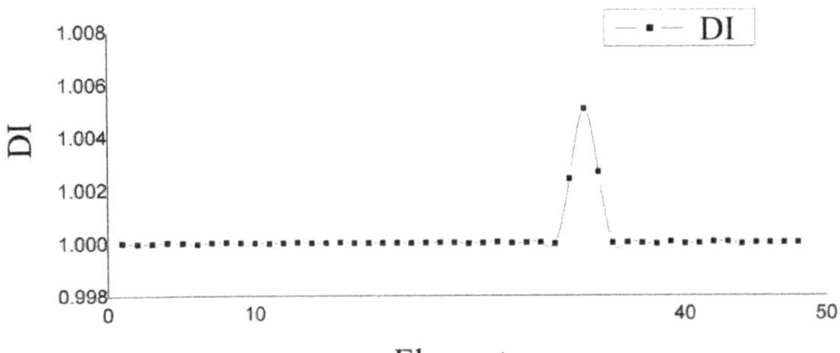

(**b**) Damage identification results of Case 2.

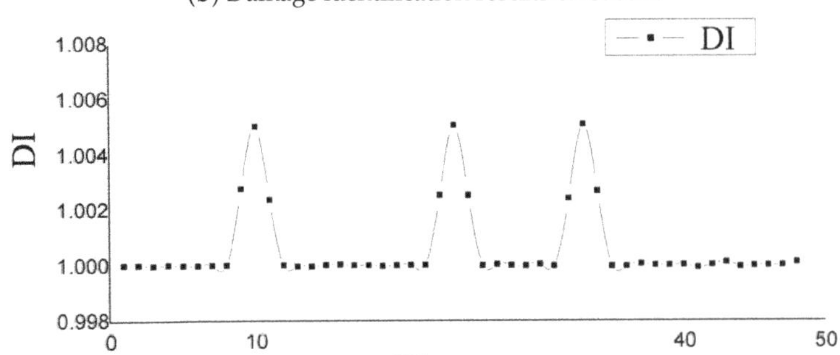

(**c**) Damage identification results of Case 3.

Figure 4. Damage identification results of simply supported beam.

4. Multiscale Damage Identification of Xinyi River Bridge

4.1. Multiscale FE Model

The Xinyi River bridge (Figure 5) is situated on the Beijing-Shanghai Expressway (within Shuyang County). The superstructure of this bridge employs a combination of partially prestressed concrete uniform cross-section box girders with a 30 m span. The overall length of the bridge is 2168.20 m, with a design load capacity of automobiles up to

Super 20 and trailers up to 120. It was completed and opened to traffic in 2001. To analyze the current damage status, a multiscale finite element model was established. Small-scale units are suitable for areas suspected of damage, allowing for further clarification and quantification of structural damage. The mid-span section employs fine small-scale Solid45 elements, while other large-scale sections utilize Beam188 elements [37]. The cross-section is presented in Figure 6.

Figure 5. Xinyi River bridge.

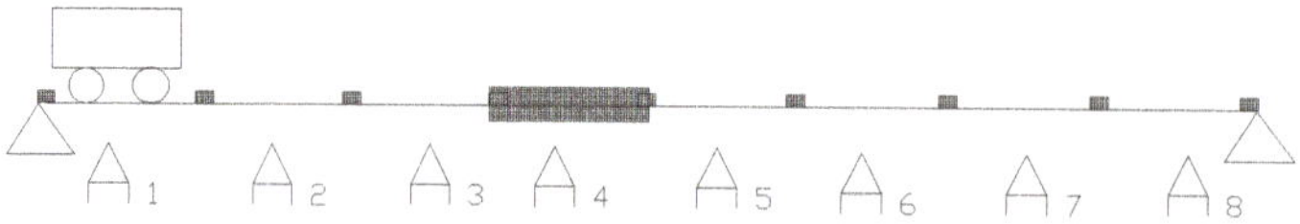

(**a**) Multiscale model.

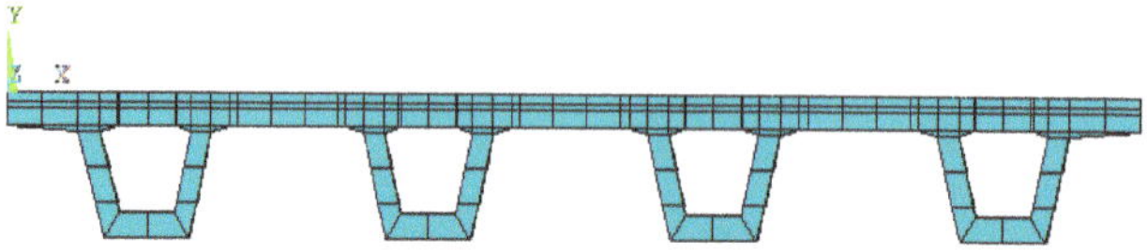

(**b**) Cross-section view of small-scale sections.

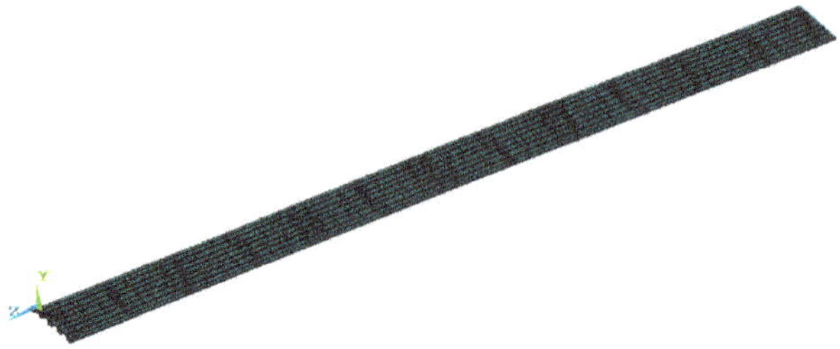

(**c**) Integral FE model.

Figure 6. Finite element model of Xinyi River bridge.

4.2. Test Case

This paper uses the modified finite element model of the Xinyi River bridge to simulate the damage to the main girder. Small-scale A4 areas and large-scale A5 areas were selected to simulate the damage condition of the main girder of the Xinyi River bridge, as follows:

Case 1: Small-scale A4 region, main beam stiffness reduced by 5%;
Case 2: Small-scale A4 region, main beam stiffness reduced by 3%;
Case 3: Small-scale A4 region, main beam stiffness reduced by 1%;
Case 4: Large-scale A5 region, main beam stiffness reduced by 5%;
Case 5: Large-scale A5 region, main beam stiffness reduced by 3%;
Case 6: Large-scale A5 region, main beam stiffness reduced by 1%;
Case 7: Large-scale A5 and small-scale A4 regions, main beam stiffness reduced by 1% each.

4.3. Damage Sensitivity Analysis of Beam-Type

Without considering noise interference, based on the finite element models of the Xinyi River bridge after different damage scenarios, a simulation was conducted to simulate the passage of a heavy vehicle (60t) at a constant speed across the bridge. Using Equations (4)–(15), the damage indicators for each condition were calculated, and the results are summarized in Table 1. The damage indicators for Conditions 3, 6, and 7 are plotted in Figure 6 for reference.

Table 1. Damage identification results.

DI	1% Noise Interference	3% Noise Interference	5% Noise Interference	10% Noise Interference
DI(A4)-1	1.091	1.091	1.091	1.086
DI(A5)-1	1.080	1.081	1.082	1.078

Without considering noise interference, from Table 1 and Figure 7, the following can be inferred:

(1) The damage identification indicators based on nodal curvatures are able to effectively identify minor damage (1%) in structures across different scales;
(2) When damage occurs in the small-scale region of the main beam, the damage identification indicators based on nodal curvatures can accurately locate the damage position;
(3) When damage occurs in the large-scale region of the main beam, the damage identification indicators based on nodal curvatures can generally determine the damage location.

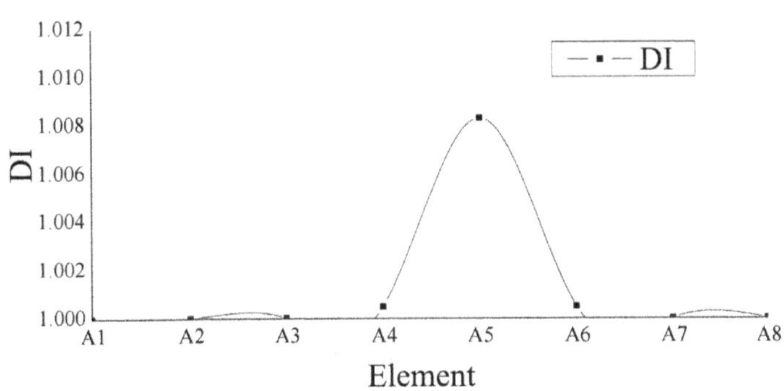

(**a**) Case 3.

Figure 7. *Cont.*

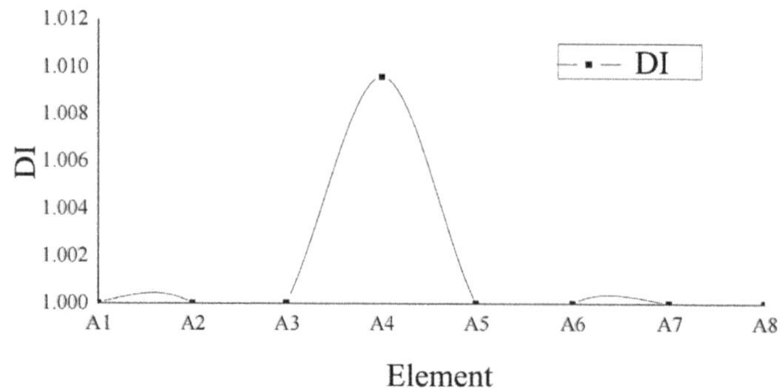

(**b**) Case 6.

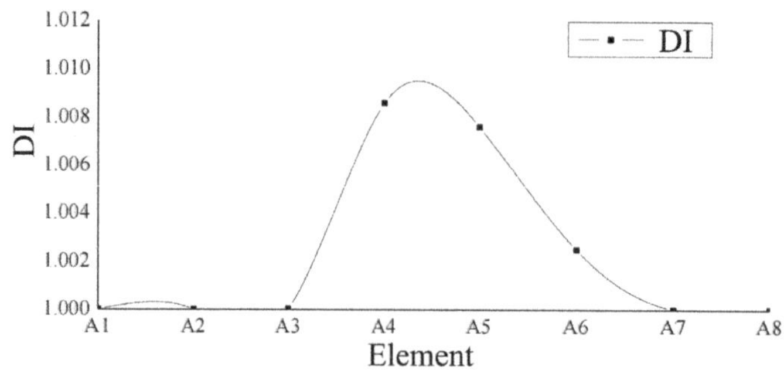

(**c**) Case 7.

Figure 7. Damage identification results of Xinyi River bridge.

4.4. Noise Analysis

Apply noise in the following equation:

$$d(t_0) = d_0(t_0)(1 + \varepsilon R) \tag{20}$$

In the formula, *d*(*t*0) and *d*0(*t*0) represent the node displacement values before and after noise addition ε at time *t*0, respectively; *N* represents the noise level; and *R* represents a white noise signal that follows a normal distribution within the range of [−1, +1].

When taking into account various levels of noise interference, specifically 1%, 3%, 5%, and 10%, firstly, based on the fundamental principle of wavelet denoising, the interference of noise on the measured signal is reduced. Then, as described above, the damage indicators under Condition 8 are calculated using Equations (4)–(15), and the results are shown in Figure 8 and Table 1.

In the presence of noise interference, the results in Figure 8 indicate the following: (1) The damage identification method based on nodal curvatures is able to achieve damage identification across different scales of the structure, and it exhibits a certain degree of noise resilience. (2) With the increase in noise level, the damage index of small-scale units changes less, while the damage index of large-scale units changes a lot; this is especially true for the A6 unit shown in Figure 8d, in which the damage index is different than the damage index obtained without noise interference. Although the damage location can still be identified, there is a large error in identifying the degree of damage in large-scale units.

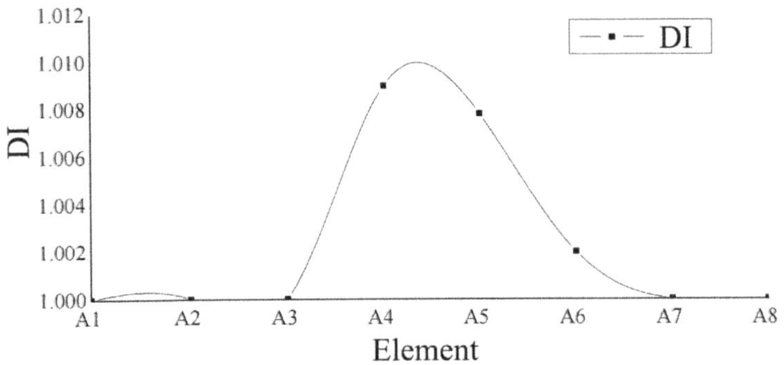

(**a**) Results of damage identification under 1% noise interference conditions.

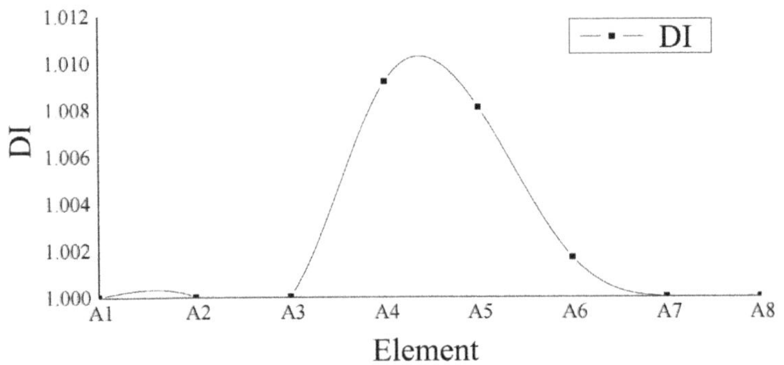

(**b**) Results of damage identification under 3% noise interference conditions.

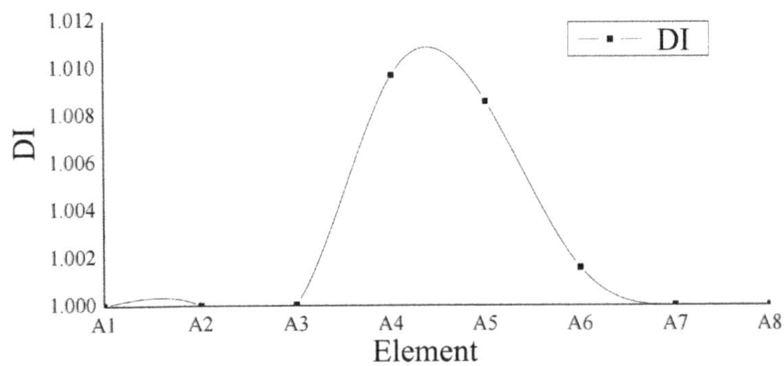

(**c**) Results of damage identification under 5% noise interference conditions.

Figure 8. *Cont.*

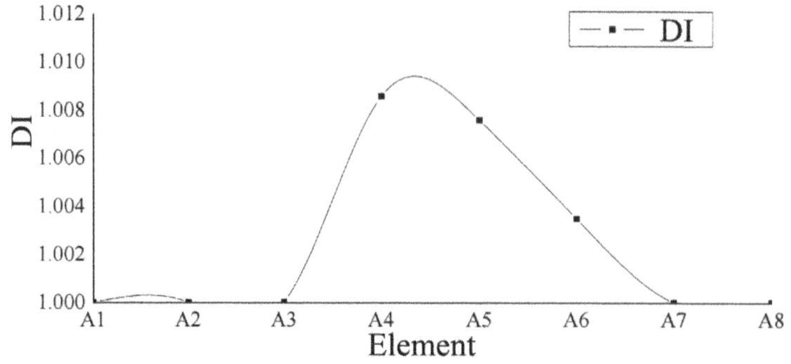

(**d**) Results of damage identification under 10% noise interference conditions.

Figure 8. Damage identification results under noise interference.

5. Conclusions

The damage identification method based on nodal curvatures discussed in this paper first utilizes wavelet transformation to perform denoising processing of time series signals. Subsequently, it develops nodal curvature damage indicators based on the fundamental relationship between structural bending moments and curvatures and the assumption that structural micro-damage has minimal impact on stress redistribution. A damage identification method is proposed by solving linear matrix equations, and its theoretical feasibility is verified through numerical simulation experiments on a continuous beam bridge. The following conclusions can be drawn:

(1) In the absence of noise interference, the damage identification indicators based on nodal curvatures can accurately locate the damage position;

(2) Considering the effect of noise interference, the damage identification method based on NCWA can basically realize the damage degree of structural small-scale units and the recognition of large-scale unit damage location;

(3) The damage identification method based on nodal curvatures has promising applications and lays a solid foundation for the health monitoring of bridge structures and their safety prognosis.

The method proposed in this paper is influenced by various factors, such as the accuracy of node displacement recognition, sensor accuracy and location, the surrounding environment, etc. The impact of these factors requires more comprehensive and systematic experiments for analysis and verification in future studies. It should be pointed out that the applicability of the method proposed in this article to strong noise interference and other scenarios in complex structures needs further research.

Author Contributions: Conceptualization, K.Y. and S.Z.; methodology, Q.Z.; software, R.Z.; validation, W.W., K.Y. and S.Z.; formal analysis, S.Z.; investigation, K.Y.; resources, K.Y.; data curation, S.Z.; writing—original draft preparation, Q.Z.; writing—review and editing, K.Y.; visualization, R.Z.; supervision, S.Z.; project administration, Q.Z.; funding acquisition, Q.Z. All authors have read and agreed to the published version of the manuscript.

Funding: This research was funded by the National Natural Science Foundation of China (No. U22A20569, 42074226, 42304046), Shenzhen High-level Talents Research Start-up Fund (No. RC2022-004), "2024 Doctoral Research Startup Project of Shenzhen Institute of Information Technology" (Grant No. SZIIT2024KJ030), Key Laboratory of Land Satellite Remote Sensing Application, Ministry of Natural Resources (Grant No. KLSMNR-G202222) and Youth Innovation Talent Project in Guangdong Provincial Universities (Grant No. 2024KQNCX174). The viewpoints in this paper represent only the author's opinion and do not represent the views of the fund committee.

Data Availability Statement: The original contributions presented in the study are included in the article, further inquiries can be directed to the corresponding authors.

Conflicts of Interest: Author Kai Ye is employed by the Shenzhen Cheng'an Softcom Technology Group Co., Ltd. The remaining authors declare that the research was conducted in the absence of any commercial or financial relationships that could be construed as a potential conflict of interest.

References

1. Ou, J.; Li, H. Structural health monitoring in main-land China: Review and future trends. *Struct. Health Monit.* **2010**, *9*, 219–231.
2. Doebling, S.W.; Farrar, C.R.; Prime, M.B.; Shevitz, D.W. *Damage Identification and Health Monitoring of Structural and Mechanical Systems from Changes in Their Vibration Characteristics: A Literature Review*; Los Alamos National Laboratory Report LA-13070-MS; Los Alamos National Laboratory: Los Alamos, NM, USA, 1996.
3. Farrar, C.R.; Lieven, N.A.J. Damage prognosis: The future of structural health monitoring. *Phil. Trans. R. Soc. A* **2007**, *365*, 623–632. [CrossRef] [PubMed]
4. Inman, D.J.; Farrar, C.R. *Damage Prognosis: Current Status and Future Needs*; Los Alamos National Laboratory Report LA-14051-MS; Los Alamos National Laboratory: Los Alamos, NM, USA, 2003.
5. Farrar, C.R.; Hemez, F. A Coupled Approach to Developing Damage Prognosis Solutions. *Eng. Mater.* **2003**, *24*, 289–306.
6. James, G.; Sharp, D. Multiscale science: A challenge for the twenty-first century. *Adv. Mech.* **1998**, *28*, 545–551.
7. Sohn, H.; Farrar, C.R.; Hunter, N.F.; Worden, K. Structural health monitoring using statistical pattern recognition techniques. *J. Struct. Eng.* **2000**, *123*, 1336–1356. [CrossRef]
8. Nair, K.K.; Kiremidjian, A.S.; Law, K.H. Time series-based damage detection and localization algorithm with application to the ASCE benchmark structures. *J. Sound Vib.* **2006**, *291*, 349–368. [CrossRef]
9. Huang, N.E.; Shen, Z. The empirical mode decomposition and the Hilbert spectrum for nonlinear and non-stationary time series analysis. *Proc. R. Soc. Lond. Ser. A* **1998**, *454*, 903–995. [CrossRef]
10. Schwabacher, M.A.; Goebel, K. A Survey of Artificial Intelligence for Prognostics. In Proceedings of the AAAI Fall Symposium, Arlington, VA, USA, 9–11 November 2007.
11. Byington, C.S.; Kalgren, P.W.; Johns, R.; Beers, R.J. Embedded diagnostic/prognostic reasoning and information continuity for improved avionics maintenance. In Proceedings of the Autotestcon 2003: IEEE Systems Readiness Technology Conference, Anaheim, CA, USA, 22–25 September 2003; pp. 320–329.
12. Przytula, K.W.; Arthur, C. Reasoning Framework for Diagnosis and Prognosis. In Proceedings of the Aerospace Conference IEEE, Big Sky, MT, USA, 3–10 March 2007; pp. 1–10.
13. Baskshi, B.R.; Stephanopoulous, G.; Wavelet:, A. Multiresolution Hierarchical Neural Network with Localized Learning. *Am. Inst. Chem. Eng. J.* **1993**, *39*, 57–81. [CrossRef]
14. Zhong, R.; Zong, Z.; Pai, P.F.; Ruan, X.; Niu, J. Instantaneous frequency tracking method for composite cable-stayed bridge. *Mech. Syst. Signal Process.* **2018**, *100*, 43–56. [CrossRef]
15. Zhang, Q. Using Wavelet Neural in nonparametric estimation. *IEEE Trans. Neural Netw.* **1997**, *8*, 227–236. [CrossRef]
16. Wang, P.; Vachtsevanos, G. Fault prognostics using dynamic wavelet neural networks. *Artif. Intell. Eng. Des. Manuf.* **2001**, *9*, 349–365. [CrossRef]
17. Chinnam, R.B.; Mohan, P. Online reliability estimation of physical systems using neural networks and wavelets. *Int. J. Smart Eng. Syst. Des.* **2002**, *4*, 253–264. [CrossRef]
18. Chinnam, R.B.; Baruah, P.A. Neuro-Fuzzy approach for estimating mean residual life in condition-based maintenance systems. *Int. J. Mater. Prod. Technol.* **2004**, *2*, 166–179. [CrossRef]
19. Zhang, G.; Lee, S.; Propes, N.; Zhao, Y.; Vachtsevanos, G.; Thakker, A.; Galie, T. A novel architecture for an integrated fault diagnostic/prognostic system. In Proceedings of the AAAI Technical Report, Palo Alto, CA, USA, 25–27 March 2002.
20. Wegerich, S.W.; Wilks, A.D.; Pipke, R.M. Nonparametric modeling of vibration signal features for equipment health monitoring. In Proceedings of the IEEE Aerospace Conference, Big Sky, MT, USA, 8–15 March 2003; pp. 3113–3121.
21. Bonissone, P.P. Knowledge and Time: A framework for soft computing applications in predictive health management (PHM). In Proceedings of the IPMU, Paris, France, 2–7 July 2006.
22. Zhong, R.; Pai, P.F. An instantaneous frequency identification method of stay cables. *J. Low Freq. Noise Vib. Act. Control* **2021**, *10*, 1–21.
23. Guan, X.; Jha, R.; Liu, Y. Probabilistic fatigue damage prognosis using maximum entropy approach. *J. Intell. Manuf.* **2012**, *23*, 163–171. [CrossRef]
24. Hu, Z.; Zhu, H.; Huang, L.; Cheng, C. Damage Identification Method and Uncertainty Analysis of Beam Structures Based on SVM and Swarm Intelligence Algorithm. *Buildings* **2022**, *12*, 1950. [CrossRef]
25. Engel, S.J.; Gilmartin, B.J.; Bongort, K.; Hess, A. Prognostics, the real issues involved with predicting life remaining. *Aerosp. Conf. Proc.* **2000**, *6*, 457–469.
26. Luchinsky, D.; Osipov, V.; Smelyanskiy, V. Model Based IVHM System for the Solid Rocket Booster. In Proceedings of the IEEE Aerospace Conference, Big Sky, MT, USA, 1–8 March 2008; pp. 1–15.
27. Bjarni, B.; Baldvin, E. Structural health modeling of the olfusá Suspension Bridge. Master's Thesis, Engineering and Natural Sciences University of Iceland, Reykjavík, Iceland, 2012.

28. Christian, B.; Dagmar, H.; Dirk, R. Advanced analysis of structural reliability using commercial FE-codes. In Proceedings of the European Congress on Computational Methods in Applied Sciences and Engineering ECCOMAS 2000, Barcelona, Spain, 11–14 September 2000; pp. 11–14.
29. Luo, Y.; Kang, Z.; Li, A. Structural reliability assessment based on probability and convex set mixed model. *Comput. Struct.* **2009**, *11*, 1408–1415.
30. Thomas, F. Modelling Uncertainty in Limit State Functions. In Proceedings of the 7th International Symposium on Imprecise Probability: Theories and Applications, Innsbruck, Austria, 25–28 July 2011.
31. Zárate, B.A.; Caicedo, J.M.; Yu, J.; Ziehl, P. Bayesian model updating and prognosis of fatigue crack growth. *Eng. Struct.* **2012**, *45*, 53–61.
32. Ling, Y.; Mahadevan, S. Integration of structural health monitoring and fatigue damage prognosis. *Mech. Syst. Signal Process.* **2012**, *28*, 89–104. [CrossRef]
33. Liu, W.K.; Qian, D.; Gonella, S.; Li, S.; Chen, W.; Chirputkar, S. Multiscale methods for mechanical science of complex material: Bridging from quantum to stachastic multiresolution continuum. *Int. J. Numer. Methods Eingineering* **2010**, *83*, 1039–1080. [CrossRef]
34. Sun, D.; Cao, Z.; Huang, C.; Wu, K.; De Schutter, G.; Zhang, L. Degradation of concrete in marine environment under coupled chloride and sulfate attack: A numerical and experimental study. *Case Stud. Constr. Mater.* **2022**, *17*, e1218. [CrossRef]
35. Bolandi, H.; Li, X.; Salem, T.; Boddeti, V.N.; Lajnef, N. Bridging finite element and deep learning: High-resolution stress distribution prediction in structural components. *Front. Struct. Civ. Eng.* **2022**, *16*, 1365–1377. [CrossRef]
36. Zhou, Y.; Li, M.; Shi, Y.; Xu, C.; Zhang, D.; Zhou, M. Research on Damage Identification of Arch Bridges Based on Deflection Influence Line Analytical Theory. *Buildings* **2024**, *14*, 6. [CrossRef]
37. Li, Z.X.; Jiang, F.F.; Tang, Y.Q. Multi-scale analyses on seismic damage and progressive failure of steel structures. *Finite Elem. Anal. Des.* **2012**, *48*, 1358–1369. [CrossRef]

buildings

MDPI

Article

Investigation on the Bearing Performance of a Single Pile in Shallow Reinforced Soft Soil Foundation under Horizontal Load

Guanglin Bai [1], Hong Zhang [1], Bo Wang [2], Feng Chen [1], Jiahao Zhao [3] and Qianjin Shu [3,*]

1 Xuzhou Survey and Design Branch of State Grid Jiangsu Electric Power Design Consulting Co., Ltd., Xuzhou 221000, China
2 State Grid Jiangsu Electric Power Co., Ltd. Economic Research Institute, Nanjing 210008, China
3 State Key Laboratory for Geomechanics & Deep Underground Engineering, School of Mechanics & Civil Engineering, China University of Mining & Technology, Xuzhou 221116, China
* Correspondence: shu-qianjin@cumt.edu.cn

Abstract: The overall reinforcement of soft soil foundation has the disadvantages of large engineering quantity and high cost. When the pile foundation bears horizontal loads in the soil, the mechanical properties of the soil near the surface have a greater impact on it compared to the deep soil. Therefore, studying the influence of shallow soil reinforcement on the horizontal bearing capacity of pile foundations has important engineering significance. Studying the influence of shallow soft soil reinforcement around piles on the horizontal bearing performance of piles is of great significance for improving the economic efficiency of pile foundation reinforcement technology in soft soil areas. In this paper, seven pile-soil finite element models are established based on ABAQUS 2022 software to study the influence of shallow reinforcement on the horizontal bearing capacity of single pile. The models were established on the basis of a field test and its validity was verified. The influence of different reinforcement degrees on the horizontal bearing capacity of piles is analyzed by taking the reinforcement width and reinforcement depth as variables. The results indicate that shallow ground improvement significantly enhances the horizontal bearing capacity of the pile. The horizontal bearing capacity of the pile is increased by 83.0%, 104.3%, and 224.4%, respectively, corresponding to a reinforcement width of 2 times, 3 times, and 4 times the diameter of the pile, respectively. With the increase of the reinforcement width, the bending moment and deformation of the pile under the same horizontal load decrease significantly, while it has no significant effect on the location of the maximum bending moment of the pile. The bearing capacity of the pile foundation gradually increases with the increase of the reinforcement depth. Compared with the unreinforced situation, the horizontal bearing capacity of the pile body is increased by 224.4%, 361.3%, and 456.8%, respectively, corresponding to a reinforcement depth of 0.1 times, 0.2 times, and 0.3 times the pile length. As the reinforcement depth increases, the corresponding increase in bearing capacity does not increase linearly, but gradually decreases. This indicates that blindly carrying out deep soil reinforcement without comprehensive evaluation is not advisable.

Keywords: pile; horizontal bearing characteristics; shallow reinforcement; soft soil

1. Introduction

With rapid economic development, many highways, bridges, and transmission towers need to be established on soft soil foundation. Soft soil, characterized by low strength and high compressibility, has a low capacity to resist lateral movement of the foundation, significantly impacting the pile foundation's ability to resist load deformation. In regions with poor soil properties, shallow ground improvement is an effective method. There have been numerous studies on soil improvement both domestically and internationally.

The research on the horizontal bearing capacity of pile foundations has been widely studied by many scholars. Kellezi et al. [1] first used the finite element software ABAQUS

2022 to study the bearing characteristics of the single pile foundation of the Horns Rev offshore wind farm in Denmark. In the modeling, the Mohr–Coulomb constitutive model is used for the soil, and the soil parameters and layer thickness are selected according to the geological survey data of the engineering site. The analysis method of this model provides a reference for analyzing the horizontal bearing characteristics of the single pile foundation. Based on the field test data, Sun et al. [2] used ABAQUS to carry out a numerical analysis of very large-diameter, horizontal loaded single pile. The effects of soil strength parameters, pile size, load, and other factors on large-diameter, horizontal loaded piles were explored. Wang Jian [3] ed ABAQUS to analyze the horizontal bearing characteristics of reinforced composite piles in clay. The research shows that the horizontal ultimate bearing capacity of reinforced composite piles is related to the strength of the cement soil, the diameter of the cement soil piles, and the undrained shear strength of the soil. A simplified algorithm for the horizontal ultimate bearing capacity of reinforced composite piles is proposed. Wang Ge et al. [4] created a beam element analysis model through the three-dimensional, finite element analysis of checking the engineering test pile, clearing the horizontal load of large-diameter, single pile resistance composition, and change rule. Zhang Xin et al. [5] found through indoor pile-soil model tests and finite element numerical simulations that the pile bucket diameter ratio has a significant impact on the horizontal bearing characteristics of composite foundations. Wang Songyang et al. [6] used a strain wedge model to obtain a calculation method for the lateral bearing performance of a single pile under three-dimensional, asymmetric local erosion conditions, and revealed the influence of different erosion sizes on the mechanical and displacement properties of single piles. Ren Meng et al. [7] analyzed the stress characteristics of bridge pile foundations under fracture zones through on-site triaxial compression tests and numerical simulations, and found that the performance of pile foundations can withstand the influence of fracture zones during their service life.

Domestic and foreign scholars have also conducted extensive research on soil reinforcement. Adsero [8] and Rollins [9] used high-pressure jet grouting piles to reinforce around pile caps to improve the horizontal bearing capacity of pile groups. Experimental results showed that, compared to traditional methods of altering pile structure, using high-pressure jet grouting piles had better economic efficiency. Faro et al. [10,11] found that short piles' horizontal load-bearing performance improved significantly after using cement to reinforce shallow ground. The horizontal bearing capacity of short piles increased with the widening of the reinforcement width, and there existed an optimal reinforcement depth. He Ben et al. [12] found that reinforcing shallow soil with 7.5 times the pile diameter had a decisive effect on the horizontal bearing capacity of flexible single piles. Static load test results showed a 100% increase in the ultimate bearing capacity of the pile foundation after reinforcement, along with a 50% reduction in the maximum pile bending moment. Taghavi et al. [13,14] compared and analyzed the horizontal bearing characteristics of pile groups before and after cement-soil reinforcement through centrifuge tests. The results showed a significant improvement in the horizontal bearing capacity of the pile group after reinforcement, and the seismic performance of the pile group was also enhanced. Wang Anhui et al. [15] discovered a 40% increase in the horizontal ultimate bearing capacity of PHC piles after cement-soil reinforcement by inserting precast concrete piles into cement-mixed piles, with significant reductions in the pile top horizontal displacement and maximum pile bending moment. Yu Guangming [16,17] and others, through field tests and numerical simulations, obtained a 43% increase in the horizontal bearing capacity of single piles reinforced with cement in gravel, proposing a suitably modified p–y curve formula based on their experiments. Zhang Dingwen et al. [18], relying on a solid engineering project for reinforcing a soft foundation with an interlayer of easily liquefiable silty soil, studied the applicability and reinforcement effects of jet grouting piles through field tests.

It can be seen that current research on soil improvement focuses on studying the effects of different improvement methods, with less emphasis on the study of the influence of the same improvement method's different improvement ranges. Additionally, most studies target a single soil layer, and research on multi-layer soil is limited. Therefore, a more in-depth study is needed on the influence of the improvement range on multi-layer soil. Based on the analysis of the existing reinforcement methods, the dynamic compaction method creates difficulties in controlling the scope of reinforcement and the strength of the soil after reinforcement. The electroosmotic method has the disadvantages of an uneven reinforcement effect, with good reinforcement effect near the cathode, poor reinforcement effect near the anode, and high fraud. Vacuum preloading also has disadvantages such as uncontrollable reinforcement range. Therefore, the cement mixing method is selected in this paper. This method can effectively control the reinforcement range, and the strength of the reinforced cement soil can be measured. Numerical simulation is used to simulate the change of horizontal bearing characteristics of reinforced concrete single pile under different reinforcement degrees, and the influence of reinforcement range on horizontal bearing characteristics of single pile is explored.

2. Finite Element Model Establishment

2.1. Model Design

This study focuses on the influence of two parameters, namely reinforcement width and reinforcement depth, on the load-bearing performance of the pile. Both parameters are simulated using a single controlled variable method based on the ABAQUS finite element model. The dimensions of the pile and the surrounding soil are introduced separately as follows.

(1) Soil around the Pile

The soil around the pile is simulated as a rectangular block, as shown in Figure 1. To eliminate the influence of boundary conditions on the load-bearing capacity of the pile foundation, the soil is set up and down to a square with a side length of 20D, and the thickness of the soil is set to 2L (where D and L are the diameter and length of the pile, respectively). The soil is divided into three parts: the shallow soil around the pile reinforced with cement, the upper part of soft soil, and the bottom sand soil. Since the model pile and the applied horizontal load in the horizontal static load test have symmetry, to improve the computational efficiency, the numerical simulation model is created as half of the actual model.

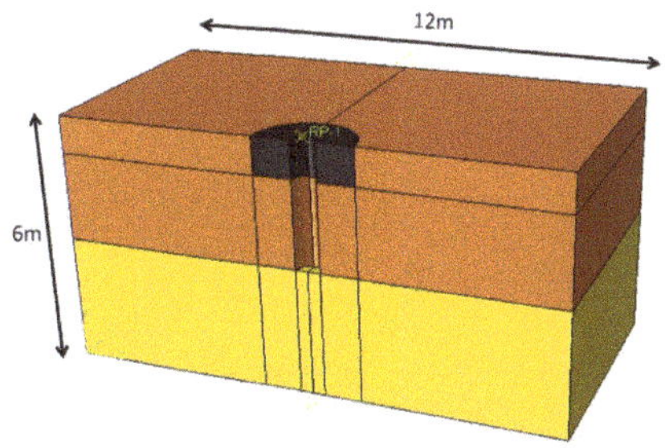

Figure 1. Soil finite element model. (The black part in the picture is cement reinforced soil, the orange part in the picture is upper layer silt soft soil, and the yellow part in the picture is sandy soil).

(2) Pile

As shown in Figure 2, the pile is a concrete pile with a length of 3 m and a diameter of 0.6 m, positioned in the middle part of the soil.

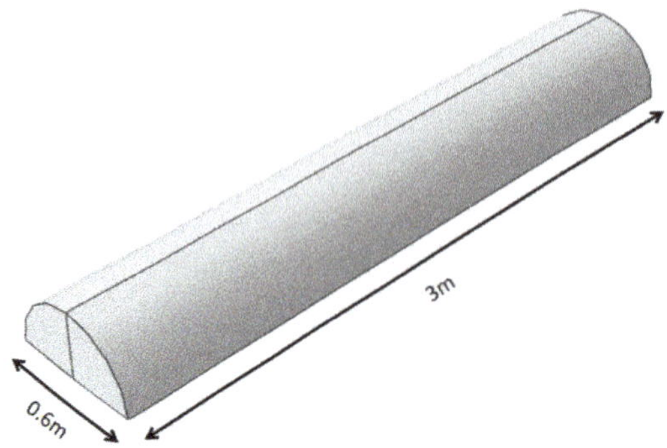

Figure 2. Pile finite element model.

2.2. Simulation Groups

As shown below, Figure 3 provides a schematic diagram of the finite element simulation. In this figure, the reinforcement depth of cement reinforced soil is the amount that needs to be changed for research. The thickness of the silty soft soil and the dimensions of the pile are held constant.

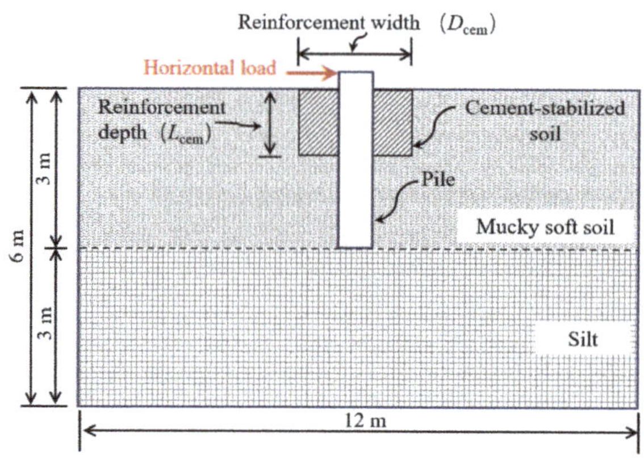

Figure 3. Finite element simulation parameters.

(1) Variation in Reinforcement Width

This section primarily investigates the influence of changes in reinforcement width on the horizontal bearing characteristics of the pile foundation. Therefore, while keeping the reinforcement depth constant, only the reinforcement width is altered. The specific groupings are outlined in Table 1, where D_{cem} is the depth of the reinforced soil and L_{cem} is the width of the reinforced soil.

Table 1. Reinforcement width variable simulation grouping.

Notation	L/m	D/m	D_{cem}/m	L_{cem}/m
1	3	0.6	1.2	0.3
2	3	0.6	1.8	0.3
3	3	0.6	2.4	0.3
4	3	0.6	1.8	0.6
5	3	0.6	2.4	0.6

(2) Variation in Reinforcement Depth

This section primarily explores the impact of changes in reinforcement depth on the horizontal bearing characteristics of the pile foundation. Therefore, while keeping the reinforcement width constant, only the reinforcement depth is altered. The specific groupings are outlined in Table 2, where D_{cem} is the depth of the reinforced soil and L_{cem} is the width of the reinforced soil.

Table 2. Reinforcement depth variable simulation grouping.

Notation	L/m	D/m	D_{cem}/m	L_{cem}/m
1	3	0.6	1.8	0.3
2	3	0.6	1.8	0.6
3	3	0.6	2.4	0.3
4	3	0.6	2.4	0.6
5	3	0.6	2.4	0.9
6	3	0.6	2.4	1.2
7	3	0.6	2.4	1.5

2.3. Element Selection and Material Definition

The soil surrounding the pile is divided into three categories: the shallow cement-stabilized soil around the pile, the shallow silt soft soil, and the sand at the bottom of the pile. Based on existing simulation studies, the silt clay adopts the Mohr–Coulomb elastic–plastic constitutive model [19]. Although the strength of the sand is higher than that of the soft clay, it is still much lower than the strength of the pile, and exhibits significant plasticity. Therefore, for the sand, the Mohr–Coulomb elastic–plastic model is also employed. For the cement-stabilized soft soil, according to the conclusions obtained by researchers like Jamsawang P. [20], the Mohr–Coulomb constitutive model is equally effective in simulating its deformation and failure under load. Therefore, all three types of soil are simulated using the Mohr–Coulomb model. In the simulation, the values for soft soil and sand are taken from a geological survey report of a location at the front edge of the Yangtze River Delta [21]. The pile is modeled using a linear elastic model. Detailed model parameters are shown in Table 3, where C is the cohesion, φ is the internal friction angle, μ is Poisson's ratio, E is the elastic model, and ρ is the density.

Table 3. Pile and soil properties.

Soil Mass	C/kPa	φ/°	μ	E/MPa	ρ/(g/cm^3)
Mucky moft soil	10.9	17.2	0.35	3.77	1.75
Silt	2.5	30.8	0.3	8.71	1.93
Cement-stabilized soil	346	38.3	0.2	130	1.76
Pile	/	/	0.2	22,000	2.50

2.4. Meshing

The rationality of meshing is a crucial step in determining the accuracy of numerical simulation results. In this study, the sweeping neutral axis algorithm is employed for meshing the soil, excluding the soil at the pile bottom. The soil close to the pile is finely meshed, while the soil farther from the pile is sparsely meshed. Structural meshing is applied to the pile bottom soil and the pile itself. Additionally, four seed points are used for refining the mesh around the pile, and three seed points are used for the internal part of the pile. The reduced integration, three-dimensional element C3D8R is used in this modeling. The horizontal loading responses for different mesh sizes are shown in Figure 4. As observed in Figure 4, when the mesh size ranges from 50 mm to 200 mm, there is minimal variation in the ultimate horizontal bearing capacity at the pile top, gradually reaching a stable state. Considering the trade-off between computational accuracy and speed, a mesh size of 100 mm is chosen for this model.

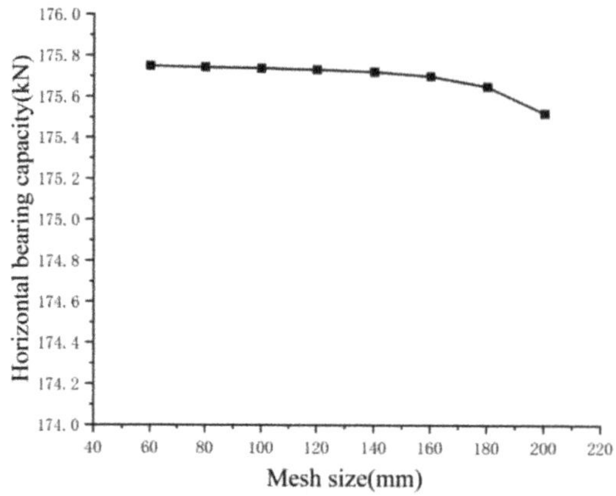

Figure 4. The ultimate bearing capacity of the pile top varies with the grid size.

2.5. Contact Definition and Boundary Conditions

Three steps of analysis are set up for this simulation. In the first step, the pile section is entirely deactivated using the birth–death element method. In the second step, the pile is activated, and the soil originally occupying the pile section is deactivated using the birth–death element method. This simulates the static press-in process of the pile and ensures a more reasonable distribution of initial ground stresses within the model. In the third step, horizontal loads are applied to the pile top.

The bottom boundary condition of the model is fully fixed, with all four sides subjected to normal constraints, and the top in a free state. The contact surfaces adopt the Mohr–Coulomb model, with the pile considered the master surface and the soil as the slave surface. The interaction on the contact surface is divided into the normal and tangential directions. In the normal direction between the pile and the soil, a "hard contact" method is applied. In the tangential direction, the Coulomb friction model is used for both soft soil and ordinary sand. The friction coefficient of soft clay to concrete is in the range of 0.3~0.5. Therefore, the tangential friction coefficient is selected as 0.3 in this range. For cement-reinforced soil, the tangential friction coefficient is in the range of 0.39~0.49, and the tangential friction coefficient is selected as 0.4 in this range [22].

2.6. Load Application and Control

The horizontal load on the top of the pile adopts the loading mode of concentrated force, and the size is set to 100 kN. The loading form is a concentrated force loading, the top surface of the pile is coupled to a point, and then a horizontal concentrated force is applied to it. According to the relevant provisions of the 2014 "Technical Code for detection of Building Foundation Piles", one of the stopping criteria for the horizontal static load test is that the horizontal displacement of the pile top reaches 30~40 mm, so this paper applies 40 mm horizontal displacement to the pile top.

2.7. Validation of Finite Element Simulation

In order to ensure the rationality of the simulation method and the reliability of the relevant results, this paper takes the field test finished by Faro et al. [10] as a reference, establishes the corresponding finite element model, and carries out the numerical simulation analysis. The comparison between the simulation results and the field test results is shown in Table 4. It should be mentioned that the code D_{cem} is the diameter of the range of the soil reinforcement plane and L_{cem} is the depth of the soil reinforcement. It can be seen that a finite element simulation can provide accurate prediction of the bearing capacities under horizontal load, with a minimum relative error of 5.53%, a maximum error of 15.05%, and an average error of 9.35%. It is indicated that the numerical simulation method proposed in Sections 2.1–2.6 is suitable for analyzing the coupling bearing mechanism of piles in marine soft foundation, considering the influence of the diameter of the range of the soil reinforcement plane and the depth of soil reinforcement.

Table 4. Comparison of horizontal bearing capacity between field measurement and numerical simulation.

Notation	D_{cem}/D	L_{cem}/D	H_{ult}/kN		Relative Error/%
			Experimental Data	Analog Data	
1	0	0	80	88.26	10.32
2	2	0.5	140	150.87	7.76
3	3	0.5	160	170.43	6.52
4	3	1	220	186.90	15.05
5	4	1	250	225.14	9.94
6	4	1.5	250	236.17	5.53
7	0	0	80	88.26	10.32

3. Results and Analysis

To explore the impact of the reinforcement range on the horizontal response of the pile, this section extracts and analyzes data, such as pile bending moment, load displacement, etc., from the simulation results. The analysis is conducted in two directions: different reinforcement widths and different reinforcement depths, investigating their respective influences on the pile.

3.1. The Influence of Reinforcement Width

In this section, with reinforcement width as the variable, we investigate the impact of changing the reinforcement width on factors such as the pile's horizontal ultimate bearing capacity, pile bending moment, displacement, etc., under the condition of constant reinforcement depth.

(1) Pile Bending Moment

Figure 5 shows the variation of the pile bending moment with the increase in reinforcement width when the pile top load is 100 kN. From Figure 5, it can be observed that when the reinforcement depth is kept constant, the reinforcement width has a significant impact on the pile bending moment. When the pile is in intact clay, the maximum bending moment of the pile is 34.60 kN·m and is located 1.2 m below the mud surface. When the reinforcement depth is 0.3 m, with the increase in reinforcement width, the maximum pile bending moment decreases to 20.52 kN·m, 12.24 kN·m, and 9.29 kN·m, representing a reduction of 40.69%, 64.62%, and 73.15%, respectively, compared to the intact clay state. However, as the reinforcement width increases, the proportion of reduction in the pile bending moment gradually decreases. This is more evident in the bending moment curve when the reinforcement depth is 0.6 m. When the reinforcement depth is 0.6 m, the maximum pile bending moment decreases to 12.13 kN·m and 11.08 kN·m with the increase in reinforcement width. It can be concluded that the impact of reinforcement width on the pile bending moment diminishes gradually as the reinforcement width increases.

Regarding the occurrence position of the maximum pile bending moment, when the reinforcement depth is kept at 0.3 m, the positions with increasing reinforcement width are 1.1 m, 1.0 m, and 0.9 m below the mud surface, showing a slight upward trend. For a reinforcement depth of 0.6 m, the positions of the maximum pile bending moment are consistently 0.5 m below the mud surface with the increase in reinforcement width. Although there is a noticeable upward shift compared to intact clay, changing the reinforcement width does not alter the occurrence position of the maximum moment. In summary, the variation in the reinforcement width has a limited impact on the occurrence position of the maximum pile bending moment.

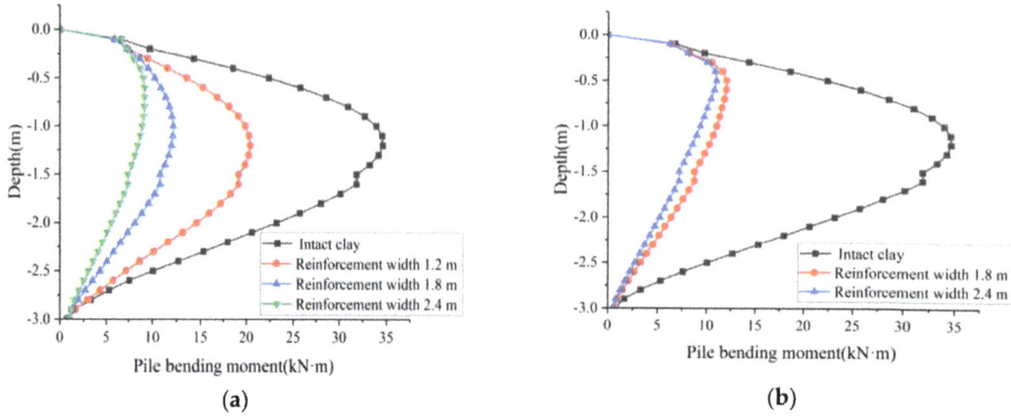

Figure 5. Bending moment distribution of pile shaft. (**a**) reinforcement depth of 0.3 m. (**b**) reinforcement depth of 0.6 m.

(2) Pile Displacement

Figure 6 displays the change curve of pile deformation with the increase of the reinforcement width, when the pile top load is 100 kN. As shown in Figure 6, when the horizontal load is 100 kN, the pile displacement curve is presented. It can be seen that shallow soil reinforcement can effectively reduce the displacement deformation of the pile. From Figure 6, it can be seen that when the reinforcement depth is kept constant, the deformation gap of the pile displacement curve gradually decreases with the increase of the reinforcement width. Without reinforcement, the pile top displacement is 55.11 mm. When the reinforcement depth is kept at 0.3 m, the pile top displacement decreases with the increase of the reinforcement width, measuring 34.11 mm, 15.62 mm, and 10.58 mm, respectively. Although the pile top displacement is gradually decreasing, this reduction rate also decreases with the increase in reinforcement width. When the reinforcement width increases from 1.8 m to 2.4 m, the displacement reduction rate is only 32%.

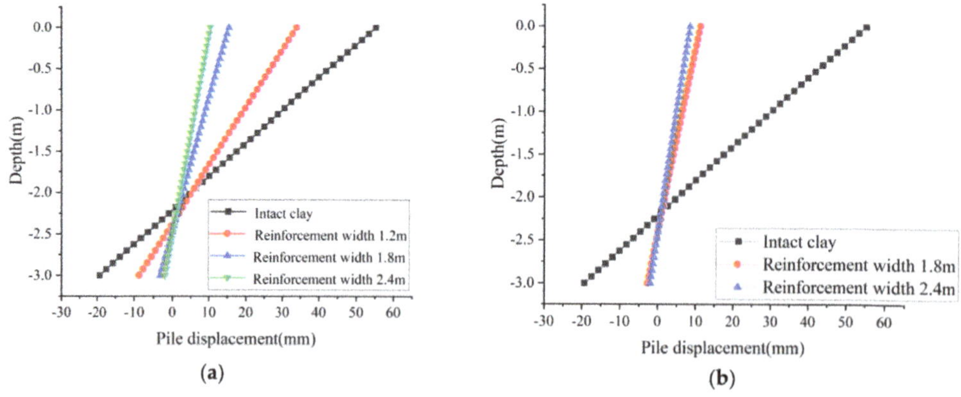

Figure 6. Pile displacement. (**a**) reinforcement depth of 0.3 m. (**b**) reinforcement depth of 0.6 m.

When the reinforcement depth is kept at 0.6 m, the pile top displacements are 11.44 mm and 8.61 mm with the increase in reinforcement width. It can be observed that while shallow soil reinforcement can significantly reduce pile deformation, the efficiency of this reinforcement gradually decreases with the increase in reinforcement width.

(3) Pile Horizontal Bearing Capacity

Figure 7 displays the displacement gradient curve obtained by changing the reinforcement width under the premise of keeping the reinforcement depth unchanged. According to the relevant provisions in the "Technical Specification for Building Pile Foundation Inspection" (JGJ106-2014) [23], the second inflection point of the curve in the figure corresponds to the horizontal ultimate load value of the single pile. According to the curve in the diagram, it can be determined that the ultimate load values of single piles under different reinforcement widths are 16.42 kN, 30.04 kN, 33.54 kN, and 53.27 kN, respectively.

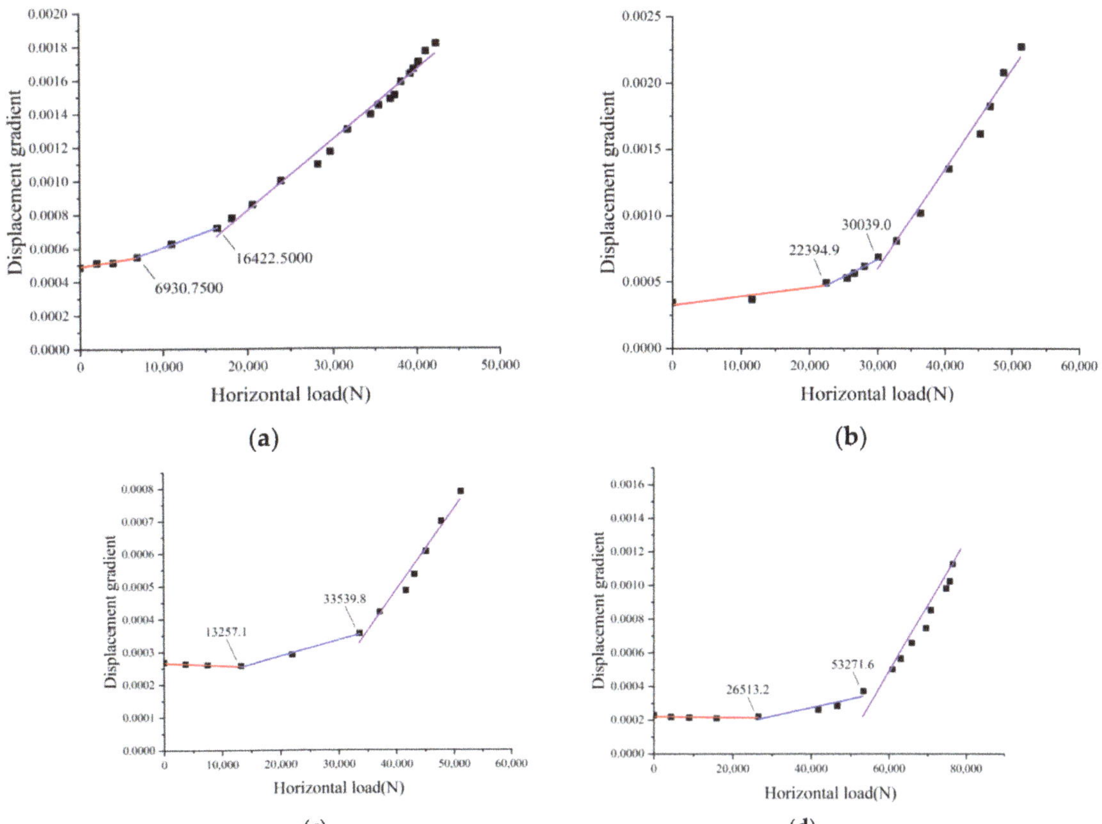

Figure 7. Displacement gradient curve (reinforcement depth of 0.3 m). (**a**) not strengthened. (**b**) reinforcement width of 1.2 m. (**c**) reinforcement width of 1.8 m. (**d**) reinforcement width of 2.4 m.

Figure 8 displays the curve of the ultimate bearing capacity of the pile with the change of the reinforcement width. Compared with the unreinforced case, the ultimate load values of the piles increased by 82.95%, 104.26%, and 224.42%, respectively.

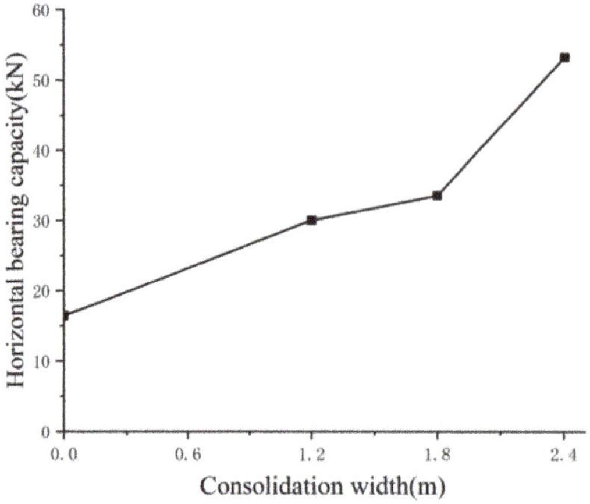

Figure 8. Relationship between the horizontal bearing capacity of single pile and reinforcement width.

3.2. The Impact of Reinforcement Depth

In this section, with the reinforcement depth as a variable, the influence of changing the reinforcement depth on the horizontal ultimate bearing capacity, pile bending moment, displacement, and other factors is investigated, while keeping the reinforcement width constant.

(1) Pile Bending Moment

Figure 9 depicts the distribution curve of the pile bending moment when changing the reinforcement depth, while keeping the reinforcement width constant. Similarly to the previous section, when the pile is in intact clay, the maximum bending moment is 34.60 kN·m, and the maximum moment occurs 1.2 m below the mud surface. From the two graphs above, it can be observed that although changing the reinforcement depth significantly reduces the pile bending moment compared to intact clay, the impact on the maximum bending moment is not significant when keeping the reinforcement width constant. When the reinforcement width is 2.4 m, with the increase in reinforcement depth, the bending moment even shows an increase. For reinforcement depths of 0.3 m, 0.6 m, and 0.9 m, the maximum bending moments are 9.29 kN·m, 11.08 kN·m, and 12.55 kN·m, respectively. It can be seen that changing the reinforcement depth does not have a significant impact on the maximum bending moment.

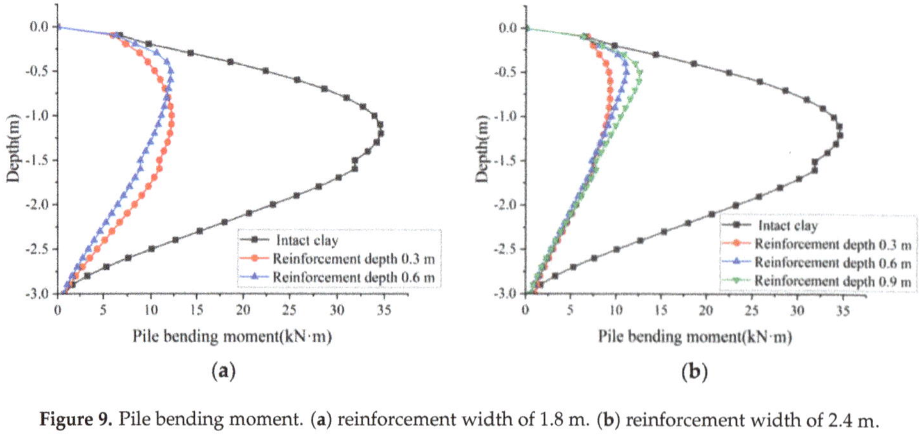

Figure 9. Pile bending moment. (**a**) reinforcement width of 1.8 m. (**b**) reinforcement width of 2.4 m.

In Figure 9a, it is evident that with the increase in reinforcement depth, the location of the maximum bending moment shifts upward. Although the shift is not very pronounced in Figure 9b, it also gradually moves upward with the increase in reinforcement depth. Therefore, there is a certain relationship between the change in reinforcement depth and the occurrence location of the maximum bending moment.

(2) Pile Displacement

Figure 10 represents the pile displacement curves obtained by changing the reinforcement depth while keeping the reinforcement width constant. Both before and after reinforcement, the pile exhibits the characteristics of a rigid pile, and shallow reinforcement significantly reduces the deformation of the pile. From the right graph, it can be observed that, with a constant reinforcement width of 2.4 m, the pile displacements increase with the reinforcement depth, measuring 10.58 mm, 8.61 mm, and 7.91 mm, respectively. Although the pile deformation is gradually decreasing, it becomes less significant when reinforcing beyond a width of 1.8 m.

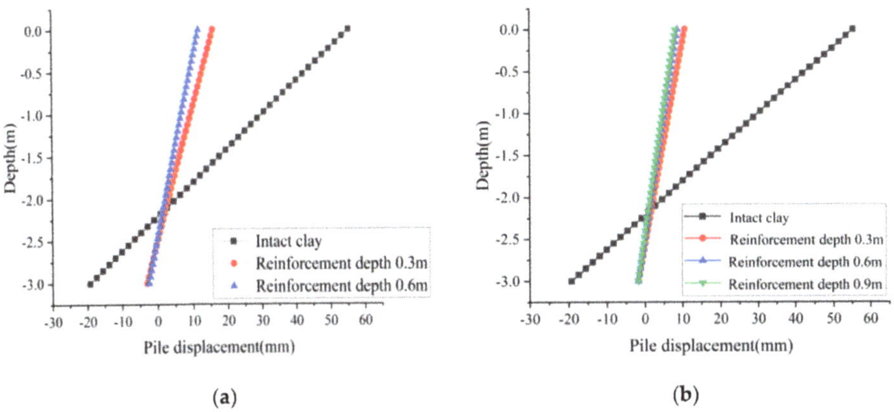

Figure 10. Pile displacement. (**a**) reinforcement width of 1.8 m. (**b**) reinforcement width of 2.4 m.

(3) Displacement Gradient

Figure 11 represents the displacement gradient curve obtained by changing the reinforcement depth under the premise of keeping the reinforcement width unchanged. According to the relevant provisions in the "Technical Specification for Building Pile Foundation Inspection" (JGJ106-2014) [24], the second inflection point of the curve in the figure corresponds to the horizontal ultimate load value of the single pile. According to the curve in the diagram, it can be determined that the ultimate load values of single pile under different reinforcement depths are 16.42 kN, 53.27 kN, 75.75 kN, 91.42 kN, 94.86 kN, and 96.09 kN, respectively.

Figure 12 displays the curve of the ultimate bearing capacity of the pile, changing with reinforcement depth. Compared to the unreinforced case, the ultimate load values of the piles increased by 224.42%, 361.33%, 456.76%, 477.71%, and 485.2%, respectively. As the reinforcement depth increases, the increase in bearing capacity does not increase linearly, but gradually decreases. From Figure 12, it can be seen that when the reinforcement depth exceeds 1.5D, which is 0.9 m, increasing the reinforcement depth again no longer significantly improves the horizontal bearing capacity of the pile. Compared to the case where the reinforcement depth is 0.9 m, when the reinforcement depth is increased to 1.2 m, the horizontal bearing capacity of the pile only increases by 3.76%. Compared to the case where the reinforcement depth is 1.2 m, when the reinforcement depth is increased to 1.5 m, the horizontal bearing capacity of the pile only increases by 1.30%. This indicates that simply increasing the reinforcement depth to improve the horizontal bearing capacity of the pile requires a comprehensive consideration of the reinforcement purpose and economy. From Figure 12, it can be seen that when the reinforcement depth exceeds 0.9 m, increasing the reinforcement depth again has little effect on the horizontal bearing performance of the pile. Considering the economy, it can be considered that the pile studied in this article has an optimal reinforcement depth of 1.5D, which is 0.9 m. Blindly carrying out deep soil reinforcement without comprehensive evaluation is not advisable.

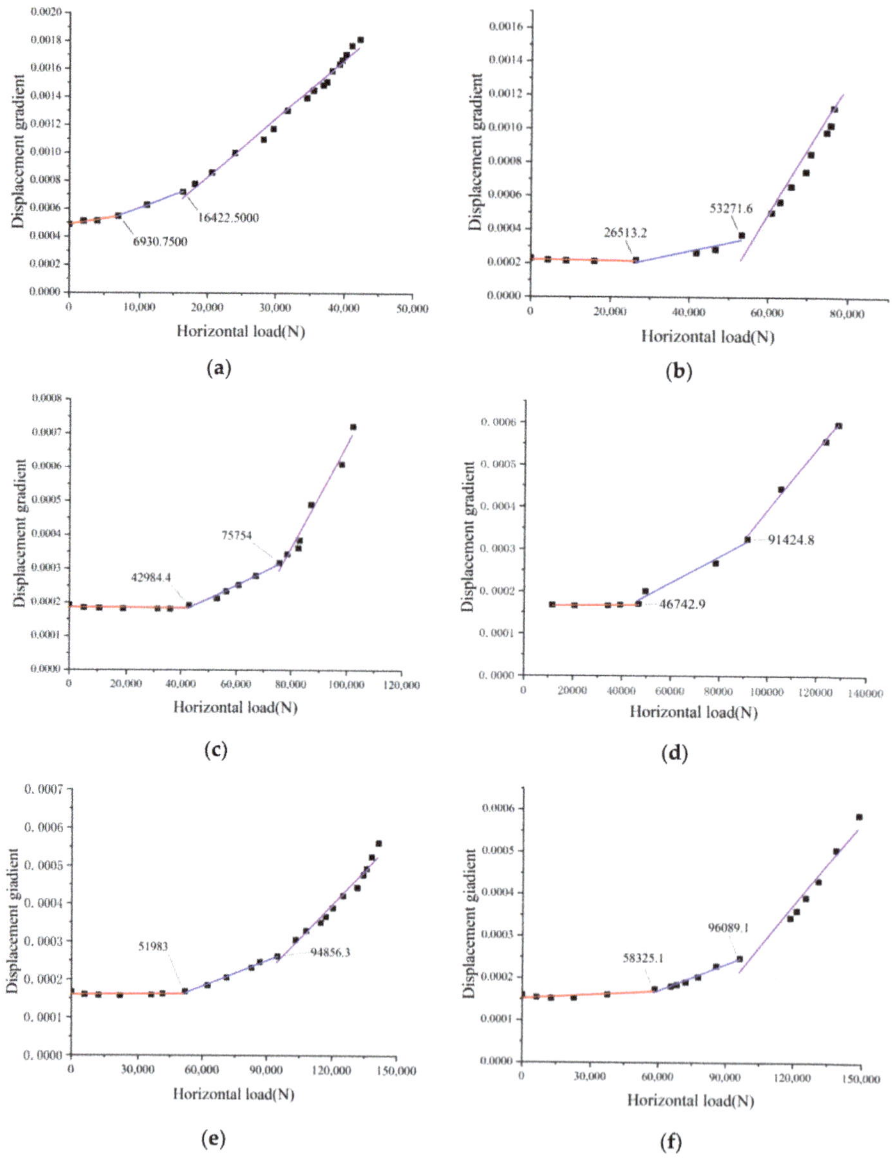

Figure 11. Displacement gradient curve (reinforcement width of 2.4 m). (**a**) not strengthened. (**b**) reinforcement depth of 0.3 m. (**c**) reinforcement depth of 0.6 m. (**d**) reinforcement depth of 0.9 m. (**e**) reinforcement depth of 1.2 m. (**f**) reinforcement depth of 1.5 m.

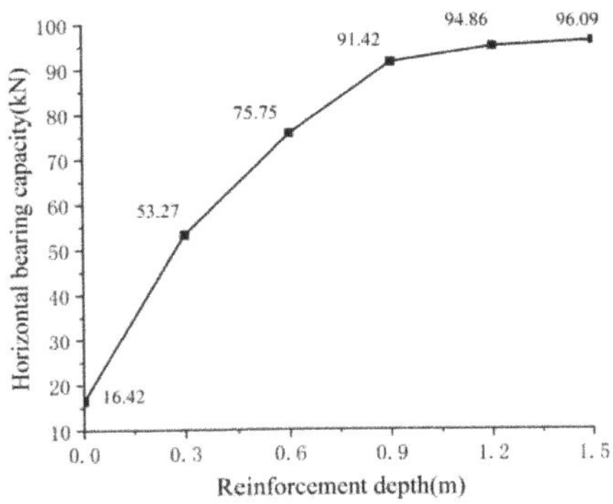

Figure 12. Relationship between horizontal bearing capacity of single pile and reinforcement depth.

4. Conclusions

In this paper, seven three-dimensional finite element models are established to study the influence of soil reinforcement width and soil reinforcement depth on the horizontal bearing capacities of single piles. The results are as follow:

(1) The comparison between the simulation carried out in this paper and the field test on the capacities of a single pile in shallow reinforcement soil showed good coincide, with a minimum relative error of 5.53%, a maximum error of 15.05%, and an average error of 9.35%. The simulation method proposed in this paper is suitable for analyzing the coupling bearing mechanism of piles in soft soil considering shallow reinforcement.

(2) Increasing the reinforcement width can significantly improve the bearing capacity of the pile foundation, and as the reinforcement width increases, the reinforcement effect will become better and better. Compared with the unreinforced case, the horizontal bearing capacity of the pile is increased by 83.0%, 104.3%, and 224.4%, respectively, corresponding to a reinforcement width of 2 times, 3 times, or 4 times the dimeter of the pile respectively. With the increase of the reinforcement width, the bending moment and deformation of the pile under the same horizontal load decrease significantly, while also having no significant effect on the location of the maximum bending moment of the pile.

(3) The bearing capacity of the pile foundation gradually increases with the increase of the reinforcement depth. However, unlike the reinforcement width, the effect of increasing the reinforcement depth on improving the bearing capacity of the pile is gradually weakening, and it can be considered that there is an optimal reinforcement depth of 1.5D. Compared with the unreinforced situation, when the reinforcement depth is 0.5 times, 1.0 times, 1.5 times, 2.0 times, and 2.5 times the pile diameter, the horizontal bearing capacity of the pile body is increased by 224.4%, 361.3%, 456.8%, 477.71%, and 485.2%, respectively.

(4) As the reinforcement depth increases, the increase in bearing capacity does not increase linearly, but gradually decreases. This indicates that simply increasing the reinforcement depth to improve the horizontal bearing capacity of the pile requires a comprehensive consideration of the reinforcement purpose and economy. Blindly carrying out deep soil reinforcement without comprehensive evaluation is not advisable.

(5) Due to space limitations, this article did not consider the combined effect of vertical and horizontal loads in the simulation analysis. Whether this has a significant impact on the relevant conclusions requires further research to confirm. In addition, further specialized research is needed to determine whether the conclusions of this article are applicable to piles with more aspect ratios and soil material properties.

Author Contributions: Methodology, Q.S.; Software, F.C. and J.Z.; Investigation, G.B., H.Z., B.W., F.C. and J.Z.; Resources, Q.S.; Data curation, G.B., H.Z. and B.W.; Writing—original draft, G.B.; Writing—review & editing, Q.S.; Funding acquisition, Q.S. All authors have read and agreed to the published version of the manuscript.

Funding: This research was supported by State Grid Jiangsu Electric Power Design Consulting Co., Ltd. (grant number JE202306).

Data Availability Statement: Data are contained within the article.

Conflicts of Interest: Authors Guanglin Bai, Hong Zhang and Feng Chen were employed by the company Xuzhou Survey and Design Branch of State Grid Jiangsu Electric Power Design Consulting Co., Ltd. Author Bo Wang was employed by State Grid Jiangsu Electric Power Co., Ltd. Economic Research Institute. The remaining authors declare that the research was conducted in the absence of any commercial or financial relationships that could be construed as a potential conflict of interest. The authors declare that this study received funding from State Grid Jiangsu Electric Power Design Consulting Co., Ltd. The funder was not involved in the study design, collection, analysis, interpretation of data, the writing of this article or the decision to submit it for publication.

References

1. Kellezi, L.; Hansen, P.B.; British Geo Technical Association (BGA). Static and dynamic analysis of an offshore mono-pile windmill foundation. In *BGA International Conference on Foundations: Innovations, Observations, Design and Practice, Proceedings of the International Conference Organised by British Geotechnical Association, Dundee, Scotland, 2–5 September 2003*; Thomas Telford Publishing: London, UK, 2003; pp. 401–410.
2. Sun, Y. Test and Numerical Analysis of Horizontal Bearing Characteristics of Super-Large Diameter Single Pile of Offshore Wind Turbine. Ph.D. Thesis, Zhejiang University: Zhejiang, China, 2016. (In Chinese).
3. Wang, J.; Zhu, Z.H.; Wang, H.Y.; Zhu, Q.H.; Fei, K. Numerical analysis of horizontal bearing capacity of rigid composite pile in clay foundation. *J. Railw. Sci. Eng.* **2020**, *17*, 1382–1389. (In Chinese)
4. Wang, G.; Liao, C.C.; Zhang, L.L.; Lin, J.Q.; Ye, G.L.; Zhang, Z.C. Resistance component analysis and application of large diameter horizontal loaded single pile. *J. Civ.Eng.* **2024**, *57*, 72–82. (In Chinese)
5. Zhang, X.; Yu, D.; Zhu, K.; Zhao, A.; Ren, M. The Horizontal Bearing Characteristics and Microscopic Soil Deformation Mechanism of Pile-Bucket Composite Foundation in Sand. *Appl. Sci.* **2024**, *14*, 907. [CrossRef]
6. Wang, S.; Ma, J.; Wang, C.; Liu, F.; Li, D. An Investigation into the Lateral Bearing Performance of a Single Pile Embedded at a Three-Dimensional Asymmetric Local Scour Site Using the Modified Strain Wedge Model. *Appl. Sci.* **2024**, *14*, 3056. [CrossRef]
7. Ren, M.; Cheng, J.; Zhang, S.; Pang, Y.; Zhu, W. Ultimate Load-Bearing Capacity and Sustainable Performance of Pile Foundations of Yanji Suspension Bridge in Fault Zone Based on Refined Geological Model. *Sustainability* **2024**, *16*, 1858. [CrossRef]
8. Adsero, M.E. Effect of Jet Grouting on the Lateral Resistance of Soil Surrounding Driven-Pile Foundations. Ph.D. Thesis, Brigham Young University, Provo, UT, USA, 2008. (In Chinese).
9. Rollins, K.M.; Adsero, M.E.; Brown, D.A. Jet grouting to increase lateral resistance of pile group in soft clay. In *Contemporary Topics in Ground Modification, Problem Soils, and Geo-Support, Proceedings of the International Foundation Congress and Equipment Expo 2009, Orlando, FL, USA, 15 March 2009*; American Society of Civil Engineers: Reston, VA, USA, 2009; pp. 265–272. [CrossRef]
10. Faro, V.P.; Consoli, N.C.; Schnaid, F.; Thomé, A.; da Silva Lopes, L. Field tests on laterally loaded rigid piles in cement treated soils. *J. Geotech. Geoenvironmental Eng.* **2015**, *141*, 06015003. [CrossRef]
11. Faro, V.P.; Schnaid, F.; Consoli, N.C. Field tests of laterally loaded flexible pile in soil with top cement-treated layers. *Proc. Inst. Civ. Eng. Ground Improv.* **2018**, *171*, 174–182. [CrossRef]
12. He, B. *Horizontal Load-Bearing Characteristics of Single Piles and Composite Pile Foundations on Soft Clay Ground*; Geo Technical Engineering, Zhejiang University: Zhejiang, China, 2016. (In Chinese)
13. Taghavi, A.; Muraleetharan, K.K.; Miller, G.A.; Cerato, A.B. Centrifuge modeling of laterally loaded pile groups in improved soft clay. *J. Geo Tech. Geoenvironmental Eng.* **2016**, *142*, 04015099. [CrossRef]
14. Taghavi, A.; Muraleetharan, K.K.; Miller, G.A. Nonlinear seismic behavior of pile groups in cement-improved soft clay. *Soil Dyn. Earthq. Eng.* **2017**, *99*, 189–202. [CrossRef]
15. Wang, A.H.; Zhang, D.W.; Liu, S.Y.; Deng, Y.G. Research on the bearing characteristics of rigid composite tubular piles under horizontal load. *J. China Univ. Min. Technol.* **2018**, *47*, 853–861. (In Chinese)
16. Wang, A.; Zhang, D.; Deng, Y. Lateral response of single piles in cement improved soil: Numerical and theoretical investigation. *Comput. Geotech.* **2018**, *102*, 164–178. [CrossRef]
17. Yu, G.; Gong, W.; Chen, M.; Dai, G.; Liu, Y. Prediction and Analysis of Behaviour of Laterally Loaded Single Piles in Improved Gravel Soil. *Int. J. Civ. Eng.* **2019**, *17*, 809–822. [CrossRef]
18. Zhang, D.W.; Liu, S.C.; Lin, W.F.; Shi, H.B.; Mao, Z.L. Field test of jet grouting piles reinforcing the liquefiable silty soil layer in soft ground. *J. Traffic Transp. Eng.* **2022**, *22*, 103–111. (In Chinese)

19. Rollins, K.M.; Brown, D.A. *Design Guidelines for Increasing the Lateral Resistance of Highway-Bridge Pile Foundations by Improving Weak Soils*; Transportation Research Board: Washington, DC, USA, 2011.
20. Jamsawang, P.; Voottipruex, P.; Jongpradist, P.; Bergado, D.T. Parameters affecting the lateral movements of compound deep cement mixing walls by numerical simulations and parametric analyses. *Acta Geotech.* **2015**, *10*, 797–812. [CrossRef]
21. Wang, Z.H. Research on the Horizontal Bearing Characteristics of Large-Diameter Cement-Soil Composite Steel Pipe Piles. Master's Thesis, Southeast University, Nanjing, China, 2021. (In Chinese).
22. Cui, Z.Z. *Research on the Stability of Borehole Walls and Vertical Bearing Performance of Drilled Cast-in-Place Piles in the Near Front Edge Area of the Yangtze River Delta*; China University of Mining and Technology: Xuzhou, China, 2020. (In Chinese)
23. *JGJ 94-2008*; Technical Specification for Building Pile Foundations. Ministry of Housing and Urban-Rural Development of the People's Republic of China: Beijing, China, 2008. (In Chinese)
24. *JGJ 106-2014*; Technical Specification for Testing of Building Foundation Piles. China Architecture Publishing & Media Co., Ltd.: Beijing, China, 2014. (In Chinese)

buildings

MDPI

Article

Numerical Parameter Analysis of High-Strength Steel Frame with Y-Eccentric Brace Using Variable Replaceable Link

Xi Chen, Shen Li, Gang Liang and Min He *

School of Civil Engineering and Construction, Xi'an University of Technology, Xi'an 710048, China; cxz1109@xaut.edu.cn (X.C.); lishen2861@xaut.edu.cn (S.L.); lg20854007@xaut.edu.cn (G.L.)
* Correspondence: hem@xaut.edu.cn

Abstract: The present study proposes a variable replaceable link for high-strength steel frames with Y-eccentric braces designed to effectively dissipate earthquake energy by confining plastic deformation to its central zone. This unique feature allows for easy post-earthquake recovery or replacement. To investigate the seismic performance of such structures, a comprehensive finite element numerical parametric analysis is conducted using ABAQUS software. Various parameters, including the length of the central zone, replaceable link length, span, and steel grade are considered to optimize the structural design. This study examines the failure modes, hysteretic behavior, bearing capacity, plastic rotation of the replaceable link, and ductility of structures under cyclic loading. The results indicate that reducing the span and utilizing high-strength steel significantly enhance the ductility and ultimate bearing capacity of the structure. This approach also reduces the cross-sectional dimensions, saves steel material, and limits the development area of plasticity, thereby facilitating post-earthquake repair of links after rare earthquakes. An optimal length of the link improves the structural stiffness and energy dissipation capacity. However, if it is too short or too long, it complicates post-earthquake repairs and impairs energy dissipation performance. The conclusions drawn from this research aim to provide valuable insights and theoretical foundations for future structural designs.

Keywords: replaceable link; eccentrically brace; high-strength steel; numerical study

1. Introduction

Steel structures have consistently demonstrated exceptional seismic performance in previous earthquakes, establishing them as the preferred choice in the construction industry. This preference has been further reinforced due to advancements in design, manufacturing, and installation technologies. Li et al. [1] conducted a remote collaborative hybrid test (RCHT) using the OpenFresco platform, demonstrating that the Y-shaped eccentric brace high-strength steel frame (YEB-HSSF) structure satisfies the seismic performance requirements specified in GB 50011-2010 [2]. The structure effectively maintained controlled inter-story drift angles and shear distribution ratios under both frequent and rare earthquake intensities. Keivan et al. [3] demonstrated that a Y-type self-centering eccentrically braced frame (SCEBF-Y) incorporating replaceable energy dissipation devices can achieve strength and stiffness comparable to conventional eccentrically braced frames (EBFs). Furthermore, the SCEBF-Y retains a recentering capability and exhibits minimal residual drift following seismic events. The performance of the SCEBF-Y can be tailored by adjusting the properties of the post-tensioning strands and cable lengths. The seismic performance of eccentrically braced hybrid frames with extremely short shear links and flush endplate bolted connections was investigated by Ma et al. [4]. Their study demonstrated that these configurations exhibit satisfactory performance metrics, including an overstrength coefficient of approximately 1.8 and an ultimate plastic rotation exceeding 0.12 rad. This research highlights the significant influence of factors such as the link length ratio on the structural behavior of these systems. The methodology proposed by Li et al. [5] presents a robust approach for evaluating the elastic stiffness and bearing capacity of eccentrically braced steel frames with K-type, D-type, and V-type configurations. Their study successfully validated the accuracy of the predicted elastic stiffness, yield strength, and ultimate bearing capacities, demonstrating consistent results within a 10% margin when compared with both finite element analysis and experimental data. Zheng et al. [6–9] proposed hybrid Bayesian copula and AI Bayesian methodologies to evaluate wind-induced risk, seismic resilience, and time-variant seismic fragility in high-rise structures by incorporating aleatory and epistemic uncertainties associated with the dynamic excitations, input loads, cost ratio, recovery time, and corrosion initiation. The study conducted by Lettieri et al. [10] investigated a solution for damage-free, self-centering eccentrically braced frames (EBFs) using seismic links known as self-centering devices (SC-links). This system incorporates post-tensioned high-strength steel bars, disk springs, and friction dampers. Through analytical and finite element models, the authors demonstrated that this innovative approach offers optimized strength, stiffness, and ductility while minimizing the residual and peak story drift angles as well as link rotations. Almasabha et al. [11] conducted an assessment on the accuracy of the AISC 341-16 equation in predicting the shear strength of short links and proposed a simple truss model (STM) which incorporates strain hardening, surpassing the predictive capability of the AISC equation. The mean overstrength values for built-up and rolled W-shape specimens were found to be 1.21 and 1.15, respectively, indicating the superior predictive accuracy of the STM and suggesting its adoption. Yin et al. [12] introduced an eccentrically buckling-restrained braced frame with web-bolted replaceable links which improves seismic performance and the ease of post-earthquake repairs, demonstrating the links' efficacy as a "structural fuse". Montuori et al. [13] proposed a theory of plastic mechanism control for eccentrically braced frames, with a specific focus on EB frames employing an inverted Y scheme in order to ensure the establishment of a robust global collapse mechanism, which is crucial for earthquake-resistant design. The methodology involves designing vertical link elements based on seismic forces, incorporating second-order effects through mechanism equilibrium curves, and assessing accuracy by means of push-over analyses. The performance of bolted and welded repairable fuse devices in earthquake-resistant composite steel frames was investigated by Valente et al. [14]. Through a combination of experimental tests and numerical analyses, they were able to identify significant differences in hysteretic behavior and failure modes between the two solutions, ultimately demonstrating their effectiveness. The seismic response of composite frames was investigated by Vasdravellis et al. [15], with a specific focus on the influence of partial composite action between the concrete slab and steel beam, as well as the impact of partial-strength connections. Eccentrically braced steel frame systems offer notable advantages, including exceptional lateral stiffness and effective dissipation of energy. These systems primarily rely on links to withstand seismic forces and absorb earthquake-induced energy release. This design ensures that the links yield first under seismic action for easy replacement after an earthquake. Consequently, the utilization of replaceable links has been widely adopted in seismic structural systems as an efficient approach to achieving commendable seismic performance and post-earthquake recoverability [16,17]. However, experimental findings on conventional Y-shaped eccentrically braced steel frames have consistently demonstrated that the uneven distribution of bending moments at both ends of the links often leads to failures in bending shear under seismic loading. Consequently, these structures exhibit suboptimal performance in terms of energy dissipation [13,18].

To address the deficiencies of conventional Y-shaped eccentrically braced steel frames, this study proposes a structural form which enables rapid post-earthquake recovery: a high-strength, steel composite Y-shaped eccentrically braced structure with replaceable links. The links and braces are constructed using traditional steel, while other components utilize high-strength steel. The variable cross-section design of the links allows for energy dissipation through a smaller energy-dissipating area during significant seismic events, ensuring shear failure and maintaining primary frame stability. Incorporating high-strength steel in non-energy-dissipating components reduces material waste, construction costs, and the overall structure weight, thereby reducing seismic forces. Due to its high lateral stiffness, an eccentrically braced structure exhibits minimal residual deformations after an earthquake, facilitating rapid structural recovery by simply replacing damaged links. In contrast to the challenging, time-consuming, and costly repairs required for traditional eccentrically braced steel frames following earthquakes, this novel structure offers quick restoration through link replacement while also preventing bending shear failure and enhancing energy dissipation effectiveness. To achieve efficient post-earthquake repair of links, a performance-based seismic design method [19] is employed in this study which demonstrates superior seismic performance levels compared with conventional approaches. Parametric analysis of high-strength steel frames with Y-shaped eccentric braces and replaceable links was conducted using finite element software. The findings from this research have significant implications for practical engineering applications of eccentrically braced structural systems and promote the utilization of high-strength steel in seismic design regions.

2. Finite Element Model Validation

To validate the suitability of replaceable links, a finite element model was employed in this study to replicate the experiment described in [20]. A half-scale model with a span of 3000 mm and a story height of 2800 mm was utilized. The dimensions and detailed measurements of the specimen are illustrated in Figure 1. H-shaped steel was used for the frame beams, frame columns, and links. ST37 steel (equivalent to Chinese Q235B carbon structural steel) was employed for the frame beams and links, while ST52 steel (equivalent to Chinese Q355 steel) was used for other components. The connection between the frame beam and link involved eight M20 high-strength bolts with a diameter of 20 mm. Stiffeners, positioned at intervals 170 mm apart from each other and having a thickness of 12 mm, were located at the flanges where the frame columns connected to the frame beams.

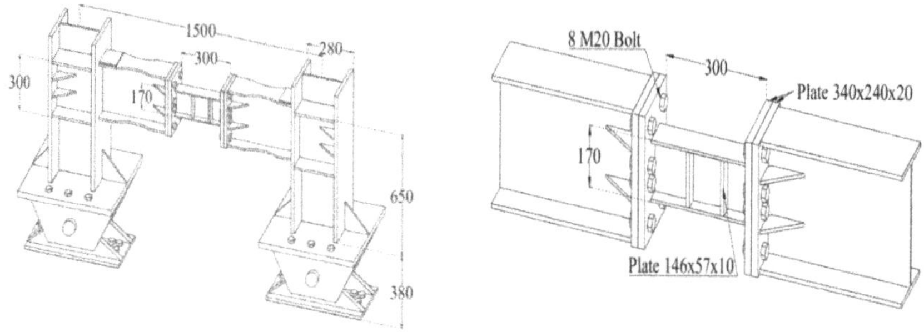

Figure 1. Schematic diagram of replaceable link (mm).

The material properties were modeled using the approach proposed by Kaufmann et al. [21], which is based on cyclic coupon tests. ST52 steel, with a nominal yield stress of 370 MPa, was utilized for the columns and associated components, such as the doubler plates, stiffeners, end plates, and continuity plates. The panel zone strength was controlled according to ANSI/AISC 341–16 [22]. The plasticity of the steel was represented by employing the Von Mises yield surface and an associated flow rule. A combined nonlinear isotropic and kinematic strain hardening approach was employed for the hardening model. Furthermore, a steel material elastic modulus of 200,000 MPa and Poisson's ratio of 0.3 were adopted. Detailed information regarding the material properties and section dimensions of all components can be found in Tables 1 and 2.

Table 1. Material properties.

Material Component	Steel Grade	Yield Strength f_y (N/mm^2)	Ultimate Strength f_u (N/mm^2)	Elastic Modulus (10^5 N/mm^2)
Frame beam	ST37	370	488	2.0
Link	ST37	302	404	2.0

Table 2. Section size of each component of model specimen.

Component	Cross Section	Steel Grade
Frame column	H280 × 280 × 12 × 20	ST52
Frame beam	H300 × 150 × 6 × 8	ST37
Link	H170 × 120 × 5 × 12	ST37
End plate	340 × 240 × 20	ST52
Stiffener at the link	146 × 57 × 10	ST52

2.1. Loading Protocol

The loading protocol for this experiment initially consisted of six cycles with an amplitude of 0.00375 rad, followed by six cycles each with amplitudes of 0.005 rad and 0.0075 rad. Subsequently, four cycles were conducted with an amplitude of 0.01 rad. Finally, additional cycles were performed sequentially in a stepwise manner as depicted in Figure 2, with increasing amplitudes of 0.015 rad, 0.02 rad, 0.03 rad, and 0.04 rad.

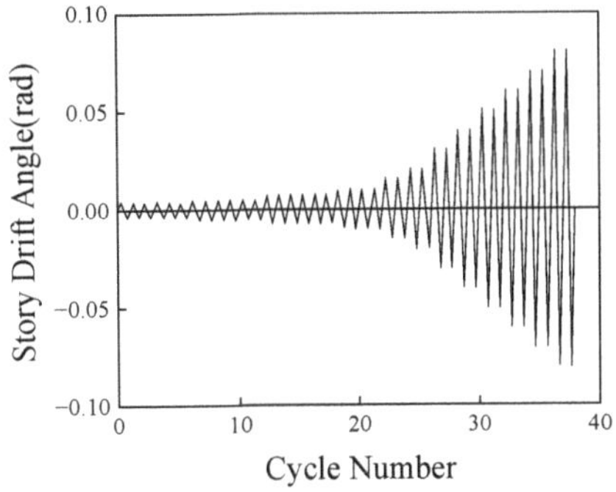

Figure 2. Loading protocol.

2.2. Finite Element Simulation

The finite element models were created using solid elements to ensure accurate representation. To enhance the efficiency of the modeling process, a simplified shape was adopted for the high-strength bolts. Linear reduced integration elements (C3D8R) with a grid size of 10 mm were utilized for both the frame beams and links, as depicted in Figure 3. The reduced integration element effectively addresses distorted deformations and yields accurate displacement results by satisfying the volume consistency constraint at a limited number of integration points. However, it is susceptible to the hourglass problem, where elements undergo deformation without generating strain, resulting in zero stiffness. Therefore, it is crucial to refine the mesh size of each component and their connections as much as possible, ensuring at least two layers of mesh at the flanges. In case of irregular components, they can be decomposed into relatively regular parts for improved meshing quality. Additionally, boundary points can be added to enhance the mesh quality of detailed sections within each component. The boundary conditions in the finite element simulation were defined based on the actual stress states of the structure. Firstly, a displacement constraint (U1 = U2 = U3 = 0) was applied at the centroid of the column base section to simulate a hinged connection. Subsequently, the translational degree of freedom along the Y axis was constrained to prevent out-of-plane instability. Lastly, horizontal loads were applied through point–surface coupling within the beam height range of the frame column.

Figure 3. Mesh generation.

The convergence of the analysis was significantly influenced by the contact settings. As illustrated in Figure 4a, the contact interactions between the high-strength bolts and the end plate primarily involved the bolt nut with the end plate, as well as the bolt shank with the holes in the end plate. In the contact characteristic settings, the tangential direction employed a "penalty" friction model. When two surfaces came into contact and relative sliding occurred, the model applied a frictional force proportional to the sliding distance to resist the movement, with the friction coefficient set to 0.3. For the normal direction, a "hard" contact behavior was used to ensure that there was no overlap within the contact area, thereby accurately simulating the interaction between the contact surfaces.

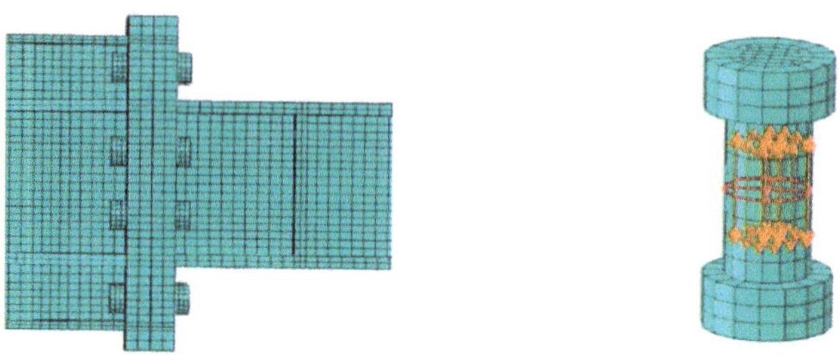

(**a**) Contact diagram (**b**) Bolt pretension

Figure 4. Contact settings.

Figure 4b demonstrates the implementation of bolt pretension. To address convergence issues arising from the instantaneous application of a large bolt load, a three-step approach was adopted for incrementally applying the bolt load. An internal surface within each bolt shank's middle section served as the application surface for exerting bolt load. Initially, a small force was applied to establish stable contact between the high-strength bolts and the connected components. This force was then gradually increased to the intended levels of bolt pretension until reaching its final value.

2.3. *Defining the Initial Defect*

The aim of the incorporation of initial imperfections was to accurately replicate the behavior of structures under realistic engineering conditions. In reality, steel materials develop initial imperfections during processing, transportation, and installation which can affect the stiffness of the structure under loading conditions. Accounting for these defects allows for a more precise reflection of the forces and deformation behaviors in practical scenarios, thereby enhancing the reliability and accuracy of the numerical simulation results. To simulate failure behavior in real-world situations and account for the impact of these geometric imperfections, it was essential to perform a buckling analysis on the samples first. Figure 5 illustrates the first five buckling modes of the model. Modifications were made to apply a scaled version of the first buckling mode, which incorporated defect factors into structural analysis using ABAQUS 6.14 software.

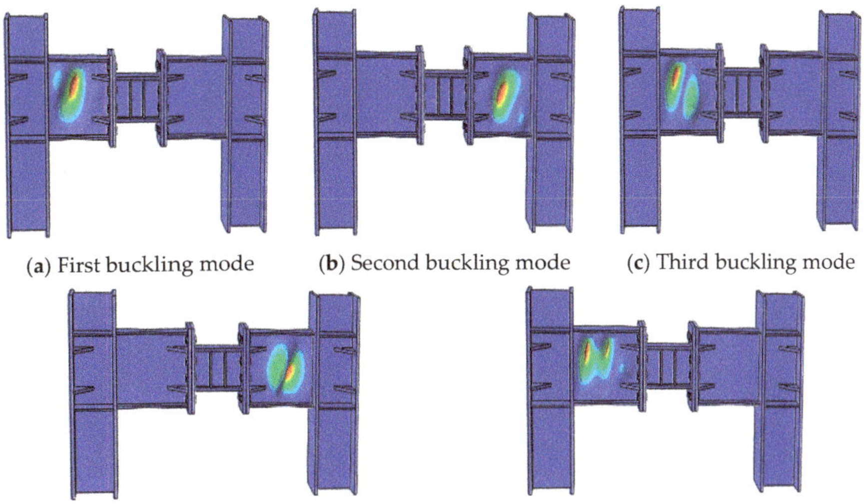

(**a**) First buckling mode (**b**) Second buckling mode (**c**) Third buckling mode

(**d**) Fourth buckling mode (**e**) Fifth buckling mode

Figure 5. Typical buckling modes of finite element mode.

2.4. Comparison of Finite Element Simulation and Test Results

The comparison between the test results and the finite element simulation of the overall structural failure is depicted in Figure 6. During testing, shear failure occurred in the link, while varying degrees of buckling were observed in the flanges and web of the frame beam. The finite element simulation results indicate that plastic failure was primarily concentrated in the link, with significant deformation also occurring in the flanges and web of the frame beam. However, due to disparities in the weakening degree between the finite element model and actual specimen for the frame beam flanges, pronounced buckling observed during experimentation was not replicated by the finite element model. Nevertheless, this disparity does not compromise the reliability of the finite element simulation. In the ultimate state conditions, complete elasticity was maintained by the frame columns, while the end plates connected via high-strength bolts exhibited no signs of failure, thus demonstrating feasibility for high-strength bolted end plate connections.

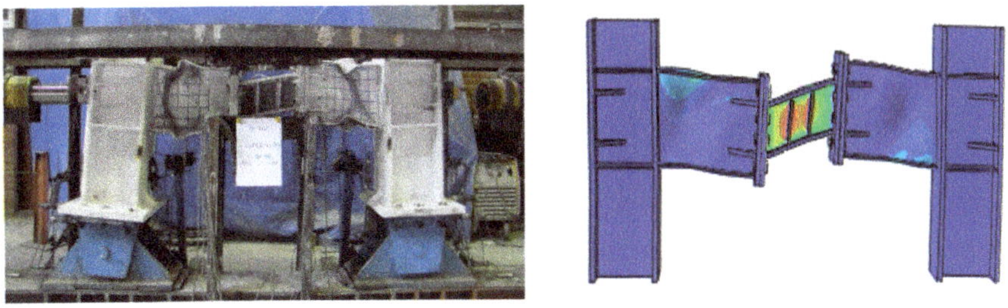

Figure 6. Failure modes for specimen.

The hysteresis curve obtained from the finite element simulation, as illustrated in Figure 7, presents a more comprehensive profile compared with the experimental curve.

Notably, the cyclic curves in the simulation exhibited remarkable coincidence without displaying any pinching phenomena under identical displacements. Compared with the experimental curves, the finite element simulation hysteresis curves are more robust and exhibit no pinching phenomena, with the cyclic curves nearly overlapping at the same displacement levels. The experimental peak loads were 473 kN and −486 kN, while the finite element simulation peak loads were 484 kN and −473 kN, resulting in errors of 2.32% and −2.67%, respectively. After reaching the peak load, the experimental bearing capacity gradually declined. In contrast, after the initial stiffness degradation in the finite element simulations, the increase in bearing capacity became gradually more stable without a subsequent decline. This discrepancy is attributed to the chosen material constitutive model in the finite element analysis, which did not incorporate load failure; the bearing capacity would only decrease when the structure transformed into a mechanism. Although this slightly deviates from real-world behavior, it does not affect the validity of the finite element analysis method.

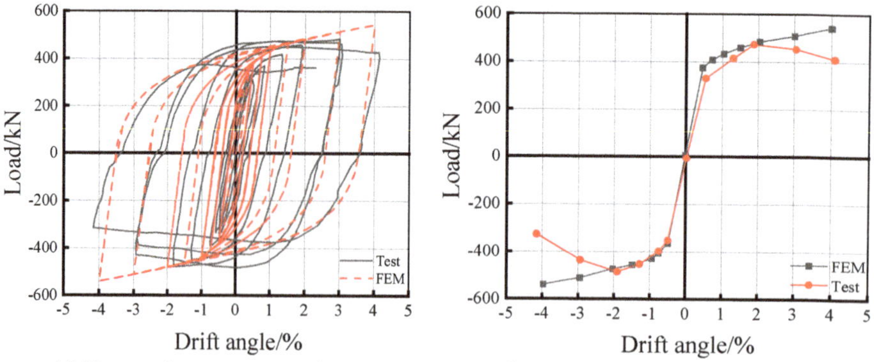

(**a**) Hysteresis curve comparison (**b**) Backbone curve comparison

Figure 7. Comparison of hysteretic loading results.

As illustrated in Figures 8 and 9 and Tables 3 and 4, both the energy dissipation and equivalent viscous damping coefficient (h_e) of the tested structures and finite element models exhibited an increasing trend with respect to the story drift angle. Upon reaching link yielding, a transition into a plastic phase occurred, resulting in a significantly expanded hysteresis curve and consequently a larger hysteresis loop area. The total energy dissipation value for the finite element simulation curve was recorded to be 13,596.96 kN·mm, while that of the experimental curve amounted to 15,769.97 kN·mm. Comparative analysis revealed that the finite element simulation demonstrated a superior energy dissipation capability by exhibiting a larger hysteresis loop area compared with the experimental observations. Notably, the high-strength bolted end plates demonstrated remarkable effectiveness in dissipating energy through link connections. Following an earthquake event, repair or replacement of damaged links can be easily accomplished by simply dismantling these high-strength bolts, thereby facilitating post-earthquake repairs while simultaneously reducing the costs associated with manpower, materials, and financial resources.

Table 3. Energy dissipation capacity of specimen.

Loading Stage	0.00375 Rad	0.005 Rad	0.0075 Rad	0.01 Rad	0.015 Rad
Energy dissipation value (kN·mm)	121.31	184.32	497.65	803.47	1252.84
Equivalent viscous damping coefficient h_e	0.13	0.19	0.26	0.30	0.34
Loading stage	0.02 rad	0.03 rad	0.04 rad	-	-
Energy dissipation value (kN·mm)	2113.04	3766.81	4857.52	-	-
Equivalent viscous damping coefficient h_e	0.38	0.40	0.40	-	-

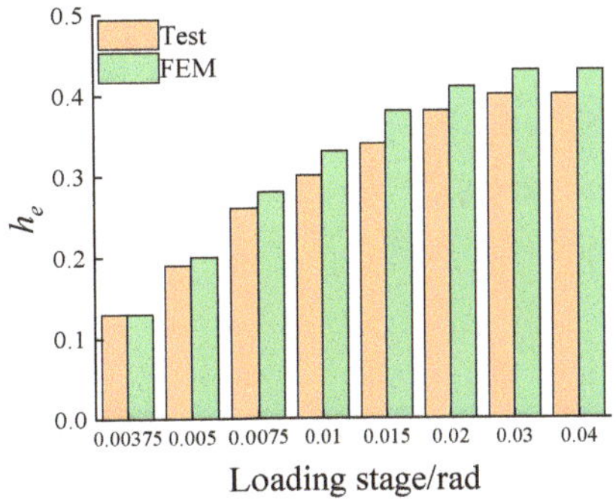

Figure 8. Comparison of equivalent viscous damping coefficients.

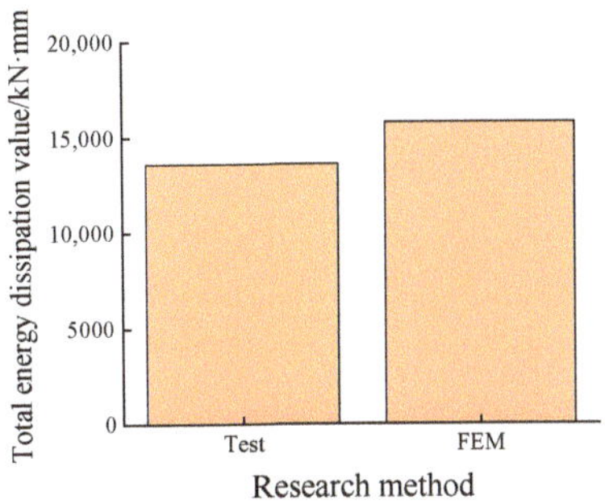

Figure 9. Comparison of total dissipated energy of specimen.

Table 4. Energy dissipation capacity of finite element model.

Loading Stage	0.00375 Rad	0.005 Rad	0.0075 Rad	0.01 Rad	0.015 Rad
Energy dissipation value (kN·mm)	105.01	238.03	543.73	881.44	1634.30
Equivalent viscous damping coefficient h_e	0.13	0.2	0.28	0.33	0.38
Loading stage	0.02 rad	0.03 rad	0.04 rad	-	-
Energy dissipation value (kN·mm)	2442.11	4116.84	5808.51	-	-
Equivalent viscous damping coefficient h_e	0.41	0.43	0.43	-	-

3. Finite Element Model Parameter Design

The components in this study were modeled using three-dimensional solid modeling techniques. The structural model employed the linear reduced-integration element C3D8R for finite element simulation. Figure 10 illustrates the geometric dimensions of the BASE specimen, which had a height of 3600 mm and a span of 7200 mm. The replaceable link measured 900 mm in length, with an energy-dissipating region spanning 550 mm. Table 5 provides details on the cross-sectional dimensions and steel grades of the BASE specimen. A total of four groups comprising nine specimens were designed for this study. A refined model was established using the finite element software ABAQUS, incorporating the relevant parameters listed in Table 6. The L series encompassed variations in the length of the energy-dissipating region among the specimens; the D series introduced alterations to the steel grade; the K series modified the span; and finally, the H series changed the length of the link.

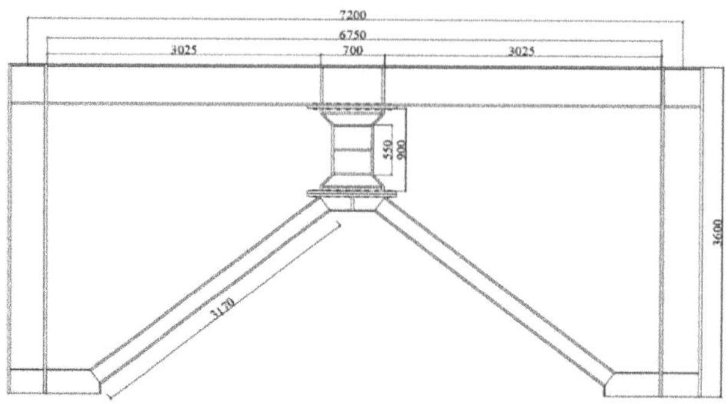

Figure 10. Dimensions of the structures (mm).

Table 5. Cross-section of the components.

Component	Cross-Section	Steel Grade
Frame column	H450 × 450 × 15 × 30	Q460
Frame beam	H440 × 300 × 10 × 20	Q460
Link	H700 × 200 × 20 × 26	Q355
Energy-dissipating region	H460 × 200 × 20 × 26	Q355
Support diagonal rods	H250 × 240 × 14 × 20	Q355
End plate	980 × 300 × 30	Q460

Table 6. Parameters of various specimens.

Specimen	Length of Link (mm)	Length of Energy-Dissipating Region (mm)	Steel for Link	Frame Beam, Frame Column	Span (mm)
BASE	900	550	Q355	Q460	7200
L-1	900	450	Q355	Q460	7200
L-2	900	350	Q355	Q460	7200
D-1	900	550	Q355	Q550	7200
D-2	900	550	Q355	Q690	7200
K-1	900	550	Q355	Q460	6600
K-2	900	550	Q355	Q460	6000
H-1	1050	550	Q355	Q460	7200
H-2	1200	550	Q355	Q460	7200

The stress–strain relationship of the steel was established based on the isotropic Mises yield criterion and kinematic hardening rule, while tie constraints were applied between the supports, frame beams, and frame columns. The high-strength bolt heads and nuts interacted with the flanges of the frame beams and end plates, the bolt rods interacted with the walls of the bolt holes, and the end plates interacted with the other end plates and flanges of the frame beams. Loading was imposed by constraining the X-direction displacement (X = 0) of the frame columns to prevent out-of-plane instability. Column constraints, vertical loads, and horizontal loads were applied using a node-to-surface coupling method. The boundary conditions are illustrated in Figure 11.

The stress–strain relationship of the steel utilized in this section was simplified to a tri-linear model, as illustrated in Figure 12. By employing the isotropic Mises yield criterion and kinematic hardening rule provided by the ABAQUS software, an isotropic, nonlinear mixed hardening elastoplastic constitutive relationship was established.

The application of loads was divided into three main categories: bolt load application, vertical load application, and horizontal load application. Prior to applying the bolt load, an "internal" surface had to be created at the location of the bolt. To prevent non-convergence due to excessive force applied all at once, a small force was initially applied to establish contact between the bolt and the connected components. This initial force was then replaced with the intended preload before adjusting the load to fix the current length of the bolt(s). The vertical load was applied as a force with the magnitude set to 0.2 N_y (where N_y represents the axial pressure when full cross-section yield occurred in the column). In order to more accurately simulate real conditions, the large deformation option was enabled.

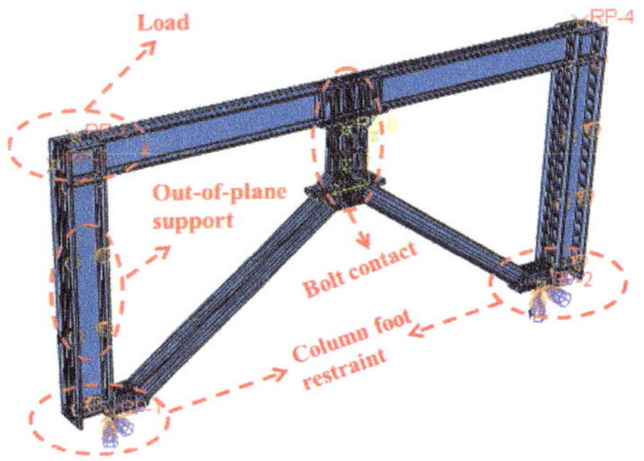

Figure 11. Boundary conditions.

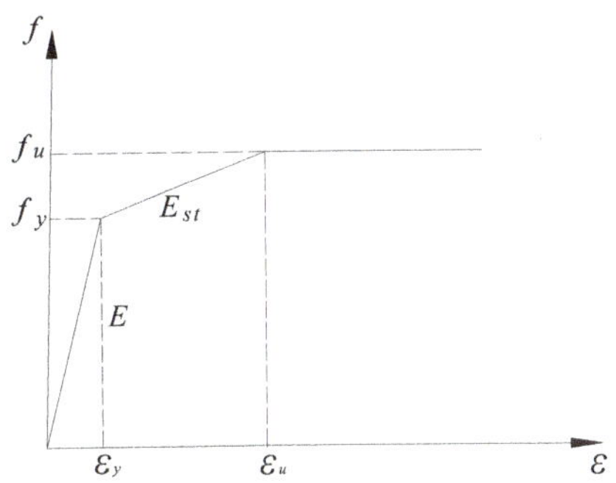

Figure 12. Stress–strain relationship of steel.

The horizontal load was applied using displacement control. For monotonic loading, the load was incrementally increased until structural failure occurred. In the case of cyclic loading, the load was systematically applied in multiples of the yield displacement, following a pattern of $\Delta_y/4$, $\Delta_y/2$, $3\Delta_y/4$, Δ_y, $2\Delta_y$, $3\Delta_y$, $4\Delta_y$, and so on. Each displacement level was cycled twice before progressing to the next level until structural failure ensued.

According to the actual loading conditions of the eccentrically braced steel frame, three failure criteria were established during the loading process. If any one of these criteria was met, then this indicated structural failure, and the loading was terminated:

(1) The stress in the link reached the ultimate stress of the material.
(2) The plastic strain in the high-strength bolts reached the ultimate strain of the material.
(3) The stress at any cross-section of the frame beams, columns, or other components reached the yield stress, resulting in plastic hinge formation.

4. Analysis of Specimen Parameters

4.1. Analysis of Finite Element Results for D-Series Specimens

4.1.1. Hysteresis Curve Analysis of D-Series Specimens

The structure underwent cyclic loading, and the resulting hysteresis curves are presented in Figure 13. Both the BASE specimen and D series specimens exhibited complete hysteresis curves with a spindle shape while completing a $4\Delta_y$ displacement cycle. This indicates that the high-strength steel frame-Y-shaped eccentrically braced frame structure with replaceable links possessed an excellent seismic energy absorption capability and strong plastic deformation ability. Prior to link yielding, the structure remained in an elastic stage, displaying a linear hysteresis curve during repeated loading cycles without significant stiffness degradation. After link yielding occurs and enters into the plastic phase, the hysteresis curve starts exhibiting loop shapes. With increasing displacement, its area enlarges along with more apparent plastic deformation. Meanwhile, the stiffness significantly decreases, while the bearing capacity gradually levels off.

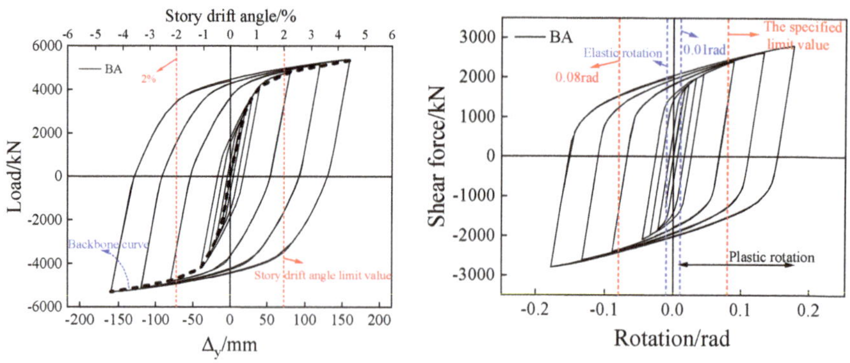

(**a**) BASE specimen

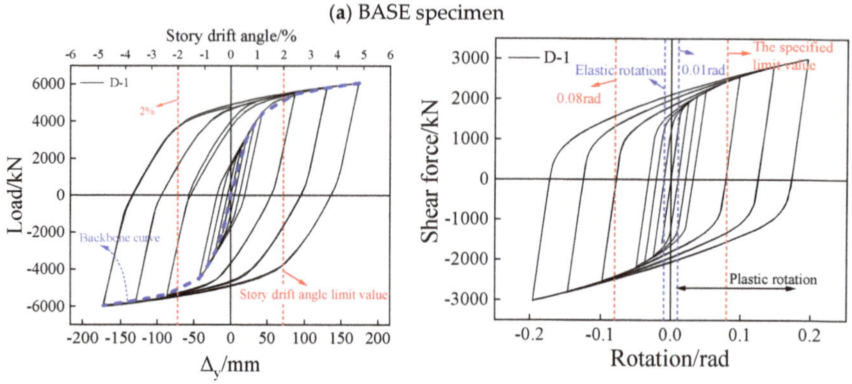

(**b**) D-1 specimen

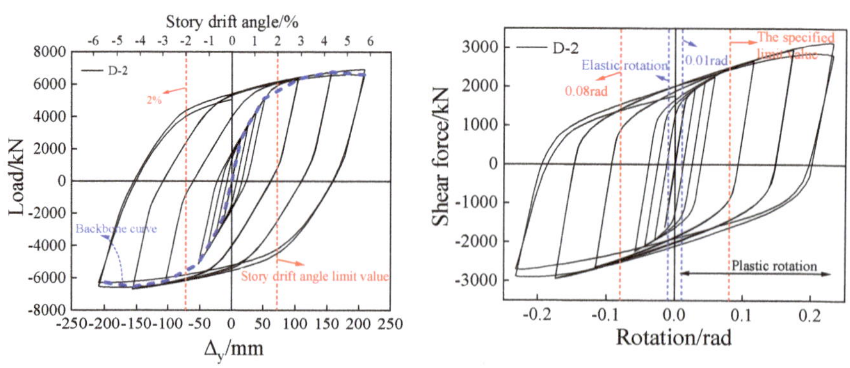

(**c**) D-2 specimen

Figure 13. Hysteretic curves of D series.

4.1.2. Analysis of Bearing Capacity, Plastic Rotation and Ductility of D-Series Specimens

The backbone and shear–rotation curves of the links are depicted in Figures 14 and 15, respectively. The ultimate bearing capacities of the BASE, D-1, and D-2 specimens were determined to be 4255.4 kN, 4470.3 kN, and 5018.5 kN, respectively. A discernible increasing trend in the ultimate bearing capacity was observed with higher steel grades. Compared with the BASE specimen, the D-1 and D-2 specimens exhibited increments in their ultimate bearing capacities of 6.9% and 20.7%, respectively. Notably, all specimens demonstrated significantly excessive plastic rotations for the links (0.18 rad for the BASE specimen; 0.19 rad for the D-1 specimen; and 0.23 rad for the D-2 specimen), surpassing the specified plastic rotation limit of 0.08 rad as stipulated by American seismic design code AISC341-16 [22] for shear-yielding links.

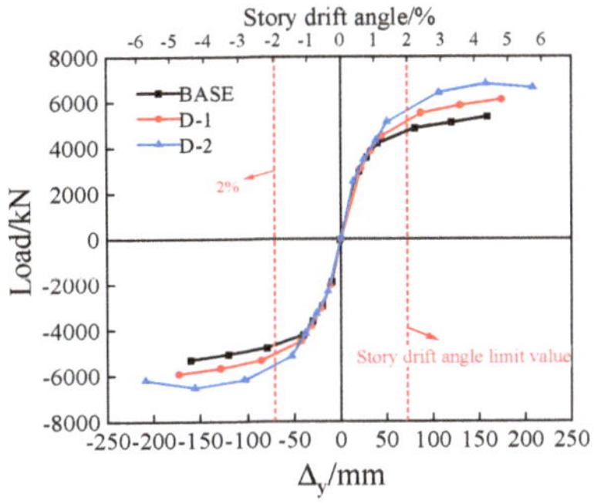

Figure 14. Backbone curves of D series.

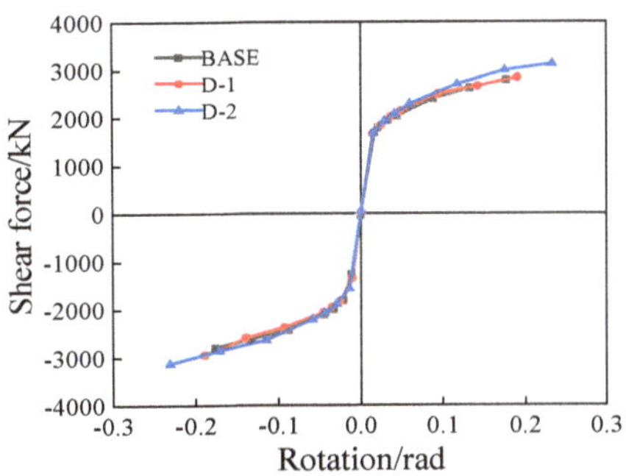

Figure 15. Shear–rotation curves of D series.

The maximum shear forces experienced by the links in each specimen were as follows: BASE = 2747.3 kN; D-l = 2954.9 kN; and D-2 = 3135.1 kN. Consistent with the overall performance, there was an increasing trend observed in the ultimate shear force with higher steel grades. In comparison with the BASE specimen, both the D-l and D-2 specimens exhibited significantly higher ultimate shear forces of 7.6% and 14.1%, respectively. Moreover, these specimens demonstrated exceptional ductility properties along with a robust capacity for plastic deformation.

4.1.3. Failure Mode of D-Series Specimens

The plastic strain cloud diagram of the BASE specimen at the ultimate state is depicted in Figure 16a. The distribution of equivalent plastic strain predominantly concentrated within the energy-dissipating region, aligning with design expectations despite relatively high stress on the upper flange. The central portion of the link effectively absorbed and dissipated seismic energy, satisfying the design requirements for variable cross-section links. At the ultimate state, the stress in the energy-dissipating region reached its maximum level. Once compromised, the link serves as a primary line of defense, while the frame beam acts as a secondary line of seismic defense.

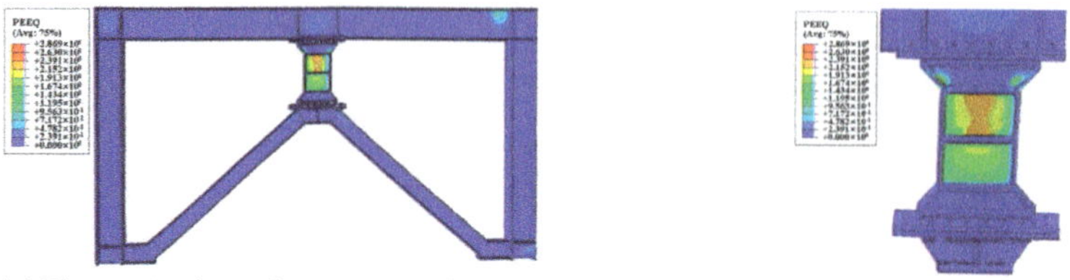

(**a**) The equivalent plastic strain diagram (**b**) The equivalent plastic strain of replaceable link

Figure 16. Failure mode of base specimen.

The plastic strain cloud diagrams of the D-series specimens at the ultimate state are depicted in Figures 17a and 18a. These diagrams exhibit a failure mode similar to that observed in the BASE specimen. Furthermore, elevated stress levels can be observed in the frame beams and supporting flanges connected to the links, as well as in the top and bottom sections of the frame columns, accompanied by plastic failure occurring within the energy-dissipating region. However, an increase in the steel grade led to a gradual reduction in the stress experienced by these areas. This suggests that higher strength grades of steel diminish the participation of frame components in energy dissipation, thereby enhancing performance within designated energy-dissipating regions.

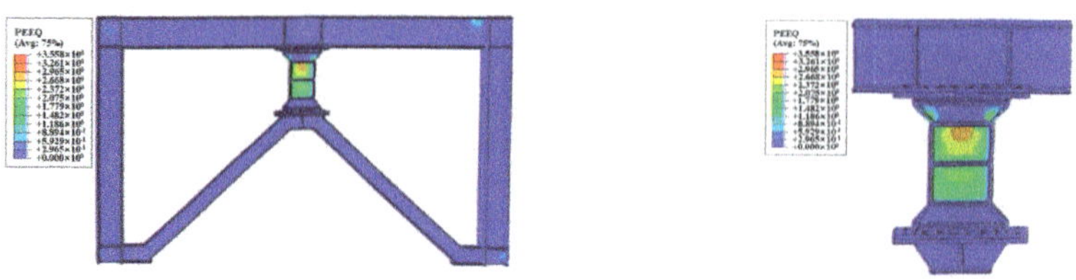

(**a**) The equivalent plastic strain diagram (**b**) The equivalent plastic strain of replaceable link

Figure 17. Failure mode of D-1 specimen.

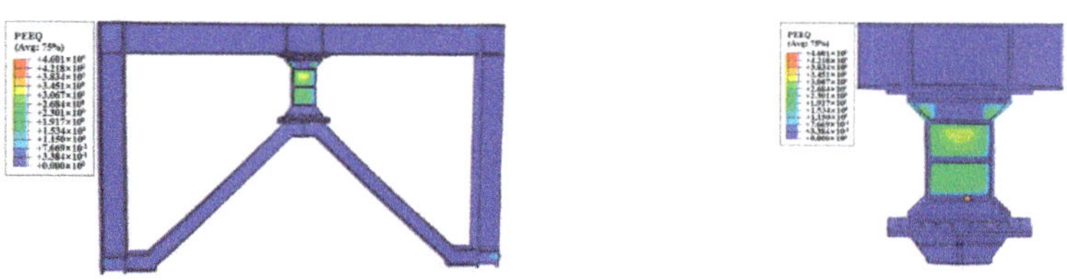

(**a**) The equivalent plastic strain diagram (**b**) The equivalent plastic strain of replaceable link

Figure 18. Failure mode of D-2 specimen.

4.2. Analysis of Finite Element Results for L-Series Specimens

4.2.1. Hysteresis Curve Analysis of L-Series Specimens

The structure was subjected to cyclic loading, and the resulting hysteresis curves are illustrated in Figure 19. The hysteresis curves of the L-series specimens exhibited a spindle shape and demonstrated significant completeness. During the elastic stage, the curves predominantly display linear behavior. As the links yield, stiffness degradation is initiated. With increasing displacement, the trend of an enhanced bearing capacity becomes less pronounced. However, upon reaching a load level of $4\Delta_y$, a decline in the bearing capacity can be observed for the second loop of the hysteresis curve. This decline can be attributed to both plastic deformations occurring within the frame and yielding of the end plates connecting the frame beams at $4\Delta_y$.

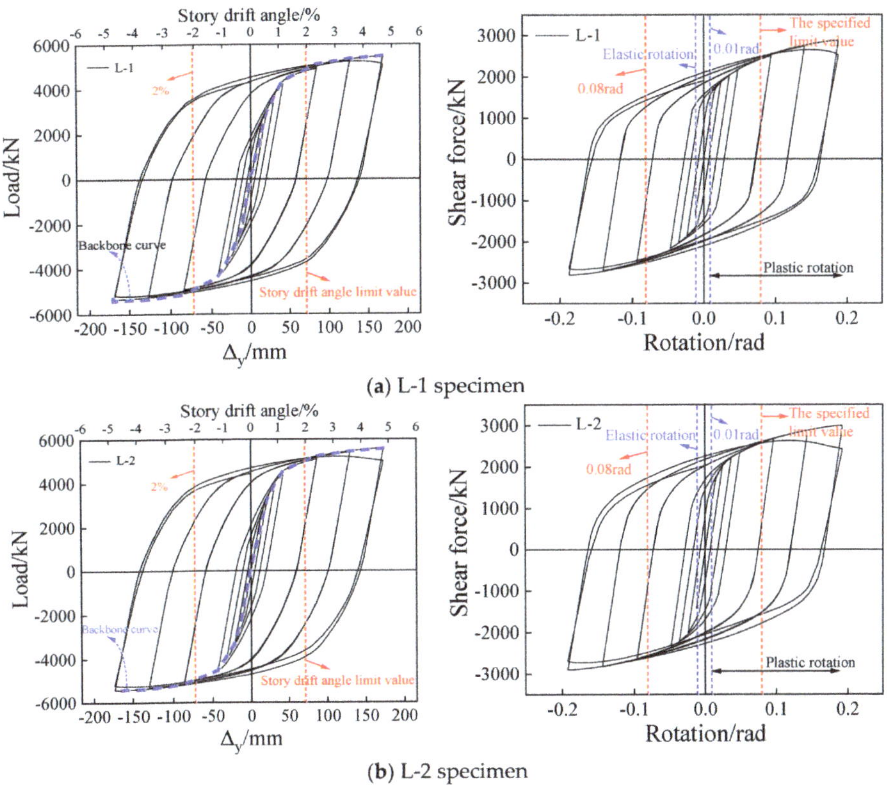

(**a**) L-1 specimen

(**b**) L-2 specimen

Figure 19. Hysteretic curves of L series.

4.2.2. Analysis of Bearing Capacity, Plastic Rotation, and Ductility of L-Series Specimens

The backbone and shear–rotation curves of the links for the specimens are illustrated in Figures 20 and 21, respectively. As shown in Figure 20, the ultimate bearing capacities of the BASE, L-1, and L-2 models were recorded to be 5393.61 kN, 5449.29 kN, and 5566.73 kN, respectively. In comparison with the BASE specimen, the L-1 and L-2 specimens demonstrated increases in their ultimate bearing capacities of 1.1% and 3.2%, respectively. It can be observed that reducing the length of the energy-dissipating region enhanced the overall structural ultimate bearing capacity.

In the finite element simulation, the plastic rotation of the link was induced by shear deformation in the energy-dissipating region. As illustrated in Figure 21, the shear–rotation curves for the links show no discrepancies in the plastic rotations during the elastic stage across all models. However, upon entering the plastic stage, it can be observed that the L-2 model exhibited a significantly accelerated development of plastic rotation and reached its maximum value at the ultimate state. The plastic rotations for the L-series links measured 0.178 rad, 0.187 rad, and 0.193 rad, significantly surpassing the limit of 0.08 rad. The ultimate shear forces for the BASE specimen and L-1 and L-2 specimens were recorded to be 2747.3 kN, 2874.9 kN, and 2975.13 kN, respectively, thus indicating increases of approximately 4.6% and 8.3% in the ultimate shear forces for the L-1 and L-2 specimens compared with the BASE specimen, respectively.

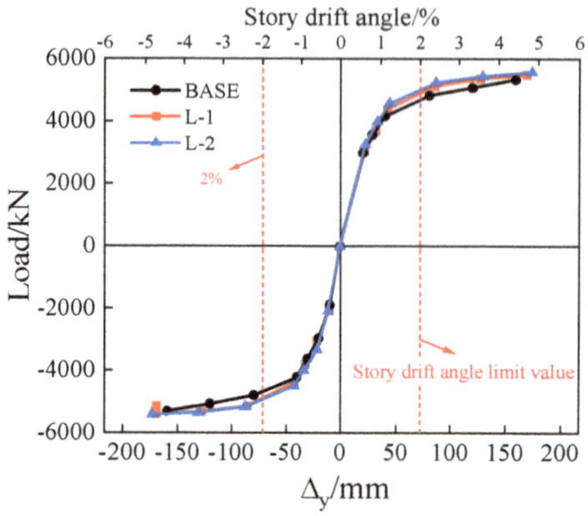

Figure 20. Backbone curves of L series.

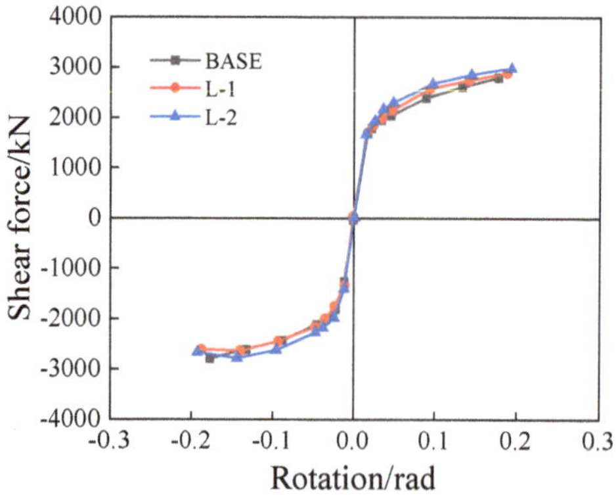

Figure 21. Shear–rotation curves of L series.

Furthermore, it can be observed that as the length of the energy-dissipating region decreased, there was a minor yet discernible increasing trend in the ultimate shear force exhibited by the links.

4.2.3. Failure Mode of L-Series Specimens

The plastic strain cloud diagrams of all of the L-series specimens at the ultimate state and the plastic strain distribution diagrams of the replaceable links are illustrated in Figures 22 and 23, respectively. In the L-series specimens, the plastic strain was predominantly concentrated in the regions associated with energy dissipation. Stress concentrations were observed on the right side of the lower flange of the frame beam connected to the link and on the supporting flange. The shorter length of these energy-dissipating regions led to more pronounced stress concentration. After completing a cyclic displacement of $4\Delta_y$, minimal deformation was observed in both end plates of the links within the L-series specimens, indicating their complete elastic behavior throughout this displacement range. Specifically, for specimen L-1, no plasticity occurred even at $4\Delta_y$, whereas for specimen L-2, elastic behavior persisted until $3\Delta_y$ before transitioning into a plastic stage at $4\Delta_y$. Despite minor plastic deformation occurring during testing, all replaceable links met their design requirements.

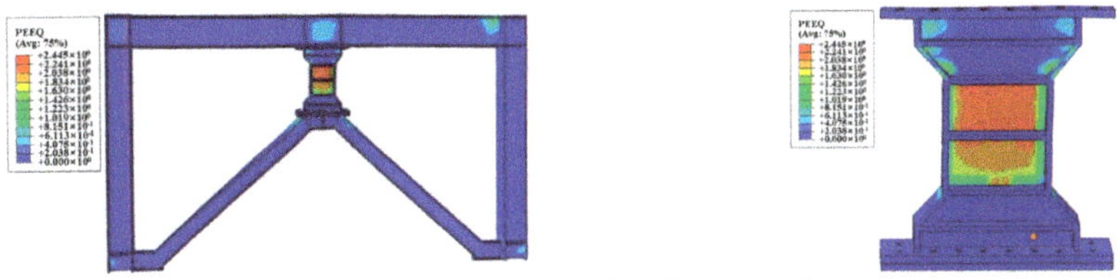

(**a**) The equivalent plastic strain diagram (**b**) The equivalent plastic strain of replaceable link

Figure 22. Failure mode of L-1.

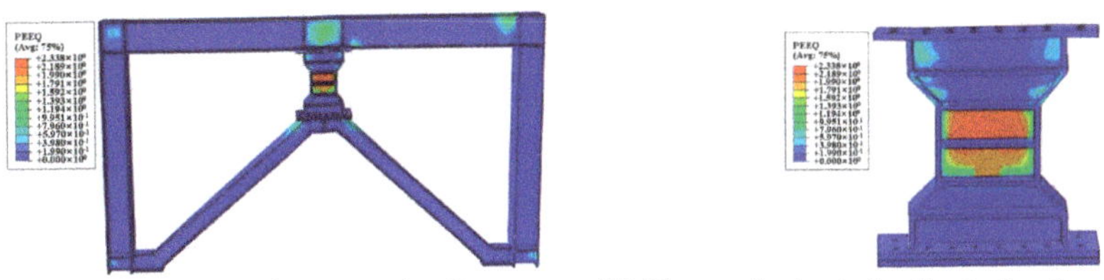

(**a**) The equivalent plastic strain diagram (**b**) The equivalent plastic strain of replaceable link

Figure 23. Failure mode of L-2.

4.3. Analysis of Finite Element Results for K-Series Specimens

4.3.1. Hysteresis Curve Analysis of K-Series Specimens

The hysteresis curves obtained from cyclic loading of the structure are depicted in Figure 24. It is evident that the hysteresis curves of the K-series specimens exhibit a spindle shape and demonstrate significant completeness. Prior to the yielding of the links, the hysteresis curves of the specimens display essentially linear behavior. Subsequently, as plastic deformation occurs in the links, loops start to form within the hysteresis curves, with their size enlarging and the area increasing with the displacement increments. The trend of an increasing bearing capacity gradually diminished as the displacement increased further. Moreover, reducing the span led to an enhancement in the structural bearing capacity while accentuating the plastic deformation.

4.3.2. Analysis of Bearing Capacity, Plastic Rotation, and Ductility of K-Series Specimens

The backbone curves and shear–rotation curves of the links are presented in Figures 25 and 26, respectively. It is evident from the figures that all specimens exhibited high ductility coefficients, indicating exceptional structural ductility and a proficient plastic deformation capability. A reduction in span led to a significant enhancement in structural ductility. As depicted in Figure 25, the backbone curves demonstrate that as the displacement increased, the trend of augmenting the structural bearing capacity gradually diminished, resulting in a decrease in the slope of the backbone curve. Furthermore, decreasing the frame span resulted in an increase in the structural bearing capacity.

According to Figure 26, the shear–rotation curves of the links indicate that during the elastic stage, there was no significant disparity in plastic rotation among the models. However, upon entering the plastic stage, it can be observed that the K-2 model exhibited a notably accelerated development of plastic rotation in the link, reaching its maximum rotation at the ultimate state. The plastic rotations of the links in the K series were significantly higher, with values of 0.178 rad, 0.188 rad, and 0.195 rad, surpassing 0.08 rad by a considerable margin. The ultimate shear forces for the links in the BASE specimen, K-1 specimen, and K-2 specimen were recorded to be 2747.3 kN, 2874.9 kN, and 2975.13 kN, respectively, representing increases of approximately 4.6% and 8.3% for specimens K-1 and K-2 compared with the BASE specimen, respectively. Moreover, it can be observed that with a reduction in the span length, there existed an ascending trend in the ultimate shear force for the links, albeit with minimal differences.

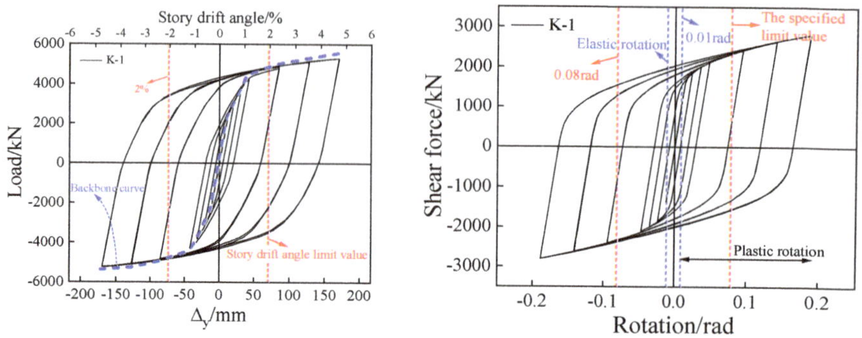

(**a**) K-1 specimen

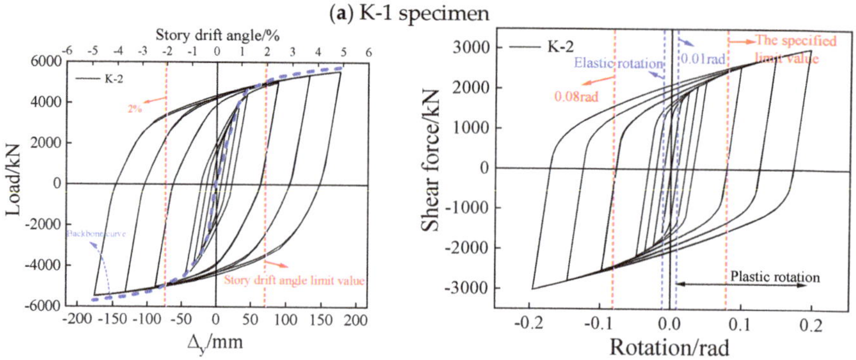

(**b**) K-2 specimen

Figure 24. Hysteretic curves of K series.

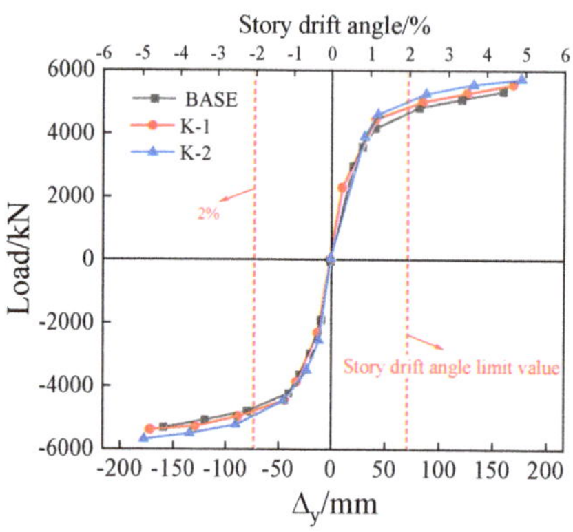

Figure 25. Backbone curves of K series.

4.3.3. Failure Mode of K-Series Specimens

The plastic strain cloud diagrams of the K-series specimens at their ultimate states are presented in Figures 27a and 28a. These diagrams exhibit a failure mode similar to that observed in the BASE specimen, where plastic failure initially occurs within the energy-dissipating region before propagating to the frame beam, indicating an ideal failure mechanism. The stress distributions in the replaceable links of both the BASE specimen and K-series specimens under different cycles are depicted in Figures 27b and 28b. It is evident from these figures that plastic strain primarily accumulated within the energy-dissipating region, with a greater concentration observed specifically for the K-2 specimen, suggesting enhanced efficiency of energy dissipation.

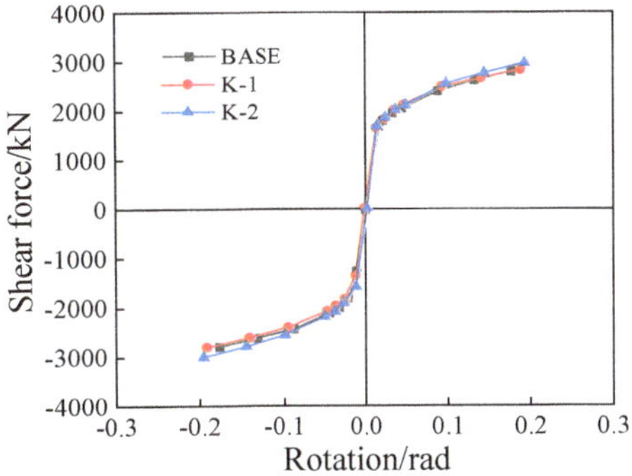

Figure 26. Shear–rotation curves of K series.

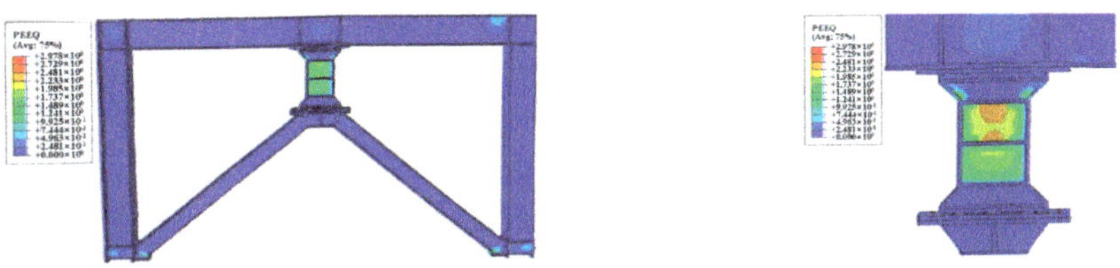

(**a**) The equivalent plastic strain diagram (**b**) The equivalent plastic strain of replaceable link

Figure 27. Failure mode of K-1.

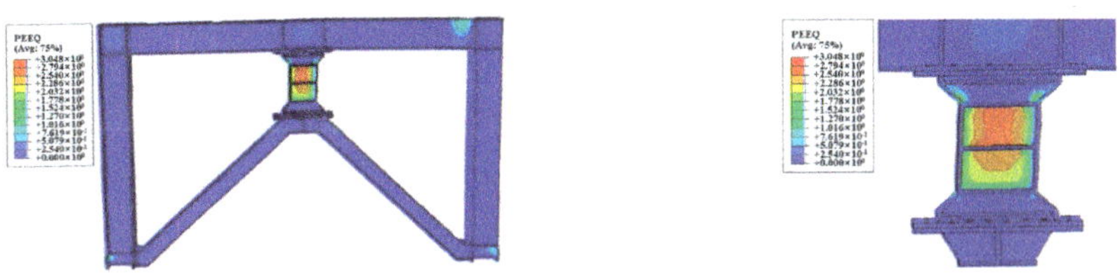

(**a**) The equivalent plastic strain diagram (**b**) The equivalent plastic strain of replaceable link

Figure 28. Failure mode of K-2.

4.4. Analysis of Finite Element Results for H-Series Specimens

4.4.1. Hysteresis Curve Analysis of H-Series Specimens

The hysteresis curves of the H-series specimens are illustrated in Figure 29, presenting a comprehensive profile and exhibiting a trend similar to that observed in the L series. With an increase in link length, there was a corresponding reduction in the structural bearing capacity. Upon reaching a story drift angle of 0.02 rad, the growth rate of the horizontal bearing capacity for each specimen tended to stabilize. In subsequent stages, the bearing capacity experienced decline during the second cycle due to frame buckling and yielding of the end plates connecting the link and frame beam, resulting in plastic deformation.

4.4.2. Analysis of Bearing Capacity, Plastic Rotation, and Ductility of H-Series Specimens

Figures 30 and 31 depict the backbone curve and shear force–rotation curve of the energy dissipation link, respectively. It can be observed from Figure 30 that the yield bearing capacities of the BASE, H-1, and H-2 specimens were 4255.4 kN, 4178.2 kN, and 4032.7 kN, respectively. The H-1 and H-2 specimens exhibited reductions of 1.81% and

5.23%, respectively, compared with the BASE specimen. In addition, the ultimate bearing capacities of the BASE, H-1, and H-2 specimens were 5393.61 kN, 5009.29 kN, and 4946.31 kN, respectively, showing differences of 7.13% and 8.29% for the H-1 and H-2 specimens compared with the BASE specimen, respectively. These results indicate that variations in the length of the energy dissipation link significantly influenced the ultimate bearing capacity. All specimens demonstrated high ductility coefficients, reflecting excellent structural plastic deformation capabilities. As the length of the energy dissipation link increased, the ductility of the structure improved. However, there was a notable decrease in the ultimate bearing capacity.

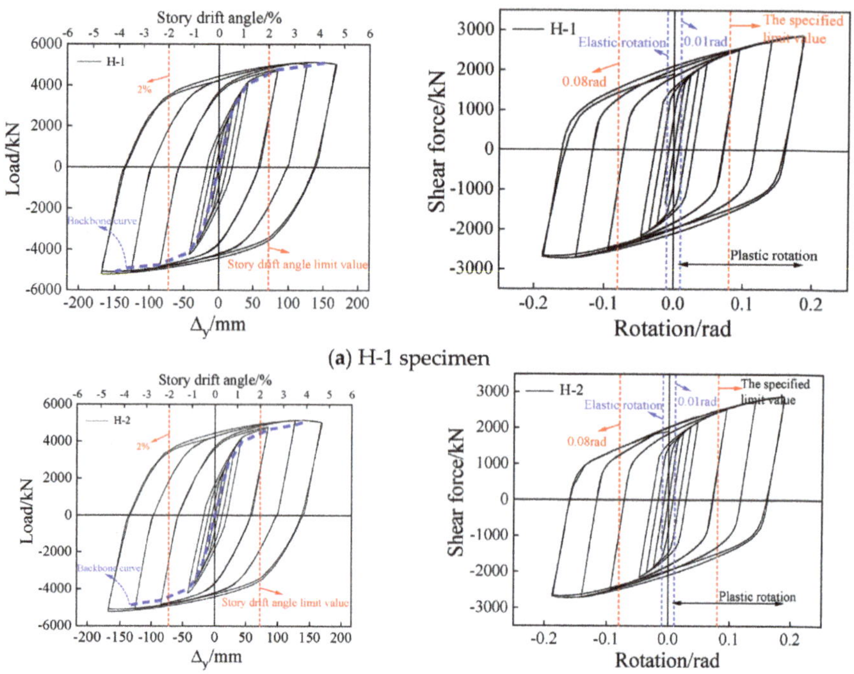

Figure 29. Hysteretic curves of H series.

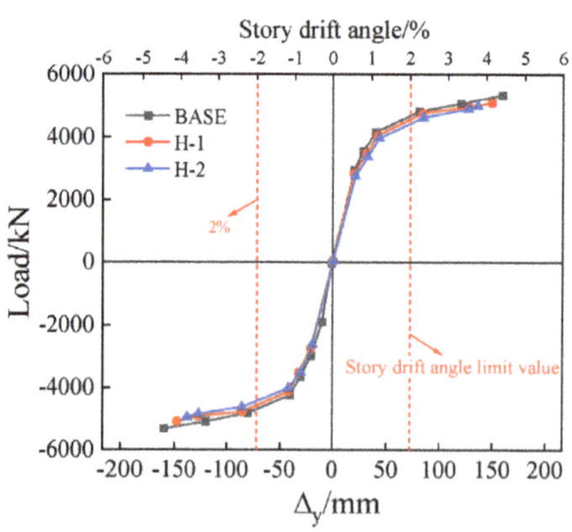

Figure 30. Backbone curves of H series.

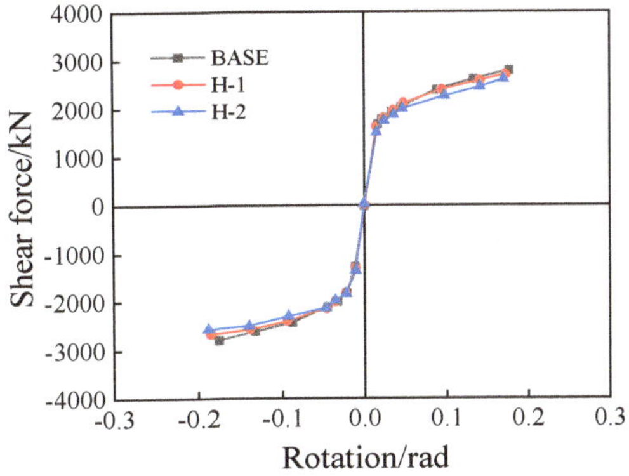

Figure 31. Shear–rotation curves of H series.

The shear force–rotation curve of the links in Figure 31 indicates that there was no difference in the plastic rotations of the links among the models during the elastic stage. However, once the plastic stage was reached, the rotation of the BASE specimen exhibited a significantly faster increase and reached its highest value at the ultimate state. The plastic rotations of the H-series links were measured to be 0.178 rad, 0.173 rad, and 0.170 rad, which significantly exceed the limit of 0.08 rad. The ultimate shear forces for the BASE, H-1, and H-2 specimens were recorded to be 2747.3 kN, 2703.96 kN, and 2664.57 kN, respectively. In comparison with the ultimate shear force value of the BASE specimen, both the H-1 and H-2 specimens exhibited decreases of approximately 1.57% and 3.1%, respectively, in their respective values for the ultimate shear forces observed. Although all specimens demonstrated plastic rotations far exceeding the limit, set to 0.08 rad, it can be concluded that longer link lengths ultimately resulted in lower bearing capacities, along with reduced levels of achieved plastic rotation. This observation suggests that excessively long links do not contribute favorably toward energy dissipation within structural systems.

4.4.3. Failure Mode of H-Series Specimens

Figures 32 and 33 illustrate the plastic strain distribution diagrams for the entire specimen and the replaceable link, respectively, under the ultimate conditions for the H-series specimens. The figures demonstrate that the plastic strain was predominantly concentrated in the energy-dissipating region. However, significant levels of plastic strain were also observed in the frame beams, supporting flanges, webs connected to the link, as well as in the top and bottom sections of the frame columns. In the case of the H-1 specimen, although the elastic behavior was retained by the flange end plate of the frame beam after completing a cyclic displacement of $4\Delta_y$, yielding occurred on both sides of this end plate from $1\Delta_y$ to $4\Delta_y$. This observation suggests that replacement might pose challenges under severe seismic conditions. Consequently, despite achieving a displacement amplitude of $4\Delta_y$ for the H-series specimens, both end plates connected to the frame beams and the support frame beams linked to the replaceable links experienced yield and entered into a plastic stage during the ultimate conditions. Such behavior failed to meet the requirements for replaceable links, and this highlights how significantly the link length influences structural hysteretic performance, with excessive lengths being detrimental to concentrated energy dissipation.

(**a**) The equivalent plastic strain diagram (**b**) The equivalent plastic strain of replaceable link

Figure 32. Failure mode of H-1.

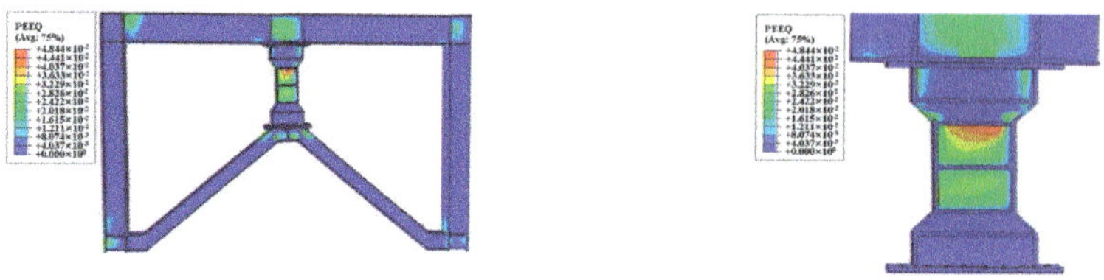

(**a**) The equivalent plastic strain diagram (**b**) The equivalent plastic strain of replaceable link

Figure 33. Failure mode of H-2.

5. Conclusions

The present study employed the BASE specimen to develop four groups of Y-shaped eccentric braced steel frame models, incorporating variations in the length of the energy-dissipating region and the combination of the steel strength, link length, and span. By conducting finite element analysis using ABAQUS software, this research investigated the influence of these parameters on the seismic performance of the models subjected to cyclic loading. The key findings are summarized as follows:

(1) The BASE specimen demonstrated that a Y-shaped eccentrically braced steel frame with high-strength steel replaceable links effectively dissipated energy. Under horizontal loading, plastic strain primarily occurred in the energy-dissipating region. In the ultimate conditions, both the beams and columns underwent plastic deformation, thereby validating the performance-based design.

(2) For the structures incorporating different steel combinations, plastic strain was predominantly localized within the energy-dissipating region. Utilizing higher-grade steel for beams, columns, and end plates enhanced the plastic strain concentration in the energy-dissipating beam segment while reducing it in the non-energy-dissipating components, thereby facilitating easier replacement of links following seismic events.

(3) The specimens demonstrated exceptional ductility, with enhanced load-bearing capacity and ductility observed as the span length decreased.

(4) The plastic strain was primarily concentrated in the energy-dissipating region, while stress was predominantly localized on the lower flange of the frame beam connected to the link. Shorter energy-dissipating regions enhanced both the ductility and load-bearing capacity. However, they also resulted in increased plastic rotation within the link. The flange end plates of the frame beam underwent limited plastic deformation, thereby allowing for potential replacement of the link.

(5) The reduction in the plastic rotation and load-carrying capacity, resulting from an increase in the link length, hampered effective energy dissipation and complicated link replacement following strong seismic events. An excessive link length can induce flange buckling, impeding efficient energy dissipation.

The proposed Y-shaped eccentrically braced steel frame structure, which integrates replaceable links and high-strength steel, is based on comprehensive research. Extensive investigations have been conducted to evaluate its seismic performance. However, further research is warranted in the following areas:

(1) This study primarily employed the finite element method to analyze the seismic behavior of the new Y-shaped eccentrically braced structure. Due to various factors influencing the computational process, there may be fluctuations in the accuracy of the simulation results. Hence, additional experimental research is necessary.
(2) This study only considered four parameters which affect structural seismic performance. Future research could explore other influential factors.

Author Contributions: Conceptualization, S.L.; methodology, X.C. and S.L.; software, M.H.; investigation, X.C.; data curation, X.C.; writing—original draft preparation, X.C.; writing—review and editing, S.L. and M.H.; project administration, G.L.; funding acquisition, G.L.; supervision, G.L. All authors have read and agreed to the published version of the manuscript.

Funding: The authors are grateful for the financial support from the Postdoctoral Research Foundation of China (2022M712562) and National Key Research and Development Plan (YS2022YFB2600177).

Data Availability Statement: The original contributions presented in the study are included in the article, further inquiries can be directed to the corresponding author.

Conflicts of Interest: The authors declare that they have no known competing financial interests or personal relationships that could have appeared to influence the work reported in this paper. All authors declare that no conflict of interest exists.

References

1. Li, T.; Su, M.; Guo, J.; Zhao, K.; Zhang, Q. Seismic performance evaluation of Y-shaped eccentric brace–high-strength steel frame structure based on remote collaborative hybrid test. *Soil Dyn. Earthq. Eng.* **2023**, *164*, 107621. [CrossRef]
2. *GB50011-2010*; Code for Seismic Design of Buildings. China Architecture and Building Press: Beijing, China, 2016.
3. Keivan, A.; Zhang, Y. Nonlinear seismic performance of Y-type self-centering steel eccentrically braced frame buildings. *Eng. Struct.* **2019**, *179*, 448–459. [CrossRef]
4. Ma, X.; Zhang, X.; Zhang, S.; Zhao, G.; Dong, B. Seismic performance of very short links in eccentrically braced substructures. *J. Constr. Steel Res.* **2024**, *216*, 108628. [CrossRef]
5. Li, S.; Xu, T.; Li, X.; Liang, G.; Xi, H. Elastic stiffness and bearing mechanism of eccentrically braced steel frames. *Structures* **2023**, *55*, 818–833. [CrossRef]
6. Zheng, X.-W.; Li, H.-N.; Gardoni, P. Hybrid Bayesian-Copula-based risk assessment for tall buildings subject to wind loads considering various uncertainties. *Reliab. Eng. Syst. Saf.* **2023**, *233*, 109100. [CrossRef]
7. Zheng, X.-W.; Li, H.-N.; Shi, Z.-Q. Hybrid AI-Bayesian-based demand models and fragility estimates for tall buildings against multi-hazard of earthquakes and winds. *Thin-Walled Struct.* **2023**, *187*, 110749. [CrossRef]
8. Zheng, X.-W.; Li, H.-N.; Lv, H.-L.; Huo, L.-S.; Zhang, Y.-Y. Bayesian-based seismic resilience assessment for high-rise buildings with the uncertainty in various variables. *J. Build. Eng.* **2022**, *51*, 104321. [CrossRef]
9. Zheng, X.-W.; Li, H.-N.; Gardoni, P. Probabilistic seismic demand models and life-cycle fragility estimates for high-rise buildings. *ASCE-J. Struct. Eng.* **2021**, *147*, 04021216. [CrossRef]
10. Lettieri, A.; de la Peña, A.; Freddi, F.; Latour, M. Damage-free self-centring links for eccentrically braced frames: Development and numerical study. *J. Constr. Steel Res.* **2023**, *01*, 107727. [CrossRef]
11. Almasabha, G.; Al-Mazaidh, R. Simple Truss Model to estimate the shear strength of short links in the ccentrically Braced Frame (EBF) steel system. *Thin-Walled Struct.* **2023**, *188*, 110811. [CrossRef]
12. Yin, Z.; Yang, B.; Zhang, X. Design of an eccentrically buckling-restrained braced steel frame with web-bolted replaceable links. *J. Constr. Steel Res.* **2022**, *192*, 107250. [CrossRef]
13. Montuori, R.; Nastri, E.; Piluso, V. Theory of plastic mechanism control for eccentrically braced frames with inverted y-scheme. *J. Constr. Steel Res.* **2014**, *92*, 122–135. [CrossRef]
14. Valente, M.; Castiglioni, C.A.; Kanyilmaz, A. Dissipative devices for earthquake resistant composite steel structures: Bolted versus welded solution. *Bull. Earthq. Eng.* **2016**, *14*, 3613–3639. [CrossRef]
15. Vasdravellis, G.; Valente, M.; Castiglioni, C.A. Dynamic response of composite frames with different shear connection degree. *J. Constr. Steel Res.* **2009**, *65*, 2050–2061. [CrossRef]
16. Tratan, A.; Dubina, D.; Dinu, F. Control of Global Performance of Seismic Resistant EBF with Removal Link. In Proceedings of the Conference on Behavior of Steel Structures in Seismic Aeas, STESSA, Naples, Italy, 9–12 June 2023; pp. 455–461.
17. Stratan, A.; Dubina, D. Bolted Links for Eccentrically Braced Steel Frames. In Proceedings of the Fifth International Workshop on Connections in Steel Structures, Delft, The Netherlands, 3–4 June 2004; Volume 3, pp. 223–232.
18. Shayanfar, M.A.; Rezaeian, A.R.; Zanganeh, A. Seismic performance of eccentrically braced frame with vertical link using PBPD method. *Struct. Des. Tall Spec. Build.* **2014**, *23*, 1–21. [CrossRef]
19. Li, S.; Su, M. A performance-based seismic design method for eccentrically braced steel frames. *Eng. Mech.* **2014**, *31*, 195–204. [CrossRef]

20. Mahmoudi, F.; Dolatshahi, K.M.; Mahsuli, M.; Nikoukalam, M.T.; Shahmohammadi, A. Experimental study of steel moment resisting frames with shear link. *J. Constr. Steel Res.* **2019**, *154*, 197–208. [CrossRef]
21. Kaufmann, E.; Metrovich, B.; Pense, A. *Characterization of Cyclic Inelastic Strain Behavior on Properties of A572 Gr. 50 and A913 Gr. 50 Rolled Sections*; ATLSS Rep, Lehigh University: Bethlehem, PA, USA, 2001.
22. *ANSI/AISC 341-16*; Seismic Provision for Structure Steel Buildings. American Institute of Steel Construction: Chicago, IL, USA, 2016.

buildings

MDPI

Article

Research on Collision Restitution Coefficient Based on the Kinetic Energy Distribution Model of the Rocking Rigid Body within the System of Mass Points

Qiuyu Mao [1,2], Tongfa Deng [1,2,*], Botan Shen [1,2] and Yuexin Wang [1,2]

1 School of Civil and Surveying & Mapping Engineering, Jiangxi University of Science and Technology, Ganzhou 341000, China; 6720211160@mail.jxust.edu.cn (Q.M.); shenbotan@jxust.edu.cn (B.S.); 6120220347@mail.jxust.edu.cn (Y.W.)

2 Jiangxi Province Key Laboratory of Environmental Geotechnical Engineering and Hazards Control, Jiangxi University of Science and Technology, Ganzhou 341000, China

* Correspondence: tfdeng@jxust.edu.cn; Tel.: +86-139-7078-3208

Abstract: Rocking structures exhibit significant collapse resistance during earthquakes. In studies of rocking rigid bodies, the collision restitution coefficient is typically determined based on the classical model of the rocking rigid bodies. However, during the rocking process, the collision restitution coefficient, influenced by the uncontrollable error in collision energy dissipation between the rigid body and the ground, indirectly impacts the final results of the equations of motion. Therefore, the rationality and reliability of the collision restitution coefficient are crucial for seismic analysis of rocking rigid bodies and self-centering members. This paper introduces a phasic energy dissipation and kinetic energy redistribution model specifically designed for the rocking rigid body within the system of mass point. This model divides the collision into three distinct stages, incorporating energy dissipation considerations in the first two stages to calculate the total kinetic energy of the rigid body. In the third stage, the remaining kinetic energy is redistributed to precisely determine the analytical solution for the collision restitution coefficient of an ideal, homogeneous rectangular rigid body during collision. Lastly, the validity and reliability of the proposed model are confirmed through comparisons with experimental data.

Keywords: rocking structure; rocking rigid body model; seismic performance; collision restitution coefficient; conservation of kinetic energy

1. Introduction

Self-centering structures in engineering and unconstrained tall objects within buildings are susceptible to rocking and collisions during seismic events. The collision process results in energy loss, influenced by numerous factors [1,2]. The aleatory and epistemic uncertainties coming from various sources have significant impacts on the accuracy of risk estimates for structures under dynamic excitations [3,4].

In 1963, Housner [5] introduced the concept of "rocking structure", drawing from his analysis of the 1952 California earthquake and the 1960 Chilean earthquake. He formulated the classical model of the rocking rigid body [6]. This model has garnered widespread recognition among researchers. Notably, within the realm of resilient seismic design, this model and its analytical methods have undergone notable advancements and enhancements since their inception [6,7].

Beck and Skinner [8] initially explored the utilization of the rocking concept in seismic bridge design in 1974 [9]. Huckelbridge and Clough [10,11] performed shaking table experiments on three- and nine-story rocking steel frame models in 1977, revealing the beneficial role of structural rocking in mitigating the need for increased structural strength and ductility [12]. In 1980, Yim et al. [13] discovered that the rocking response of rigid bodies during earthquakes was highly susceptible to variations in size, slenderness ratio (h/b), and ground motion characteristics. In the same year, Aslam et al. [14] examined the rocking behavior of rigid bodies under earthquake conditions through experimental investigations, concluding that the overturning tendency of rigid bodies is significantly influenced by base parameters, the collision restitution coefficient, and changes in ground motion. In 1981, the South Rangitikei railway bridge in New Zealand became the first to incorporate rocking piers, successfully implementing the rocking structure concept in bridge engineering [15]. Apostolou et al. [16] conducted an analysis of rocking vibration parameters for rigid bodies resting on rigid and linear elastic foundations in 2007, emphasizing the significance of rigid body size, slenderness ratio, and foundation vibration parameters as critical factors influencing overturning [6]. As the concept of "resilient structures" gained popularity, more researchers began exploring the field of rocking structures. In 2009, Mariott et al. [17] introduced a novel method utilizing external energy dissipation devices, as opposed to internal reinforcement, to address the challenge of post-earthquake reinforcement replacement. They validated the effectiveness of these rocking piers through quasi-static and quasi-dynamic testing, demonstrating their superior energy dissipation and self-centering capabilities. Addressing concerns regarding the durability of prestressed bundles and external energy dissipation devices, Guo et al. [18], in 2015, explored the use of basalt fiber-reinforced composite materials, modified aluminum alloys, and glass fiber-reinforced polymer materials for the fabrication of prestressed reinforcements, external dissipation devices, and pier sleeves. They subsequently evaluated the seismic performance of these rocking piers through quasi-static testing. In 2016, the Huangxulu overpass bridge, a self-centering rocking bridge project in China spanning the Jing-Tai Expressway [19,20], implemented a seismic design method based on the displacement of its rocking bridge structure. This development ushered in the era of rocking bridge construction in the 21st century, paralleled by the contemporary construction of the Wigram-Magdala overpass bridge in New Zealand [21,22]. With the increasing application of rocking structures in engineering, researchers have increasingly focused on effectively incorporating seismic reduction principles of rocking structures into practical engineering implementations [23–28].

Prior research indicates that the magnitude of the collision restitution coefficient significantly impacts the shape of the overturning region of a rocking rigid body [6], subsequently influencing pertinent studies. Hence, accurate and rational estimation of the collision restitution coefficient holds crucial significance for seismic investigations into rocking structures. Schau et al. [29] conducted a study in 2014 that investigated the rocking motion of rigid bodies subjected to earthquake excitation, utilizing numerical integration of equations of motion and finite element methods. Highlighting that any variations in ideal geometric shapes or contact conditions would result in an increase in the collision restitution coefficient, they proposed a collision restitution coefficient value of 0.8 for rigid bodies exhibiting a slenderness ratio of 2. This value theoretically surpasses the theoretical value of 0.7 employed in classical models. In 2016, Kallionizis et al. [30] compiled a collision restitution coefficient based on previous experiments conducted on rocking rigid bodies. Their findings revealed that for rigid bodies with a small slenderness ratio, the classical model's collision restitution coefficient exhibits significant discrepancies compared to experimental outcomes. Notably, experimental values consistently exceeded the predictions of the classical theoretical model. Based on their observations, an improved model was proposed, assuming that collisions between the rigid body and the base generate impulses acting specifically between the two bottom corners of the rigid body. While the revised model aligned closely with experimental data, it was noteworthy that the length of the rotational surface was determined empirically [6]. In 2018, Čeh et al. [31] conducted free rocking tests on single rigid bodies, excluding slipping and bouncing. Their findings supported the previous research conclusions of Kallionizis et al. Additionally, they developed an algorithm capable of simulating rigid body rocking. In 2022, Jia et al. [7] conducted shaking table tests on rocking rigid bodies, aiming to investigate their seismic response characteristics. Their measurements revealed an actual collision restitution coefficient of 0.901, exceeding the theoretical value of 0.873 predicted by the classical model of the rocking rigid body.

Considering the present research landscape and its outcomes, the classical model of the rocking rigid body exhibits a significantly high margin of error in its collision restitution coefficient. The primary source of the substantial error lies in the fact that the classical model of the rocking rigid body, while utilizing the conservation of angular momentum for its solutions, disregards the system satisfying conservation of angular momentum as a comprehensive entity encompassing both the rigid body and the ground. Conversely, the advanced models incorporate empirically determined parameters, which lack a theoretical foundation. This paper aims to further refine the rocking structure rigid body model by meticulously analyzing the energy dissipation mechanisms during collisions and introducing a kinetic energy distribution model tailored for rocking rigid bodies within the mass point system. This model divides the collision into three distinct stages, incorporating energy dissipation considerations in the first two stages to calculate the total kinetic energy of the rigid body. In the third stage, the remaining kinetic energy is redistributed under the action of stresses. This approach yields an analytical solution specific to homogeneous regular rectangular rigid bodies. Compared to both experimental data and the classical model, the theoretical values derived from this model exhibit significantly reduced errors, with a correlation coefficient exceeding 0.9. This offers a novel perspective for exploring energy dissipation mechanisms in rocking rigid bodies.

2. Model and Collision Restitution Coefficient

2.1. Classical Model of the Rocking Rigid Body

The classical model of the rocking rigid body was first proposed by Housner [5] in 1963, as shown in Figure 1. The center of gravity of the rigid body is located at point C, where h is half the height of the rigid body, and b is half the width of the rigid body. After the initiation of rocking, the rigid body rotates about the pivot points O or O' during different time periods.

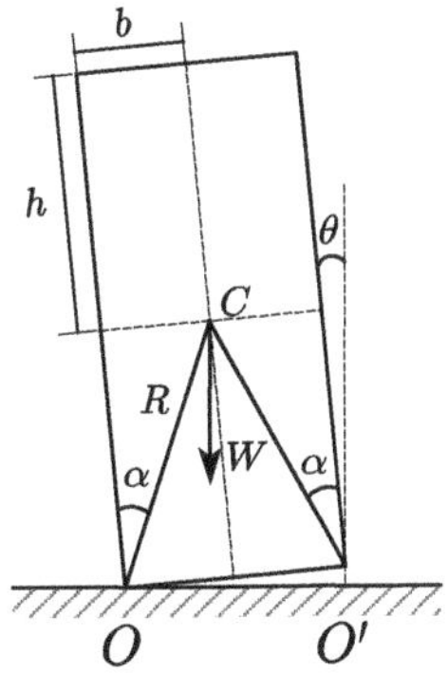

Figure 1. Classical model of the rocking rigid body [5].

Here are the assumptions for the classical model of the rocking rigid body [32]:

1. The model is simplified for a 2-D geometry;
2. The block and the base are rigid;
3. The rigid body is a homogeneous rectangular block;
4. There is no bouncing or complete lifting of the body;
5. Blocks are assumed not to be sliding on the ground or relative to each other;
6. Material stiffness and damping are not considered, i.e., the blocks cannot be compressed nor do they store/dissipate elastic energy;
7. The collision lasts for an infinitely small time interval;
8. The rigid body is not damaged during the collision.

Since the rigid body is a homogeneous and symmetrical structure, the derivation of the collision restitution coefficient in the classical model of the rocking rigid body only considers the instantaneous collision when the rigid body transitions from rotating about point O to rotating about point O'. The derivation process for the collision restitution coefficient in the classical model of the rocking rigid body is given as follows.

Based on Konig's theorem of angular momentum, before the collision, the angular momentum of the rigid body about O' is the sum of the angular momentum of the rigid body about its center of mass and the angular momentum obtained by concentrating the mass of the rigid body at the center of mass about O' [7]. Therefore, the angular momentum of the rigid body about O' before the collision is

$$L^{-} = I_C\dot{\theta}^{-} + WR\dot{\theta}^{-}\sin(\frac{\pi}{2} - 2\alpha)R \tag{1}$$

where I_C represents the angular momentum of the rigid body about its center of mass, $\dot{\theta}^{-}$ is the angular velocity of the rigid body before the collision, W is the total mass of the rigid body, R is the semi-diagonal of the rigid body, and α is the angle between the semi-diagonal and the height side of the rigid body.

By applying the equation for $I_O = I_C + WR^2$ and trigonometric relationships, the angular momentum of the rigid body about O' before the collision can be simplified to

$$L^{-} = I_O\dot{\theta}^{-} - 2Wb^2\dot{\theta}^{-} \tag{2}$$

where I_O is the angular momentum of the rigid body about point O.

The angular momentum after the collision is

$$L^{+} = I_{O'}\dot{\theta}^{+} \tag{3}$$

where $I_{O'}$ is the angular momentum of the rigid body about point O', and $\dot{\theta}^{+}$ is the angular velocity of the rigid body after the collision.

According to the conservation of angular momentum, at the instant of collision, the collision restitution coefficient at the instant of collision when the rigid body transitions from rotating about point O to rotating about point O' can be obtained as

$$r = \frac{\dot{\theta}^{+}}{\dot{\theta}^{-}} = \frac{I_O - 2Wb^2}{I_{O'}} \tag{4}$$

Since the rigid body is a homogeneous rectangular rigid body, $I_O = I_{O'} = \frac{4}{3}WR^2$, the collision restitution coefficient in the classical model of the rocking rigid body can be simplified to

$$r = 1 - \frac{3}{2}\sin^2\alpha \tag{5}$$

The classical model of the rocking rigid body has the following issues.

1. The conservation of angular momentum presupposes that the net external torque on the system is zero. However, during the collision, the rigid body is subjected to a net external torque from combined forces such as gravity, impact, and friction, which is not zero. Hence, the entire system, consisting of the rigid body and the ground, satisfies the conditions for the conservation of angular momentum, rather than the rigid body system itself, meaning that the angular momentum of the rigid body system does not conserve during the collision process.
2. Before the collision, the center of rotation for the rigid body system is at point O. At this time, the angular momentum of the system about point O' is not the entirety of the angular momentum possessed by the rigid body system but rather the relative angular momentum generated at point O' by the velocity of the system rotating about

point O. It is also unreasonable to equate the relative angular momentum of the rigid body system before the collision about O' with the entire angular momentum possessed by the rigid body after the collision.

3. Although the model assumes that the angular momentum of the rigid body system about point O' is conserved before and after the collision and calculates the collision restitution coefficient for the rocking rigid body based on this assumption, the model implicitly uses the difference in the magnitude of the angular momentum of the rigid body about points O and O' before the collision, denoted as $|L_O| - |L_{O'}|$, to represent the change in the magnitude of the angular momentum of the rigid body system during the collision, i.e., the reduction in angular momentum of the rigid body system during the collision. The collision described by this model is a complex process, and using $|L_O| - |L_{O'}|$ to represent the reduction in angular momentum of the rigid body system does not clarify its actual significance from a mechanical standpoint.
4. In previous studies, the collision restitution coefficient for rocking rigid bodies calculated by this theory has shown significant discrepancies compared to experimental values.

2.2. Kinetic Energy Distribution Model of the Rocking Rigid Body within the System of Mass Point

To avoid the issues encountered with the classical model of the rocking rigid body, this study establishes the kinetic energy distribution model within the system of mass points within a rocking rigid body in a Cartesian coordinate system, based on the assumptions of the classical model, as shown in Figure 2. Furthermore, the kinetic energy redistribution equation for the rocking rigid body model is derived. The variation in kinetic energy distribution of this model is divided into the following three stages.

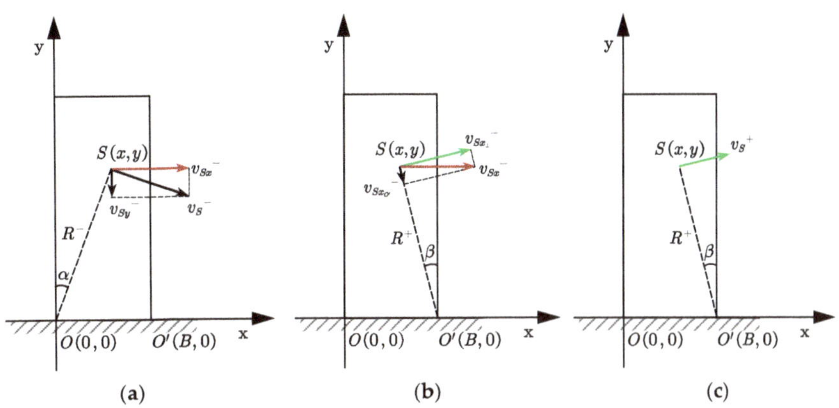

Figure 2. The kinetic energy distribution model of the rocking rigid body within the system of mass point: (**a**) moment before collision; (**b**) moment of collision; and (**c**) moment after collision.

1. Moment of rotation around point O before collision. The kinetic energy at point S of the rigid body is decomposed into kinetic energy $T_{Sx}{}^-$, which is parallel to the ground, and kinetic energy $T_{Sy}{}^-$, which is perpendicular to the ground. Due to the collision, the kinetic energy $T_{Sy}{}^-$, which is perpendicular to the ground, is completely dissipated.
2. Moment of collision. The kinetic energy $T_{Sx}{}^-$, which is parallel to the ground, is further decomposed into kinetic energy $T_{Sx_{O'}}{}^-$ along the direction $O'S$ and kinetic energy $T_{Sx_\perp}{}^-$ perpendicular to direction $O'S$. The center of rotation shifts from point O to point O', and the kinetic energy $T_{Sx_\perp}{}^-$ along direction $O'S$ is completely dissipated under the combined action of the vertical reaction force from the ground and the horizontal frictional force.

3. Moment of rotation around point O' after collision. The finite element analysis conducted by Kouris et al. [25] indicates that the variation of stress within the rigid block during the collision process is quite complex, and it is not appropriate to deduce the changes in the kinetic energy distribution of the rigid body within the system of mass points based on the alteration of stresses within the rigid body. Based on the assumption that the rigid body remains undamaged during the collision, the velocity of each point within the rigid body system is redistributed under the influence of the stresses. The total sum of kinetic energy at each point of the rigid body does not increase or decrease during this process; hence, the kinetic energy ${T_{Sx_\perp}}^-$ along the perpendicular direction $O'S$ is redistributed into kinetic energy ${T_S}^+$ rotating around O'.

It is clear from the kinetic energy distribution model of the particle system rocking rigid body that the first two stages are characterized by energy dissipation, where the rigid body decomposes and calculates the loss of kinetic energy at any point S; in contrast, during the third stage, the kinetic energy at each point of the rigid body undergoes redistribution due to internal forces. The process of kinetic energy decomposition and loss, as well as the derivation of the kinetic energy redistribution equation for this model, are described as follows.

The linear velocity of the rigid body at point S immediately before the collision is

$$ {v_S}^- = \dot{\theta}^- R^- \tag{6} $$

where R^- represents the distance from point S to point O.

The velocity of this point parallel to the ground is

$$ {v_{Sx}}^- = \dot{\theta}^- R^- \cos\alpha \tag{7} $$

where α is the angle between OS and the vertical side.

At the moment of collision, the velocity of this point perpendicular to $O'S$ is

$$ {v_{Sx_\perp}}^- = \dot{\theta}^- R^- \cos\alpha\cos\beta \tag{8} $$

where β is the angle between SO' and the vertical side.

Within the coordinate system, point O serves as the origin $(0,0)$, the coordinates of point O' are $(B,0)$, and the coordinates of point S are (x,y); according to trigonometric relationships, Equation (8) can be rewritten as

$$ {v_{Sx_\perp}}^- = \dot{\theta}^- \frac{y^2}{\sqrt{y^2+(B-x)^2}} \tag{9} $$

At this time, the kinetic energy of the rigid body at point S is

$$ {T_{Sx_\perp}}^- = \frac{1}{2}m\left(\dot{\theta}^-\right)^2 \frac{y^4}{y^2+(B-x)^2} \tag{10} $$

where $m = \rho dxdy$ is the mass of the rigid body at point S.

The total kinetic energy possessed by the rigid body at this moment is

$$ T^- = \frac{1}{2}\rho\left(\dot{\theta}^-\right)^2 \int_0^H\int_0^B \frac{y^4}{y^2+(B-x)^2}dxdy \tag{11} $$

where T^- is the total kinetic energy before redistribution of the rigid body, B is the width of the rigid body, H is the height of the rigid body, and ρ is the density of the rigid body.

Immediately after the collision, the kinetic energy of the rigid body is redistributed such that the velocity at any point on the rigid body conforms to the velocity of the rigid body rotating around point O'. The linear velocity at point S of the rigid body at this moment is

$$v_S{}^+ = \dot{\theta}^+ R^+ \tag{12}$$

where R^+ is the distance from point S to point O'.

According to trigonometric relationships, Equation (12) can be rewritten as

$$v_S{}^+ = \dot{\theta}^+ \sqrt{y^2 + (B-x)^2} \tag{13}$$

The kinetic energy of the rigid body at point S at this time is

$$T_S{}^+ = \frac{1}{2} m \left(\dot{\theta}^+\right)^2 \left[y^2 + (B-x)^2\right] \tag{14}$$

The total kinetic energy possessed by the rigid body at this time is

$$T^+ = \frac{1}{2} \rho \left(\dot{\theta}^+\right)^2 \int_0^H \int_0^B \left[y^2 + (B-x)^2\right] dxdy \tag{15}$$

The redistribution of kinetic energy alters the kinetic energy at each point of the rigid body without changing the total kinetic energy, that is, $T^- = T^+$. Thus by combining Equation (11) with Equation (15), the kinetic energy redistribution equation for the kinetic energy distribution model of the rocking rigid body within the system of mass points can be derived as

$$\left(\frac{\dot{\theta}^+}{\dot{\theta}^-}\right)^2 = \frac{\int_0^H \int_0^B \frac{y^4}{y^2 + (B-x)^2} dxdy}{\int_0^H \int_0^B \left[y^2 + (B-x)^2\right] dxdy} \tag{16}$$

Therefore, the collision restitution coefficient of the rocking rigid body kinetic energy distribution model within the system of mass points can be represented by the following equation:

$$r^2 = \frac{\int_0^H \int_0^B \frac{y^4}{y^2 + (B-x)^2} dxdy}{\int_0^H \int_0^B \left[y^2 + (B-x)^2\right] dxdy} \tag{17}$$

3. Validation of the Results

To validate the rationality of the kinetic energy distribution model of the rocking rigid body within the system of mass point, this paper collects data from previous studies and derives analytical solutions for the collision restitution coefficient of both the classical model of the rocking rigid body and the kinetic energy distribution model of the rocking rigid body within the system of mass point. These are then compared with experimental values, and an error analysis is conducted.

3.1. Verification of Analytical Solution under Various Slenderness Ratios

Kalliontzi et al. [30] compiled statistics on the collision restitution coefficient of the rocking rigid body and concluded that the experimental values obtained were all higher than those of the classical model of the rocking rigid body, as shown in Table 1.

Table 1. Statistics on the test results of the collision restitution coefficient for rocking rigid body [30].

Slenderness Ratio (h/b)	r_e	r_H	Relative Error	Block and Base (Materials at Contact Interface)
2	0.787401	0.7	0.110999	wood/steel
2	0.87178	0.7	0.197045	concrete/aluminum
2.85	0.927362	0.836	0.097806	granite/granite
3	0.877496	0.85	0.033012	wood/steel
4	0.938083	0.912	0.028825	wood/steel
4	0.927362	0.912	0.017597	concrete/steel
4	0.938083	0.912	0.028825	granite/granite
4	0.948683	0.912	0.039676	wood/aluminum
4	0.921954	0.912	0.011835	steel/steel
4.33	0.959166	0.924	0.016439	steel/wood
5.88	0.974679	0.958	0.015916	granite/granite
8.33	0.979796	0.979	0	granite/granite

Note: r_e represents the experimental value of the collision restitution coefficient, and r_H is the collision restitution coefficient from Housner's classical model.

In reality, the collision restitution coefficient measured in collision experiments varies slightly with different materials; however, since the materials used in the compiled experimental data are not prone to deformation and the collision time is extremely short, the error introduced by different materials is considered negligible. Owing to errors inherent in the testing process, for slenderness ratios that have been tested multiple times, an average value is adopted to represent the collision restitution coefficient for that specific aspect ratio.

Based on the formula for the collision restitution coefficient derived by the kinetic energy distribution model of the rocking rigid body within the system of mass point, a computation program was developed using MATLAB, as shown in Appendix A. The processed data were imported into this program to calculate the corresponding collision restitution coefficient r_r for the kinetic energy distribution model of the rocking rigid body within the system of mass point, as shown in Table 2.

Table 2. Comparison of collision restitution coefficients for rocking rigid body with statistical data.

Slenderness Ratio (h/b)	r_e	r_H	r_r
2	0.8295905	0.7	0.83030915
2.85	0.927362	0.83666	0.90454519
3	0.877496	0.848528	0.91256288
4	0.934833	0.911043	0.94736684
4.33	0.959166	0.943398	0.95441345
5.88	0.974679	0.959166	0.97412902
8.33	0.979796	0.979796	0.98664748

Note: r_e represents the experimental value of the collision restitution coefficient, r_H is the collision restitution coefficient from Housner's classical model, and r_r is the collision restitution coefficient for the kinetic energy distribution model of the rocking rigid body within the system of mass point.

Based on the data, it is evident that the collision restitution coefficient of the kinetic energy distribution model of the rocking rigid body within the system of mass points aligns more closely with the experimental values. To visualize the relationship between the classical model of the rocking rigid body, the kinetic energy distribution model of the rocking rigid body within the system of mass points, and the experimental data, a relationship of the slenderness ratio and collision restitution coefficient was constructed, as shown in Figure 3. This figure uses the error between the collision restitution coefficient from the kinetic energy distribution model of the rocking rigid body within the system of mass points and the experimental values as the length of the error bar. The curve for the classical model of the rocking rigid body falls within the error bar range only at slenderness ratios of 3 and 8.33, whereas the error is greater than that of the kinetic energy distribution model of the rocking rigid body within the system of mass points for all other slenderness ratios.

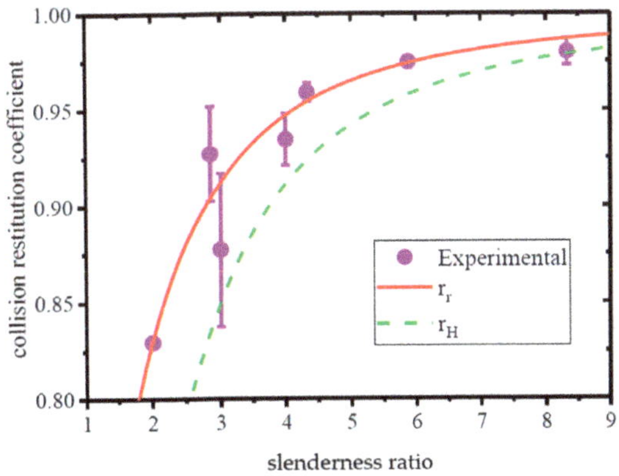

Figure 3. Slenderness ratio versus collision restitution coefficient curves for statistical data.

Čeh et al. [31] conducted research on the energy dissipation of oscillating aluminum blocks using three sets (S—small, M—medium, L—large) of blocks. Each set contained 10 aluminum blocks with different slenderness ratios. To prevent slipping and bouncing, tape was applied between the block and the base, as shown in Figure 4.

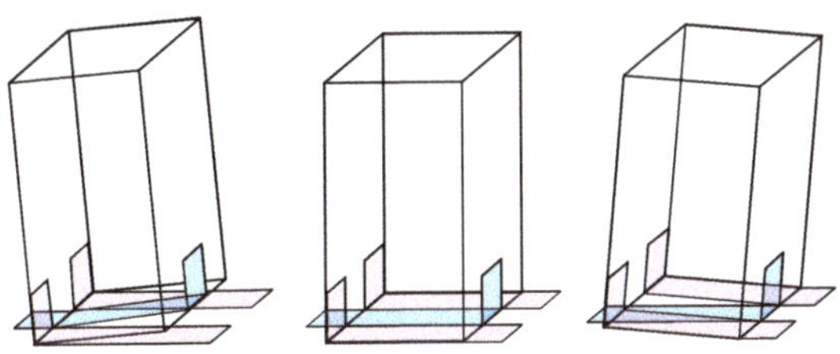

Figure 4. Rigid body-tape systems to avoid slipping and bouncing [31].

Using the computational program developed in MATLAB, the collision restitution coefficient r_H for the classical model of the rocking rigid body and the collision restitution coefficient r_r for the kinetic energy distribution model of the rocking rigid body within the system of mass points were calculated based on the experimental data. The specimen parameters, experimental data, and calculation results are presented in Table 3.

Similarly, Table 3 shows that the collision restitution coefficient derived from the kinetic energy distribution model of the rocking rigid body within the system of mass points is in better agreement with the experimental values. To more clearly visualize the relationship between the classical model of the rocking rigid body, the kinetic energy distribution model of the rocking rigid body within the system of mass points, and the experimental data, a relationship between the slenderness ratio and collision restitution coefficient was created, as depicted in Figure 5.

Figure 5 clearly demonstrates that the collision restitution coefficient for the aluminum blocks in Group L is closer to the theoretical values predicted by the kinetic energy distribution model of the rocking rigid body within the system of mass points. Additionally, the experimental data shows a relatively stable deviation from the theoretical predictions of the kinetic energy distribution model of the rocking rigid body within the system of mass point. The deviation between the experimental data and the classical model of the rocking rigid body increases gradually as the aspect ratio decreases.

Table 3. The experimental results for the collision restitution coefficient of the rocking rigid body by Čeh et al. [31].

Block	m (g)	Slenderness Ratio (*h*/*b*)	r_e	r_H	r_r
S1	113.3	1.5	0.69	0.538	0.743
S2	161.2	2.25	0.825	0.753	0.859
S3	226.6	3	0.89	0.850	0.913
S4	274.5	3.75	0.919	0.900	0.941
S5	339.6	4.5	0.948	0.929	0.958
S6	453.2	6	0.958	0.959	0.975
S7	500.8	6.75	0.96	0.968	0.980
S8	614.1	8.25	0.964	0.978	0.986
S9	727.4	9.75	0.966	0.984	0.990
M1	363.6	1.5	0.67	0.538	0.743
M2	544.4	2.25	0.84	0.753	0.859
M3	727.2	3	0.882	0.850	0.913
M4	907.7	3.75	0.914	0.900	0.941
M5	1089.6	4.5	0.941	0.929	0.958
M6	1453.2	6	0.959	0.959	0.975
M7	1634	6.75	0.966	0.968	0.980
M8	1997.6	8.25	0.969	0.978	0.986
M9	2361.2	9.75	0.974	0.984	0.990
L1	856.6	1.5	0.708	0.538	0.743
L2	1284.3	2.25	0.852	0.753	0.859
L3	1713.2	3	0.904	0.850	0.913
L4	2140.9	3.75	0.937	0.900	0.941
L5	2569.2	4.5	0.952	0.929	0.958
L6	3425.8	6	0.97	0.959	0.975
L7	3853.5	6.75	0.972	0.968	0.980
L8	4710.1	8.25	0.978	0.978	0.986
L9	5566.7	9.75	0.979	0.984	0.990

Note: r_e represents the experimental value of the collision restitution coefficient, r_H is the collision restitution coefficient from Housner's classical model, and r_r is the collision restitution coefficient for the kinetic energy distribution model of the rocking rigid body within the system of mass point.

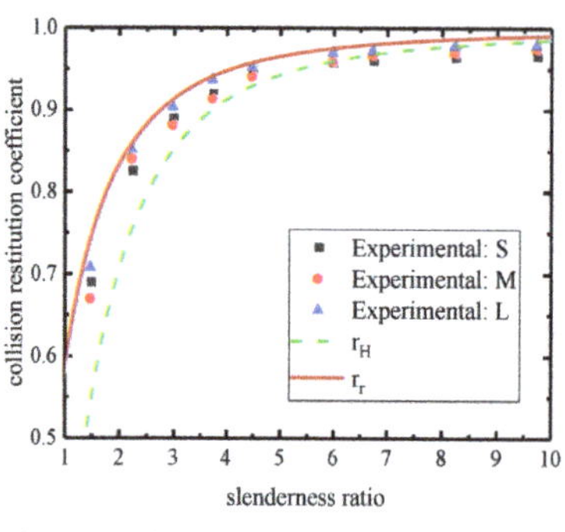

Figure 5. Slenderness ratio versus collision restitution coefficient relationships for the three sets of data.

3.2. Correlation and Error Analysis

Analysis reveals a correlation coefficient of 0.946 between the theoretical values of the kinetic energy distribution model of the rocking rigid body within the system of mass points and the statistical data. The correlation coefficients with the experimental values for Groups S, M, and L are 0.998, 0.996, and 0.998, respectively. To facilitate further analysis, scatter plots depicting the relative errors between the theoretical values of the two models and the four data sets are provided in Figures 6 and 7.

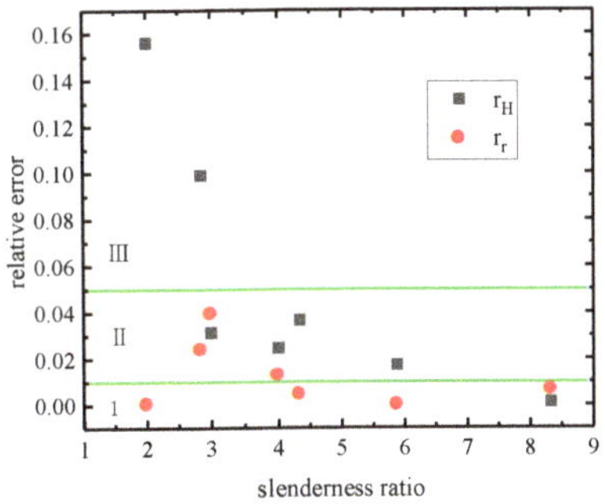

Figure 6. Relative errors between the theoretical and statistical data.

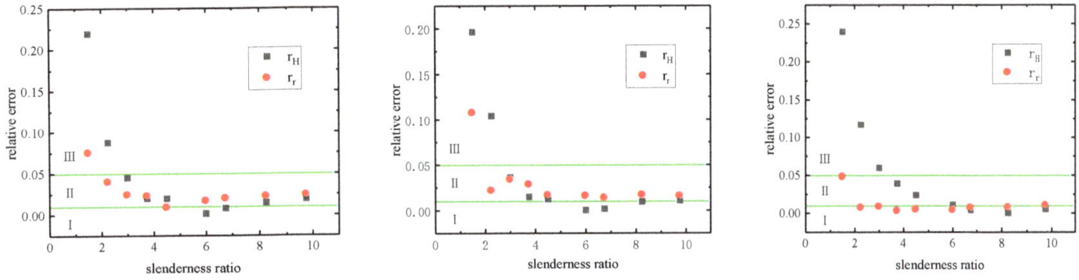

(a) (b) (c)

Figure 7. Relative errors between the theoretical values and the experimental data measured by Čeh [31]: (**a**) the errors between the theoretical values and the data from Group S; (**b**) the errors between the theoretical values and the data from Group M; (**c**) the errors between the theoretical values and the data from Group L.

In the figures, the relative errors are divided into three zones. Zone I represents errors less than 1%, Zone II represents errors greater than 1% but less than 5%, and Zone III represents errors greater than 5%. As can be seen from Figure 6, the relative errors for the classical model of the rocking rigid body are primarily distributed in Zones II and III, while the relative errors for the kinetic energy distribution model of the rocking rigid body within the system of mass points are all within Zones I and II.

Figure 7 reveals that for the three sets of data, when the slenderness ratio is less than 6, the theoretical values from the kinetic energy distribution model of the rocking rigid body within the system of mass points are more consistent with the experimental data. For Groups S and M, the advantage of the kinetic energy distribution model of the rocking rigid body within the system of mass points over the classical model of the rocking rigid body is not significant, but for Group L, the superiority of the kinetic energy distribution model of the rocking rigid body within the system of mass points is more pronounced. Furthermore, the figure illustrates that the error associated with the kinetic energy distribution model of the rocking rigid body under the system of mass points remains largely constrained within 10% across the three test groups. Only one exception occurred in the Group M experiment. The analysis shows that it is easier to produce errors in smaller components because the experiment is in a non-ideal state. Consequently, it is necessary to further analyze the influence of quality.

The experiments were not conducted under ideal conditions: collisions were not instantaneous, and some horizontal kinetic energy was consumed due to friction, with additional energy loss attributed to the tape affixed to the bottom of the blocks. To facilitate analysis, a scatter plot of the absolute error between theoretical and experimental values was created using the mass of the blocks as the independent variable (since the experiments fall between two theoretical values, for ease of comparison, the absolute error for r_r is represented by $r_r - r_e$, and the absolute error for r_H is represented by $r_e - r_H$), as shown in Figure 8. The relative errors between the theoretical values and experimental data for the same slenderness ratio are plotted on the same line chart, as depicted in Figure 9.

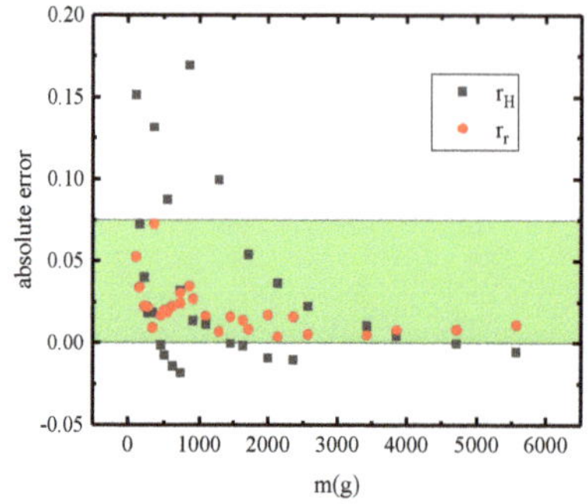

Figure 8. Absolute errors between the theoretical values and experimental data measured by Čeh [31].

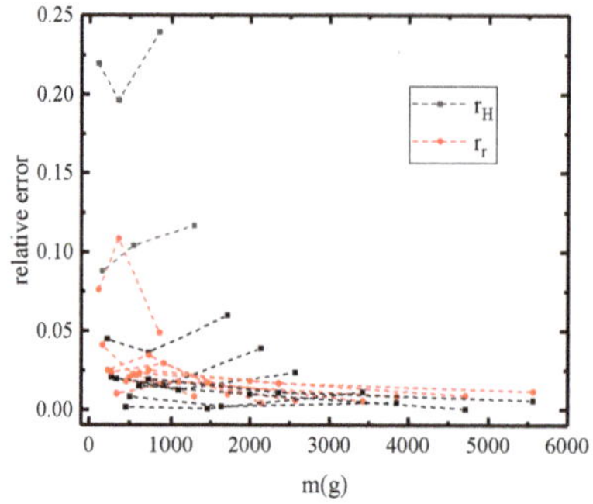

Figure 9. Relative errors between the theoretical and experimental data for the same slenderness ratio.

Figure 8 indicates that the absolute errors of the kinetic energy distribution model of the rocking rigid body within the system of mass points decrease gradually with an increase in mass and, unlike the classical model of the rocking rigid body, exhibit a relatively stable absolute error.

Figure 9 shows that at the same slenderness ratio, the relative errors of the kinetic energy distribution model of the rocking rigid body within the system of mass points tend to decrease with an increase in mass, whereas the relative error for the classical model of the rocking rigid body increases as the mass increases.

Based on the above analysis, it can be seen that the model proposed in this paper has a strong correlation with the test data, and the correlation coefficient is greater than 0.9. The error is more stable than the classical model, and the relative errors are basically controlled within 5%. In addition, since the larger the mass of the component is, the less affected by the environment, the model shows a stronger adaptability to large components (smaller error), indicating that the model can be applied to components with larger mass in engineering. Further research is needed to make it also applicable to components with smaller mass.

4. Conclusions

To refine the model of the rocking rigid body, this study proposes a phasic energy dissipation and kinetic energy redistribution model for the rocking rigid body within the system of mass points. This research conducts a step-by-step analysis of the collision process within the model and redistributes the kinetic energy post-collision, deriving an analytic solution for the collision restitution coefficient under this model. By analyzing the correlation coefficients and relative errors between the collision restitution coefficient from the proposed model and the classical model of the rocking rigid body against experimental data, the results indicate that the proposed model outperforms the classical model of the rocking rigid body. This confirms the validity of the presented model, demonstrating that the numerical solutions are accurate and reliable, leading to the following conclusions and recommendations.

1. The kinetic energy distribution model of the rocking rigid body within the system of mass points, starting from the kinetic energy and based on the energy dissipation mechanism during collisions, calculates the energy loss and conservation phases separately, which aligns more closely with reality and shows a higher degree of agreement with experimental data. Since small components initially have lower kinetic energy and are subject to more influencing factors, the advantages of this theory are more prominent in larger structures.
2. The analytic solution presented in this study is derived under the ideal conditions of extremely short collision times and negligible horizontal kinetic energy loss. However, actual collisions involve a brief contact time, primarily related to the materials involved in the collision and the instantaneous velocity. Therefore, the precision of the corresponding solutions varies with different materials and masses of the rigid bodies. If the collision time could be determined and the loss of horizontal kinetic energy taken into account, the accuracy of the solutions could be further improved.
3. In order to further improve the proposed model to explore the energy loss caused by non-instantaneous collision and friction in the swaying rigid body, it is necessary to analyze the contact between the rigid body and the ground during the collision process. The collision surface of the rigid body can be divided into a certain number of points. By introducing parameters such as the mass and elastic modulus of the rigid body, the force at each point in the collision process and the contact time with the ground can be calculated, and the kinetic energy lost during the collision process can be calculated more accurately.
4. The rigid bodies in the kinetic energy distribution model of the rocking rigid body within the system of mass points are established as homogeneous regular rectangular blocks in a two-dimensional plane. To better align with practical engineering applications, research into the collision restitution coefficient for three-dimensional-space rocking rigid bodies is necessary. This model provides new insights for the analysis of rocking in non-homogeneous and irregularly shaped rigid bodies. Further research can be conducted on non-homogeneous or irregularly shaped rigid bodies to refine the existing model of the rocking rigid body.
5. In order to expand the proposed model, the idea of finite element analysis can be incorporated into subsequent research, and the rigid body can be divided into multiple units to calculate its kinetic energy loss and conversion, so as to analyze the collision of non-homogeneous and irregular rigid bodies in three-dimensional space, making its application more practical.

Author Contributions: Conceptualization, Q.M. and T.D.; formal analysis, Q.M.; funding acquisition, T.D. and B.S.; investigation, Q.M., T.D. and Y.W.; methodology, Q.M.; software, Q.M.; validation, T.D.; writing—original draft, Q.M.; writing—review and editing, T.D. and B.S. All authors have read and agreed to the published version of the manuscript.

Funding: This work was supported by the Thousand Talents Plan of Jiangxi Province (Grant No. JXSQ2022017), Jiangxi Provincial Department of Education (Grant No. GJJ2200858), and the Research Start-Up Fund of Jiangxi University of Science and Technology (Grant No. 205200100637).

Data Availability Statement: No new data were created or analyzed in this study. Data sharing is not applicable to this article.

Conflicts of Interest: The authors declare no conflict of interest.

Appendix A

```
clear
clc
syms x y
B=1;
e=[ ];%Slenderness ratio
H=e*B;
s1=(y^4)/((y^2)+(B-x)^2);
s11=int(s1,x,0,B);
s12=int(s11,y,0,H);
s2=((y^2)+(B-x)^2);
s21=int(s2,x,0,B);
s22=int(s21,y,0,H);
S=s12/s22;
S=double(S);
r=sqrt(S)%collision restitution coefficient
```

References

1. Jeong, M.Y.; Lee, H.; Kim, J.H.; Yang, I.Y. Chaotic behavior on rocking vibration of rigid body block structure under two—dimensional sinusoidal excitation (in the case of no sliding). *KSME Int. J.* **2003**, *17*, 1249–1260. [CrossRef]
2. Jeong, M.Y.; Yang, I.Y. Characterization on the rocking vibration of rigid blocks under horizontal harmonic excitations. *Int. J. Precis. Eng. Manuf.* **2012**, *13*, 229–236. [CrossRef]
3. Zheng, X.W.; Li, H.N.; Gardoni, P. Hybrid Bayesian-Copula-based risk assessment for tall buildings subject to wind loads considering various uncertainties. *Reliab. Eng. Syst. Saf.* **2023**, *233*, 109100. [CrossRef]
4. Zheng, X.W.; Li, H.N.; Shi, Z.Q. Hybrid AI-Bayesian-based demand models and fragility estimates for tall buildings against multi-hazard of earthquakes and winds. *Thin-Walled Struct.* **2023**, *187*, 110749. [CrossRef]
5. Housner, G.W. The behavior of inverted pendulum structures during earthquakes. *Bull. Seismol. Soc. Am.* **1963**, *53*, 403–417. [CrossRef]
6. Zhao, Z.; Su, X. Literature review of researches on rigid body model of rocking structure. *Eng. Mech.* **2019**, *36*, 12–24.
7. Jia, C.; Pan, J.; Li, J.; Jia, C.; Pan, J.; Li, J.; Ma, L. Seismic analysis of eccentric single-rigid-body considering the influence of colliding-to-wall. *J. Vib. Shock.* **2022**, *41*, 116–123.
8. Beck, J.L.; Skinner, R.I. The seismic response of a reinforced concrete bridge pier designed to step. *Earthq. Eng. Struct. Dynamics* **1973**, *2*, 343–358. [CrossRef]
9. Du, X.-L.; Zhou, Y.-L.; Han, Q.; Wang, Z. State-of-the-art on rocking piers. *Earthq. Eng. Eng. Dyn.* **2018**, *38*, 1–11. Available online: https://dzgc.paperonce.org/#/digest?ArticleID=2426 (accessed on 27 December 2023).
10. Hukelbridge, A.A.; Clough, R.W. *Preliminary Experimental Study of Seismic Uplift of a Steel Frame*; Report No. UCB/EERC-77/22; University of California: Berkely, CA, USA, 1977.
11. Hukelbridge, A.A. *Earthquake Simulation Tests of a Nine Story Steel Frame with Columns Allowed to Uplift*; Report No. UCB/EERC-77/23; University of California: Berkely, CA, USA, 1977.
12. Zhou, Y.; Lü, X. State-of-the-art on rocking and self-centering structures. *J. Build. Struct.* **2011**, *32*, 1–10.
13. Yim, C.; Chopra, A.K.; Joseph, P. Rocking response of rigid blocks to earthquake. *Earthq. Eng. Struct. Dyn.* **1980**, *8*, 565–587. [CrossRef]
14. Aslam, M.; Scalise, D.T.; Godden, W.G. Earthquake rocking response of rigid bodies. *J. Struct. Div.* **1980**, *106*, 377–392. [CrossRef]
15. Cormack, L.G. The design and construction of the major bridges on the mangaweka rail deviation. *Trans. Inst. Prof. Eng. New Zealand* **1988**, *15*, 16–23.
16. Apostolou, M.; Gazetas, G.; Garini, E. Seismic response of slender rigid structures with foundation uplifting. *Soil Dyn. Earthq. Eng.* **2007**, *27*, 642–654. [CrossRef]
17. Marriott, D.; Pampanin, S.; Palermo, A. Quasi-static and pseudo-dynamic testing of unbonded post-tensioned rocking bridge piers with external replaceable dissipaters. *Earthq. Eng. Struct. Dyn.* **2009**, *38*, 331–354. [CrossRef]
18. Guo, T.; Cao, Z.; Xu, Z.; Lu, S. Cyclic load tests on self-centering concrete pier with external dissipators and enhanced durability [J]. *J. Struct. Eng.* **2015**, *142*, 04015088. [CrossRef]
19. Han, Q.; Jia, Z.L.; Xu, K.; Zhou, Y.; Du, X. Hysteretic behavior investigation of self-centering double-column rocking piers for seismic resilience. *Eng. Struct.* **2019**, *188*, 218–232. [CrossRef]

20. Han, Q.; Jia, Z.; He, W.; Xiao, Y.; Jia, J.; Du, X. Seismic design method and its engineering application of self-centering double-column rocking bridge. *China J. Highw. Transp.* **2017**, *30*, 169–177.
21. Mashal, M.; Palermo, A. Low-damage seismic design for accelerated bridge construction. *J. Bridge Eng.* **2019**, *24*, 04019066. [CrossRef]
22. Zhou, Y.; Zhang, J.; Han, Q.; Zhou, Y.; Zhang, J.; Han, Q.; Cheng, S.; He, H. Effect of viscous dampers on seismic response of rocking double-column bents under near-field ground motions with strong pulses. *China Civ. Eng. J.* **2020**, *53*, 288–293.
23. Kalliontzis, D.; Schultz, A.E.; Sritharan, S. Generalized dynamic analysis of structural single rocking walls (SRWs). *Earthq. Eng. Struct. Dyn.* **2020**, *49*, 633–656. [CrossRef]
24. Avgenakis, E.; Psycharis, I. An integrated macroelement formulation for the dynamic response of inelastic deformable rocking bodies. *Earthq. Eng. Struct. Dyn.* **2020**, *49*, 1072–1094. [CrossRef]
25. Kouris, E.-G.S.; Kouris, L.-A.S.; Konstantinidis, A.A.; Kourkoulis, S.K.; Karayannis, C.G.; Aifantis, E.C. Stochastic dynamic analysis of cultural heritage towers up to collapse. *Buildings* **2021**, *11*, 296. [CrossRef]
26. Kazantzi, A.K.; Lachanas, C.G.; Vamvatsikos, D. Seismic response distribution expressions for on-ground rigid rocking blocks under ordinary ground motions. *Earthq. Eng. Struct. Dyn.* **2021**, *50*, 3311–3331. [CrossRef]
27. Várkonyi, P.L.; Kocsis, M.; Ther, T. Rigid impacts of three-dimensional rocking structures. *Nonlinear Dyn.* **2022**, *107*, 1839–1858. [CrossRef]
28. Thomaidis, I.M.; Camara, A.; Kappos, A.J. Dynamics and seismic performance of asymmetric rocking bridges. *J. Eng. Mech.* **2022**, *148*, 04022003. [CrossRef]
29. Schau, H.; Johannes, M. Rocking and sliding of unanchored bodies subjected to seismic load according to conventional and nuclear rules. In Proceedings of the 4th International Conference on Computational Methods in Structural Dynamics and Earthquake Engineering, Kos Island, Greece, 12–14 June 2013.
30. Kalliontzis, D.; Sritharan, S.; Schultz, A. Improved coefficient of restitution estimation for free rocking members. *J. Struct. Eng.* **2016**, *142*, 06016002. [CrossRef]
31. Čeh, N.; Jelenić, G.; Bićanić, N. Analysis of restitution in rocking of single rigid blocks. *Acta Mech.* **2018**, *229*, 4623–4642. [CrossRef]
32. Anagnostopoulos, S.; Norman, J.; Mylonakis, G. Fractal-like overturning maps for stacked rocking blocks with numerical and experimental validation. *Soil Dyn. Earthq. Eng.* **2019**, *125*, 105659. [CrossRef]

buildings

MDPI

Article

Response Modification Factor of High-Strength Steel Frames with D-Eccentric Brace Using the IDA Method

Yan Ma *, Jun Yang and Xiaotong Ma

School of Civil Engineering, North Minzu University, Yinchuan 750021, China; 2016046@nmu.edu.cn (J.Y.); 2015039@nun.edu.cn (X.M.)
* Correspondence: 2016044@nmu.edu.cn

Abstract: The design innovation of high-strength steel frames paired with D-eccentric bracing exhibits remarkable resistance to plastic deformation during seismic events. This method strategically combines regular steel connections (with yield strengths below 345 MPa) and high-strength steel beams and columns (such as Q460 or Q690, with yield strengths over 460 MPa), effectively reducing cross-sectional sizes while preserving the elasticity of non-energy-dissipating members. This configuration results in substantial ductility and superior energy dissipation capabilities. The response modification factor (R) is vital for achieving both effective and economical seismic resilience, particularly in the development of efficient and cost-effective seismic designs. However, the 2016 edition of the Code for Seismic Design of Buildings (GB50011-2010) fails to incorporate the concept of R, opting instead to apply a uniform value to all structural systems. This oversight is fundamentally flawed, necessitating a comprehensive investigation into the R value specifically for the high-strength steel frame with a D-eccentric brace. This research primarily aims to improve structural performance design, provide guidance for future projects, and encourage the adoption of this advanced seismic performance structure in earthquake-prone areas. To achieve these objectives, a performance-based seismic design approach is employed. This method involves designing structures with varying numbers of stories (4, 8, and 12), different link lengths (900, 1000, and 1100 mm), and various steel strengths (Q460 and Q690). This study uses the Incremental Dynamic Analysis (IDA) method to determine the R values for each prototype. The derived performance coefficients act as crucial references for the development of future innovative structural designs. This research greatly enhances seismic design practices and facilitates the wider adoption of high-strength steel frames with D-eccentric braces due to their outstanding seismic performance.

Keywords: response modification factor; D-eccentric brace; high-strength steel; IDA method; performance-based seismic design; seismic performance

1. Introduction

Determining the structural response modification factor (R) is crucial for establishing prescribed seismic design actions for buildings. Notably, R values vary across different countries' codes and standards, enabling designers to employ a linear elastic force-based design methodology that accounts for nonlinear behavior and deformation constraints. Furthermore, the response modification factor is essential for developing equal-ductility inelastic response spectra within the context of performance-based seismic design.

In the seismic design approach adopted in China in 1978 [1], seismic design forces were determined by applying a seismic influence coefficient, which is the reciprocal of the response modification factor ($1/R$), to the elastic earthquake forces corresponding to the design seismic intensity. However, the Code for Seismic Design of Buildings (GBJ 11-89) [2], introduced in 1989, abandoned the concept of a response modification factor. Instead, seismic design forces were established directly using the elastic design response spectrum for minor earthquakes. This method determines seismic design forces through a singular structural response modification factor, aiming to reduce the elastic response at the design seismic intensity without considering the diverse inelastic energy dissipation capacities of different structures. To introduce performance-based seismic design, the General Rule for Performance-based Seismic Design of Buildings (CECS 160:2004) [3] provided values for structural seismic influence coefficients and displacement amplification factors for 25 structural systems. These values were derived from the Code for Seismic Design of Industrial and Civil Buildings (TJ11-78) [1] and data from buildings that experienced seismic ground motions. However, this approach lacks sufficient theoretical research and experimental data to substantiate its validity. Consequently, numerous researchers in China have reevaluated the current seismic design methodology. From the authors' perspective, quantifying and evaluating R would enhance the scientific, rational, and economical aspects of seismic design while facilitating the implementation of performance-based seismic design (PBSD) in China.

Seismic design of structures has traditionally relied on elastic analysis as the primary tool. However, recent earthquake events have revealed that elastic analysis alone is insufficient to capture the actual behavior of structures. Consequently, there has been a fundamental shift in seismic design provisions within design and material codes, such as ASCE, UBC, and NEHRP codes, following the Northridge earthquake in 1994 [4–6]. The prevalent approach, known as the equivalent static method, used in many seismic design codes, involves the utilization of the response modification factor (R), also referred to as the force reduction factor. Table 1 lists the response modification factor for different structural types in Chinese and American standards. Essentially, the design loads are determined by dividing the earthquake loads by the R factor, causing the structure to enter the inelastic range. Consequently, the structure must undergo significant inelastic deformations to dissipate the earthquake's energy. The ability of a structure to endure earthquake loads is directly linked to its capacity to deform within the inelastic range, commonly referred to as its ductility capacity. For systems displaying idealized bilinear behavior (as depicted in Figure 1), structural ductility, denoted as μ, is defined as the ratio of the maximum displacement to the displacement at the yielding point. Structures with higher values of the force reduction factor R require greater ductility capacity μ. Thus, the factors R and μ are interconnected and play a pivotal role in the energy dissipation mechanism of structures.

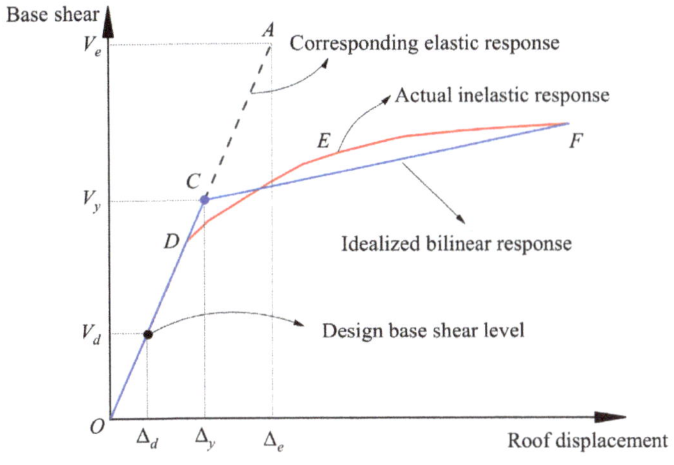

Figure 1. Performance curve of the structure.

Table 1. Seismic response modification factor in individual national specifications.

Canonical Name	Structure Type	Recommended Value
Code for Seismic Design of Industrial and Civil Buildings (TJ11-78)	Steel frame structure	4
General Rule for Performance-based Seismic Design of Buildings (CECS 160:2004)	Steel frame structure Steel eccentrically braced frame Steel concentrically braced frame	4 3.33 3.70
Code for Seismic Design of Buildings (GBJ 11-89)	Steel eccentrically braced frame	2.8125
Uniform Building Code 1997 of America (UBC97)	Steel eccentrically braced frame Steel ordinary braced frame	7.0 5.6
American Society of Civil Engineers (ASCE7-05)	Steel eccentrically braced frame (Beam column rigid connection)	8
	Steel eccentrically braced frame (Hinge joint of beam and column)	7
National Earthquake Hazard Reduction Program (NEHRP 2000)	Steel eccentrically braced frame	8

The response modification factor (R) and displacement amplification factor (C_d) can be determined using the structural response curve shown in Figure 1 (Uang, 1991) [7]. This figure shows the relationship between the roof displacements of the structure and the base shear. The roof displacements are plotted on the horizontal axis, while the base shear is represented on the vertical axis. The line OA illustrates the structure's performance curve under perfect elastic conditions. However, when the structure encounters earthquake excitation, its actual elastic response is depicted by the curve ODEF. The anticipated response in an ideal scenario is represented by the idealized performance curve labeled as ODCF. The actual nonlinear response of the structure can be approximated by a bilinear elastic–plastic relationship, transitioning from curve ODEF to ODCF in Figure 1. The initial linear response of the structure is represented by the bilinear line ODCF, with the area under this line corresponding to the area under the actual curve.

The response modification factor R is defined as follows:

$$R = \frac{V_e}{V_d} = \frac{V_e}{V_y} \times \frac{V_y}{V_d} = R_\mu \times R_\Omega \tag{1}$$

The ductility reduction factor, represented as R_μ, is essential for assessing the reduction in strength caused by inelastic behavior. The definition of R_μ involves the ratio of the maximum seismic demand for elastic response (V_e) to the base shear at the maximum inelastic displacement (V_y), which corresponds to the structure's yield force. In this context, another crucial parameter is denoted as R_Ω, the overstrength factor, which indicates the ratio of V_y to the design base shear (V_d). Following this, the displacement amplification factor (C_d) is calculated using the provided equation.

$$C_d = \frac{\Delta_{max}}{\Delta_d} \tag{2}$$

In seismic design considerations, two key parameters related to structural displacement stand out: Δ_{max}, which represents the maximum displacement during moderate or severe seismic ground motion, and Δ_d, the designated roof displacement. Δ_{max} indicates the largest displacement magnitude a structure may undergo during an earthquake, while Δ_d represents the intended displacement level that the structure is designed to withstand. Δ_{max} and Δ_d, these parameters, play a crucial role in the comprehensive evaluation of a structure's seismic performance. They are frequently combined with other metrics to assess structural integrity and safety under seismic loads.

Four Y-shaped eccentrically braced steel frames were designed using high-strength steel combinations for multiple high-rise buildings in compliance with China's seismic code. The response modification factor for these frames, determined through the Incremental Dynamic Analysis method, was found to be 4.889 [8]. A method to calculate the elastic stiffness and bearing capacity of K-type, D-type, and V-type eccentrically braced steel frames was proposed. Using the virtual work principle and the ideal failure mode of these frames, the yield and ultimate bearing capacities were determined based on frame drift angle and verified through experimental results [9]. Based on previous studies, pushover analysis and nonlinear dynamic analysis of 4-, 8-, 12-, and 16-story eccentrically braced high-strength steel frame structures with different bracing forms were conducted using performance-based seismic design methods. The results indicated that D-shaped eccentrically braced steel frames had the greatest stiffness and bearing capacity [10]. The seismic performance of three chevron-braced steel frames (CBSFs) of 3-, 6-, and 9-story Performance-based Plastic Design (PBPD) designs were compared, leading to the proposal of a PBPD framework for CBSFs considering mainshock–aftershock sequences [11]. Nonlinear time history analyses and performance-based seismic design were conducted on steel MRFs, EBFs, and BRBFs. A simple expression linking the peak seismic layer velocity to the peak ground velocity was provided, with its efficiency demonstrated through numerical examples [12]. Additionally, steel plate shear wall structures, consisting of 4, 8, and 12 stories, were developed following the Code for Seismic Design of Buildings (GB 50011-2010) [13]. The response modification factors for both single-frame planes and integral steel shear wall structures were obtained under uniform and inverted triangle loads using the pushover analytical method in Midas/Gen. A recommended response modification factor of 3.25 was proposed for steel plate shear wall structures with a maximum of 12 floors [14].

Furthermore, a high-strength steel fabricated framed-tube structure with replaceable shear links (HSS-SFT) was introduced. To investigate the response modification factor of HSS-SFTs, eight models with ideal yield mechanisms were established. Pushover analyses were conducted using the stepwise lateral force adjustment method. Based on the analysis results, a recommended response modification factor (R) of 3.65 is suggested for HSS-SFT designs under seismic action [15].

To assess the response modification factors of buckling restrained braced frames (BRBFs) used in the rehabilitation of steel frames, static nonlinear analyses were conducted on building models with single and double bracing bays, multi-floors, and various brace configurations including chevron V and inverted V. The response modification factors for different types of BRBFs with a single bracing bay ranged from 7 to 16, whereas for double bracing bays, the range extended from 8 to 22 [16].

Additionally, linear time history analyses were performed using multiple earthquake records on 4-, 8-, 12-, and 16-story steel frame buildings with different concentrically braced systems, namely X brace, V brace, and inverted V brace. The results showed an increase in the response modification factor with the number of stories for all systems except the inverted V-braced system [17].

Furthermore, the research group conducted studies on the response modification factors of eccentrically braced structures. A K-type eccentrically braced structure with varying numbers of stories (5, 10, 15, and 20 stories) and link lengths (900, 1000, 1100, and 1200 mm) was designed using the performance-based seismic design method. Static nonlinear pushover analysis and Incremental Dynamic Analysis (IDA) were carried out. The average R value, obtained through both methods of analysis, was determined to be 7.194 for the 20-story structure. Additionally, a linear relationship between the response modification factor (R) and the first period (T_1) of structures with fewer than 20 stories is established as $R = 4.275T_1 + 4.1274$ [18].

In summary, our research delivers key insights into the seismic performance and response modification factors of various structural systems, greatly aiding in the seismic design and rehabilitation of high-rise buildings. Eccentrically braced steel frames (EBFs) integrate the advantages of both moment-resisting and concentrically braced steel frames, providing superior ductility and energy dissipation. Illustrated in Figure 2 are four common EBF configurations. One key characteristic of these frames is a component known as a "fuse", which is intended to experience extensive plastic deformation, keeping the remaining structural elements elastic. Utilizing high-strength steel in this design enables the use of smaller beam and column cross-sections due to its enhanced elastic strength. Our earlier studies concentrated on identifying the response modification factors for high-strength steel frames equipped with Y-type and K-type braces, employing both Incremental Dynamic Analysis (IDA) and pushover analysis methods. Our research team thoroughly examined the parameters R and C_d for high-strength steel frames with D-type braces using an enhanced pushover analysis method. In this current study, we utilized the IDA method to determine the response modification factor specifically for high-strength steel frames with D-type braces (D-HSS-EBFs).

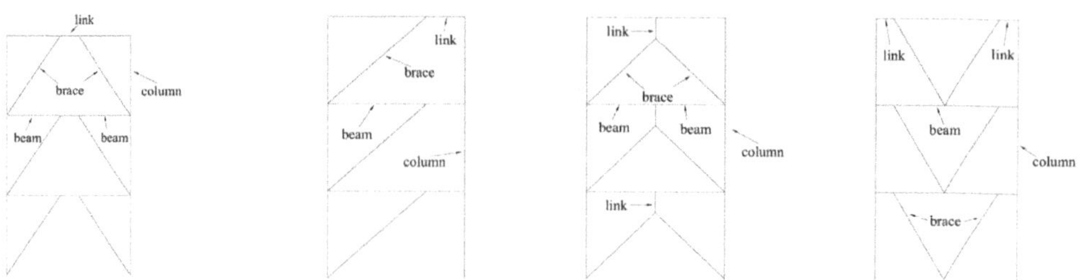

(a) K-type (b) D-type (c) Y-type (d) V-type

Figure 2. Types of eccentrically braced frames.

2. Structural Design and Finite Element Model

2.1. The Design Principle

According to the Code for Seismic Design of Buildings, Figure 3 illustrates the plan layout of the D-HSS-EBF. Designed to endure a seismic acceleration of 0.30 g, this prototype structure corresponds to a peak ground acceleration of 0.3 g, with a 10% chance of exceedance over 50 years. The building stands 3 m tall with columns 6 m apart. It features a plane dimension of 18 m × 30 m, comprising 3 spans along the Y direction and 5 spans along the X direction.

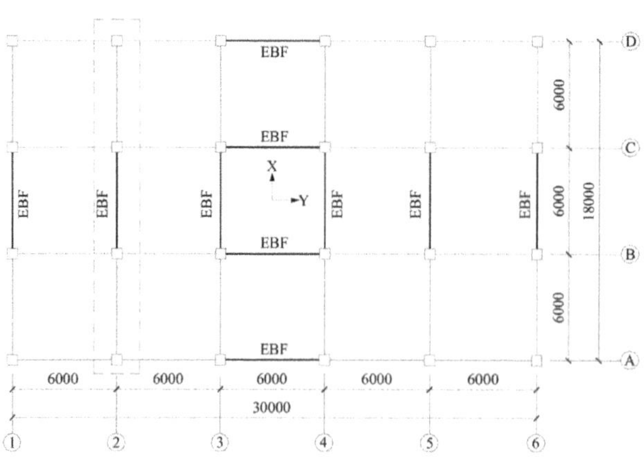

Figure 3. D-8-0.9 structural plan layout.

The frame column utilizes a square steel tube, while welded H-shaped sections form the other components. Constructed from 120 mm thick cast-in-place C30 concrete, the floor is designed to handle several load considerations: a floor load of 5.0 kN/m^2, a live load of 2.0 kN/m^2, a roof load of 5.5 kN/m^2, a roof live load of 2.0 kN/m^2, and a basic wind pressure of 0.35 kN/m^2.

A schematic diagram of the frame structure after calculation is presented in Figure 4. As depicted in Figure 5, the steel stress–strain curve in the SAP2000 model remains at its default value [19].

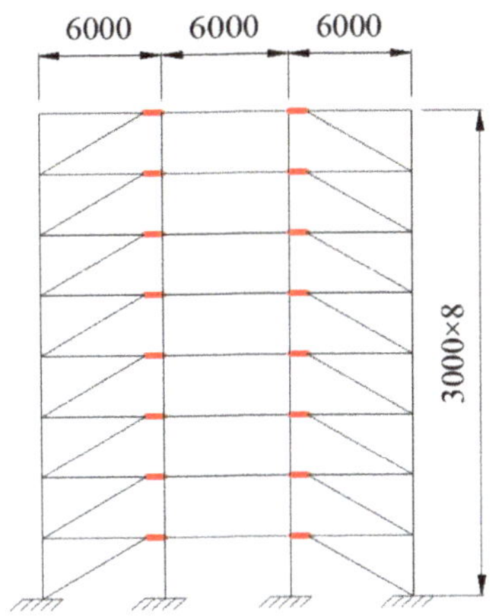

Figure 4. Computing frame element.

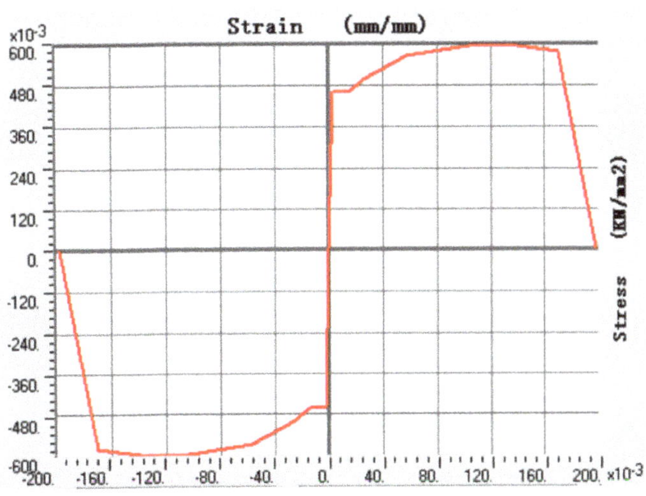

Figure 5. Stress–strain curves of steel.

A standardized naming system is employed to classify prototype structures, indicating steel materials and their properties. For instance, the designation "D1-8-0.9" represents an 8-story D-type eccentrically braced frame incorporating Q345 and Q460 steel, with a link length of 0.9 m. Using the same pattern, the structure combining Q345 and Q690 steel follows suit, yet the prefix "D2" denotes a distinct steel blend. Hence, the prototype structure, composed of Q345 and Q690 steel, comprises 8 layers and boasts a link length of 0.9 m, denoted as "D2-8-0.9".

In accordance with the basic principles of performance-based seismic design (PBDS), as outlined in the academic work of Li and Tian [20], the design approach for D-HSS-EBFs is established. A detailed depiction of the design process is provided in the design flow diagram, illustrated in Figure 6. As specified in the Load Code for the Design of Building Structures (GB50009-2012) [21], the fundamental period of high-rise steel structures is $T = 0.1N$, with N indicating the number of stories. In the case of an eight-story D-HSS-EBF, this period calculates to 0.8 s. Table 2 provides a detailed summary of the PBSD design parameters relevant to D-HSS-EBF.

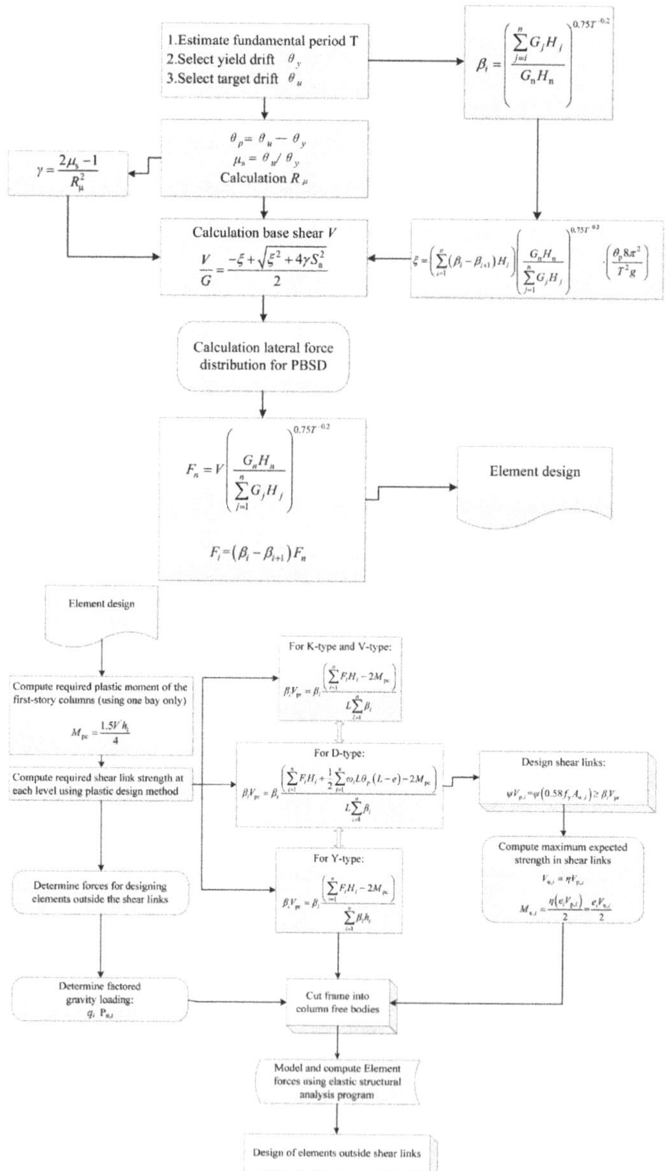

θ_y	yield drift	A_c	area of the column cross section
θ_u	ultimate drift	ε	average strain in columns
θ_{ys}	shear drift	H	structure height
θ_{yf}	bending drift	F	lateral force
h	storey height	Δ_{vert}	vertical axial deformation
a	beam length outside link	Δ_{horiz}	horizontal deformation
e	link length	H_j	the height of level j from the base
L_b	the length of brace	G_n	the weight at the top level
L	the span of the EBFs	H_n	the height of roof level from base
θ	the angle of the brace	T	the fundamental period
ε_{yb}	the strain of brace	F_i	the lateral force at level i
f_{yb}	yield strength of the braces	V	the total design base shear
E	modulus of elasticity	γ	the energy modification factor
σ	average axial stress in columns owing to overturning moment	μ_s	the structural ductility factor
M	overturning moment caused by the lateral loads	R_μ	the ductility reduction factor
f_{yc}	yield strength of the columns	S_v	the design spectral pseudo velocity
γ_R	material partial coefficients	E_e	the elastic components of the energy
θ_p	plastic drift	E_p	the plastic components of the energy
γ_p	plastic rotation of link	ψ	strength reduction factor
β_i	the shear distribution factor at level i	f_y	the yielding strength of links
V_i	the story shear forces at level i	$A_{w,i}$	the web area of ith story link
V_n	the story shear forces at the top (nth) level	V'	one-bay base shear
G_j	the seismic weight at level j	h_1	the bottom story height
ξ	a dimensionless parameter	V_u	ultimate shear capacity of link
ω_i	the vertical force in ith story beams	λ_i	the lateral force distribution coefficient
M_{pc}	the plastic moment of column base	$M_{pb,i}$	the plastic moment of beams with no braces
V_{pr}	plastic shear strength of top story link	$V_{pb,i}$	the corresponding shear force with $M_{pb,i}$
$\beta_i V_{pr}$	the plastic shear strength of ith story link.	P_{brace}	the axial force of brace
I	moment of inertia of the EBFs		

Figure 6. Flow chart of PBSD method.

Table 2. The relationship of ductility reduction factor and period.

Design Parameters	Rare Earthquake
Seismic influence coefficient α (m/s^2)	0.643
S_a (m/s^2)	1.537
Fundamental period T/s	0.8 s
Yield drift θ_y	0.5156%
Target drift θ_u	1.7156%
Plastic drift θ_p	1.2%
Ductility μ_s	3.33
Ductility reduction factor R_μ	3.33
Energy correction factor γ	0.51
V/G	0.39
Design base shear V/kN	2861.43

2.2. Shear Link Model

The performance of eccentrically braced structures during earthquakes relies primarily on the shear yield behavior of the links, which is essential for both energy dissipation and maintaining structural integrity. The specimen and finite element model are illustrated in Figure 7. To accurately describe this linkage behavior, the analytical framework incorporates the ShearV2 plastic hinge model introduced by Li and Gao [22]. The SAP2000 software was used to determine the force–displacement correction parameters for the model, with the shear plastic hinge reflecting the force–deformation relationship as depicted in Figure 8. To accurately capture the nonlinear behavior of the shear link, the analysis incorporated empirical data from Tables 5 and 6 of FEMA-356 [22]. The link's ultimate shear force (V_u) was set at 1.5 times the yield shear force (V_p). Deformation standards for immediate occupancy (Δ_{IO}), life safety (Δ_{LS}), and collapse prevention (Δ_{CP}) were evaluated based on the parameters in Tables 5 and 6 of FEMA-356. Figure 8 illustrates this analysis.

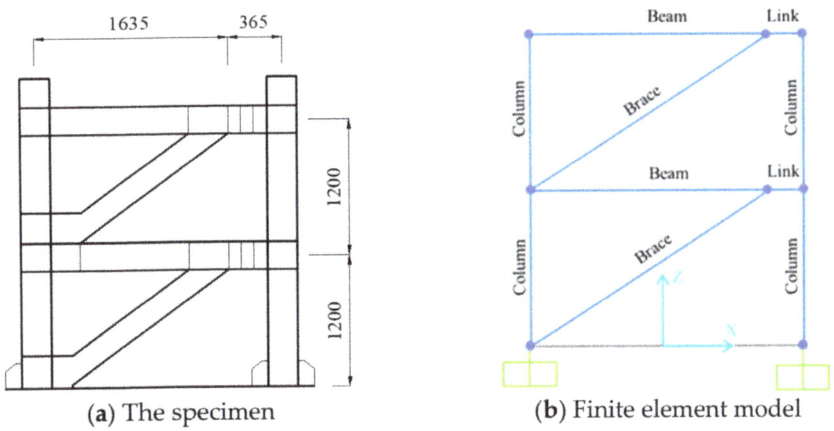

Figure 7. Comparison diagram of test specimen and finite element model.

Ref. Richards [23] drew from the experimental studies by Popov and Ramadan [24,25] to propose a shear–deformation relationship for the shear link, overlooking the intricate interplay between bending moment and shear force. The pertinent test results are presented in Figure 8b.

A spring model, integrating four nodes, was devised to simulate the diverse behaviors of the shear link, shown in Figure 9. This model ensures elasticity is maintained under various loading conditions, preventing yielding at its ends. The two external nodes mirror the arrangement of the two internal nodes, ensuring a distance of zero between each external node and its corresponding internal node. At the link's end, the critical moment hinge emerges, while the shear hinge extends across its web, constrained to primary and secondary nodes. The length of this pivotal structural unit, denoted "e", precisely matches that of the shear link. By confining the inelastic response to the ends of the link, this arrangement ensures that the rest of the link stays elastic, thereby maintaining structural stability.

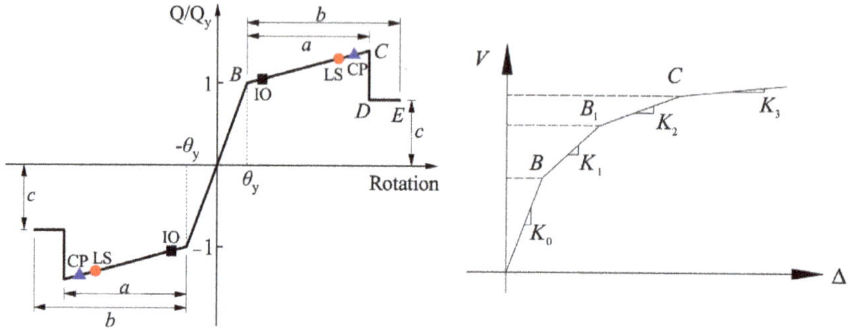

(a) Cyclic behavior (b) Shear-deformation relationship

Figure 8. Generalized force–deformation relation for shear link (FEMA-356).

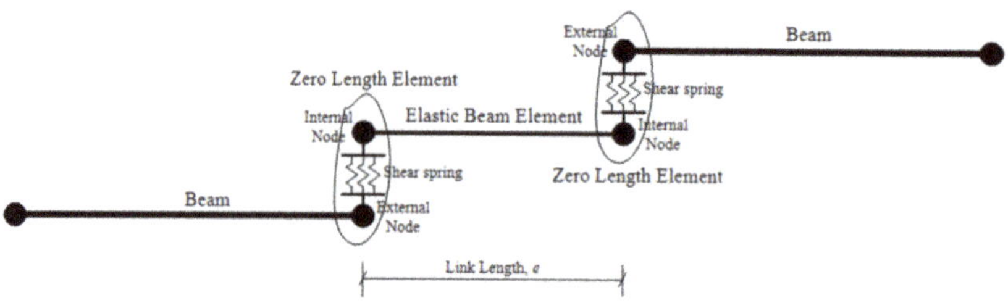

Figure 9. Shear link springs model.

An amplification factor λ is introduced to address the enhancement effect of flange shear on the shear resistance of the plastic link. This factor seamlessly integrates into the shear control formula, resulting in a refined expression, as follows:

$$V'_{\mathrm{P}} = \lambda V_{\mathrm{P}} \tag{3}$$

$$\lambda = 1 + \frac{A_{\mathrm{f}}^2 - 0.0625(et_{\mathrm{w}})^2}{0.58A_{\mathrm{w}}b_{\mathrm{f}}e} \tag{4}$$

Equations (3) and (4) distinctly establish the fundamental parameters needed for assessing the shear of a plastic link. The width of the flange is represented by b_f in these equations, while the cross-sectional area of an individual flange is denoted by A_f. The connection length is signified by e, whereas the thickness and area of the web are, respectively, indicated by t_w and A_w. The noticeable aspect is that within the SAP2000 software, the plastic hinges display a clear lack of elasticity, confining all plastic deformation exclusively to the hinge region. Meanwhile, the remaining structure consistently demonstrates its inherent elastic response.

Point B in Figure 8 vividly illustrates the yield of the hinge, showing a critical bearing capacity ($V_B = 1.1\alpha V_P$). Subsequently, at point C, the ultimate bearing capacity of the hinge is marked, with a gradual decrease in capacity beyond point C ($V_C = 1.5\alpha V_P$), and the associated angle $\theta_C = 6.12\alpha f_y/G$. Point D signifies the residual strength of the hinge, set at 0.6 for beams with a flange height-to-thickness ratio $h_0/t_w \leq 418/{f_y}^{1/2}$, and at 0.2 for those with $h_0/t_w \geq 640/f_y{}^{1/2}$. For other flange ratios, the values are linearly interpolated between 0.2 and 0.6. The angle θ_D matches θ_C. Point E marks the hinge's total failure. The specific parameters related to these points are listed in Table 3.

Table 3. Parameter for shear–displacement relation of link.

V_B	V_{B1}	V_C	K_0	K_1	K_2	K_3
$1.1V_P$	$1.3V_P$	$1.5V_P$	$2GA_w/e$	$0.03K_0$	$0.015K_0$	$0.002K_0$

In this research, each link is designed to undergo shear yield across the entire cross-sectional area simultaneously. For each link, three shear hinges are strategically positioned at the two ends and the center of the link.

2.3. Plastic Hinge of Non-Yielding Member

The model for plastic hinges in frame beams and columns incorporates the default M3 and P-M-M plastic hinges, positioned at both ends of the structural elements. Specifying these plastic hinge models allows for determining the plastic moment bearing capacity of the beams and columns.

$$Q_{CE} = M_{CE} = ZF_{ye} \tag{5}$$

$$Q_{CE} = M_{CE} = 1.18ZF_{ye}\left(1 - \frac{p}{p_{ye}}\right) \leq ZF_{ye} \tag{6}$$

D-HSS-EBFs use M_{CE} for design bending moment, with Z representing section modulus and F_{ye} indicating steel yield strength. Columns bear axial force P and have compressive capacity P_{ye}. The midpoint support's axial force is termed P hinge, and SAP2000 software automatically assigns plastic hinge characteristics.

3. IDA Method

3.1. The IDA Analysis Procedure

The primary distinction between the Incremental Dynamic Analysis (IDA) method and pushover analysis in the computation of structural performance coefficients lies in the derivation of the structural performance curve. The steps involved in acquiring structural performance coefficients through IDA analysis are outlined below:

(1) Selection of Ground Motion Records: Initiate the process by judiciously selecting a set of ground motion records, typically ranging from 10 to 20 records.

(2) Scaling of Ground Motion Records: Proceed by amplifying the chosen ground motion records into multiple sets. This amplification involves applying appropriate scaling factors to each record, yielding a series of ground motion sets, with each set corresponding to distinct maximum acceleration levels, as specified in Table 4.

Table 4. Maximum acceleration value for time history analysis under four-level ground motions.

Earthquake Hazard Level	6 Degrees	7 Degrees	8 Degrees	9 Degrees
Frequent earthquakes	18	35(55)	70(110)	140
Fortification earthquakes	49	98(147)	196(294)	392
Rare earthquake	125	220(310)	400(510)	620
Extremely rare earthquake	147	294(441)	588(882)	1176

(3) Time History Analysis: Conduct time history analyses of the structure employing the various sets of ground motion records. During this phase, capture and record pertinent data points characterizing the base shear and roof displacement responses for each ground motion record. Subsequently, we employed Origin (2021) software to conduct a data fitting process, resulting in the derivation of the structural performance curve.

(4) Target Displacement Determination: Employ the spectral method to ascertain the target displacement point within the structural system.

(5) Performance Coefficient Computation: Utilize the formulas delineated in Formulas (7)–(10) to compute a spectrum of performance coefficients for the structure, reflecting its behavior under different seismic hazard levels.

$$R = \frac{V_e}{V_d} = \frac{K_e \Delta}{V_d} \tag{7}$$

$$C_d = \frac{\Delta}{\Delta_d} \tag{8}$$

$$R_\mu = \frac{V_e}{V_y} \tag{9}$$

$$R_\Omega = \frac{V_y}{V_d} \tag{10}$$

In summary, the IDA method for determining structural performance coefficients involves a systematic procedure encompassing ground motion record selection, record scaling, time history analysis, structural performance curve derivation, target displacement determination, and subsequent computation of performance coefficients. This approach provides a comprehensive insight into the structure's behavior across varying seismic intensities.

3.2. The Calculation of Factors by IDA

The specific steps for obtaining response modification factors through the capacity spectrum method, as shown in Figure 10 within the IDA framework are elucidated as follows:

- Nonlinear Dynamic Analysis: Perform nonlinear dynamic analyses under various ground motions, resulting in envelope curves for roof displacement and base shear force. Similar curves are obtained for other ground motion groups.
- Curve Fitting: Fit all the acquired Δ-V envelope curves to generate the IDA response curve, which bears resemblance to the pushover curve.
- Yield Point Determination: Determine the yield shear force (V_y) and yield displacement (Δ_y) by introducing bilinear characteristics into the fitted IDA response curve.
- Spectrum Transformation: Convert the fitted IDA response curve into the spectrum acceleration–displacement curve ($S_{ai} - S_{di}$).
- Capacity Point Determination: Plot the response curve obtained in step 4 alongside the plastic–elastic demand spectrum for different ground motions within the same coordinate system. Iteratively identify the capacity points for different ground motions.
- Factor Calculation: Utilize the provided Formulas (7) and (10) to calculate the response modification factor (R) and displacement amplification factor (C_d) for the structure.

In summary, the IDA methodology involves a comprehensive sequence of steps encompassing structural modeling, ground motion selection and amplification, dynamic analysis, curve fitting, and subsequent factor computation. This rigorous approach offers valuable insights into the structural behavior under varying seismic conditions, facilitating the assessment of performance and safety.

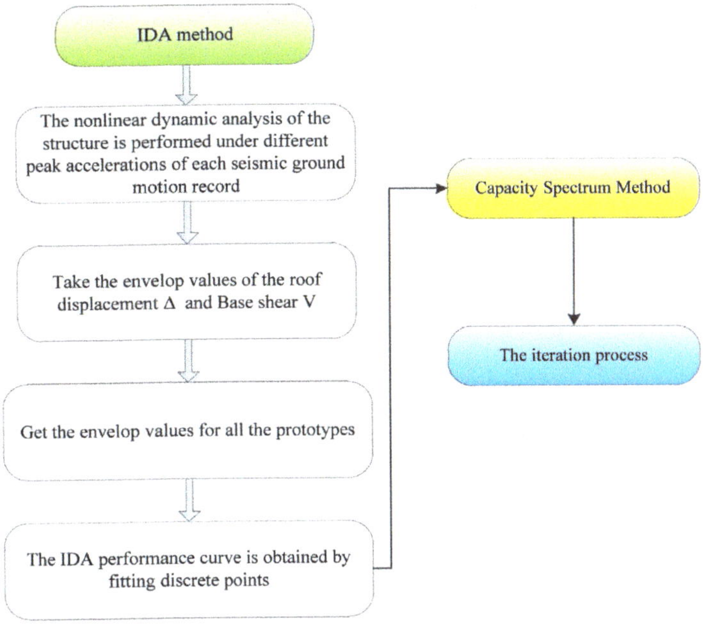

Figure 10. The IDA method.

4. Calculation Results for Each Case

4.1. IDA Method Process

The results of the Incremental Dynamic Analysis (IDA) for a single ground motion record, specifically the RSN184 seismic record, are outlined below. This analysis involves conducting elastic–plastic time history analyses while adjusting the peak ground acceleration of the seismic motion. The base shear and roof displacement values for each prototype structure are meticulously calculated and presented in Table 5. Additionally, the IDA analysis produces envelope points for each load case and presents pushover curves using both inverted triangle and uniform loading patterns, as depicted in Figure 11. These results provide a comprehensive understanding of the structural response under varying seismic conditions, enabling a robust assessment of structural performance and behavior.

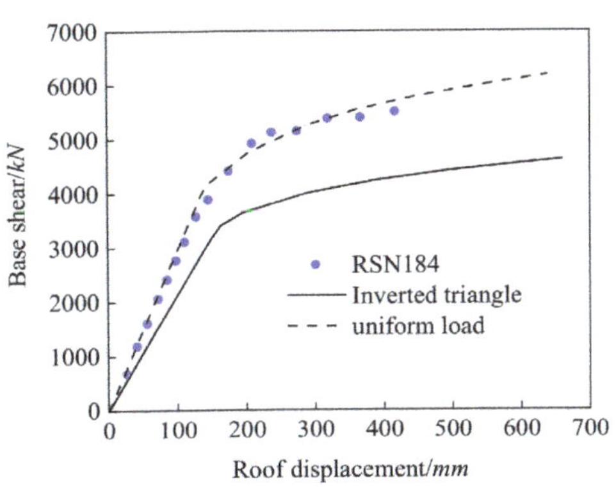

Figure 11. IDA analysis results of example D1-12-0.9 under seismic wave RSN184.

Table 5. The results of model D1-12-0.9 under ground motion RSN184.

Number	Acceleration (cm/s^2)	Roof Displacement Δ (mm)	Base Shear *V* (kN)
1	110 (Frequency earthquake)	28.503	684.572
2	170	43.528	1193.839
3	230	58.553	1615.191
4	294 (Moderate earthquake)	74.582	2064.633
5	344	87.102	2415.764
6	394	99.623	2766.893
7	444	112.144	3118.025
8	510 (High earthquake)	128.672	3581.38
9	580	146.322	3892.323
10	690	175.751	4420.492
11	800	209.99	4931.886
12	882 (Extreme earthquake)	238.591	5129.263
13	980	275.121	5155.457
14	1090	320.271	5387.287
15	1200	367.952	5408.085
16	1310	417.785	5509.309
17	1420	270.218	5235.821

Upon examination of Table 5, it becomes evident that the enveloped values of base shear and roof displacement for the structure exhibit a gradual increase with the progressive augmentation of acceleration amplitude. However, it is noteworthy that an intriguing observation arises when the peak acceleration is elevated from 1310 cm/s^2 to 1420 cm/s^2: both the base shear and roof displacement experience a decrease. This phenomenon can be primarily attributed to a critical structural condition. Specifically, when the peak acceleration is adjusted to 1420 cm/s^2, the inter-story displacement of the fifth floor attains a significant value of 65.589 mm, corresponding to an inter-story drift angle of 2.19%. Importantly, this drift angle exceeds the specified limit of 2% for elastic–plastic inter-story drift angle. Consequently, the structure reaches its ultimate state, and the analysis can be rightfully terminated at this point. This observation underscores the importance of adherence to specified performance criteria in assessing the structural response under varying seismic conditions.

From Figure 11, it can be observed that the general trend of the results obtained through the Incremental Dynamic Analysis (IDA) for the single ground motion record in model D1-12-0.9 aligns consistently with the observed variations in the pushover analysis curve.

The IDA analysis results for model D1-12-0.9 under the influence of 10 seismic ground motions are graphically depicted in Figure 12. An examination of this graph reveals a notable trend: most data points closely align with the uniformly distributed loading pushover curve. This suggests that for model D1-12-0.9, the performance curve derived through the uniformly distributed loading mode in the pushover analysis more accurately represents the actual structural response during seismic events. This finding underscores the importance of selecting an appropriate loading mode when conducting structural assessments in seismic scenarios.

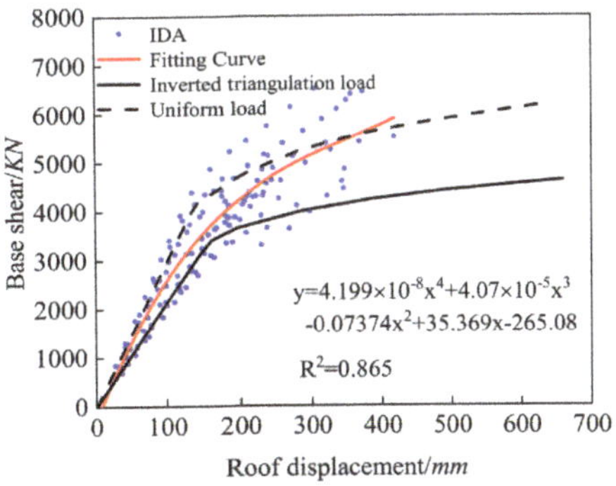

Figure 12. IDA fitting curve of D1-12-0.9.

4.2. Calculation of Structural Performance Factors

(1) Determination of Significant Yielding Point

Utilizing the nonlinear least squares method, a polynomial fitting procedure is applied to the IDA analysis data presented in Figure 13. The resultant performance curve equation for case D1-12-0.9 is expressed as follows:

$$y = 4.199 \times 10^{-8}x^4 + 4.07 \times 10^{-5}x^3 - 0.07374x^2 + 35.369x - 265.08 \quad (0 \le x \le 417.78) \tag{11}$$

Here, 'x' denotes the roof displacement of the structure, while 'y' represents the base shear of the structure. By substituting the maximum roof displacement value, Δ_m = 417.78 mm, into the equation, the maximum yield shear force, V_m, is computed as 5893.6 kN. Additionally, the structural elastic stiffness, K_e, is determined as the average value obtained from the time history analysis conducted under various seismic ground motions.

$$K_e = \frac{829 + 802 + 684 + 1188 + 844 + 1013 + 1229 + 689 + 880 + 779}{29 + 37 + 28 + 47 + 26 + 42 + 39 + 36 + 44 + 36} = 24.552\ (\text{kN/mm}) \tag{12}$$

Using Matlab (2018a) software, the integral of the equation is solved to determine the area enclosed by the performance curve and the horizontal axis. Subsequently, based on the formulas provided in this chapter, the significant yielding point for the given case is calculated.

$$V_y = \frac{2AK_e - V_m\Delta_m K_e}{\Delta_m K_e - V_m} = 4238.76\text{kN} \tag{13}$$

$$\Delta_y = \frac{V_y}{K_e} = 174.96\text{mm} \tag{14}$$

It is noteworthy that there are discrepancies between the yielding points obtained from pushover analysis for the same case. V_y and Δ_y were determined as 3784.9 kN and 191.73 mm, respectively, using the pushover analysis method. The primary reason for this disparity can be attributed to the differences in the initial stiffness and yield shear force values derived from the IDA analysis of the structural performance curve, which are relatively larger than those obtained from the pushover analysis. Furthermore, it is essential to acknowledge that the IDA fitting results exhibit a certain degree of dispersion, with an R-squared value (R^2) of 0.865.

(2) Determination of Structural Target Displacement

Using the formulas outlined in this study ((11) and (12)), the fitted performance curve is transformed into the capacity spectrum for the first three modes. Additionally, a demand spectrum curve is established to represent seismic excitations acting upon the structure. Through a series of iterative calculations, a ductility demand of μ = 1.945 and a structural target displacement of Δ = 340.5 mm are determined. The intersections between the capacity spectrum and the demand spectrum for each mode are visually depicted in Figure 14. These intersections provide critical insights into the structural response under seismic loading conditions and inform the assessment of structural performance and safety.

$$S_{ai} = \frac{V/G}{\alpha_i} \tag{15}$$

$$S_{di} = \frac{\Delta}{\gamma_i X_{i,roof}} \tag{16}$$

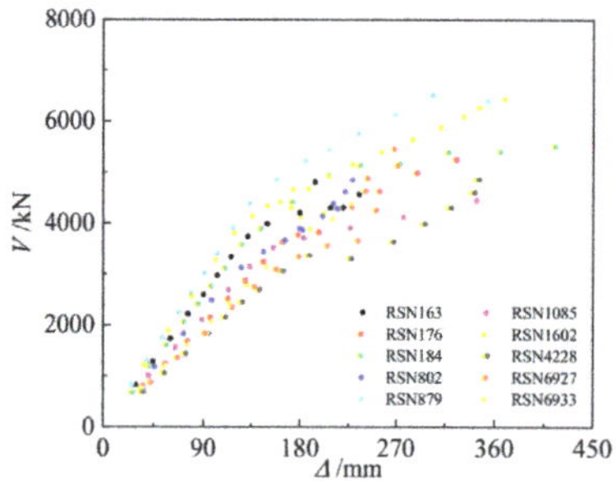

Figure 13. Example D1-12-0.9 of IDA calculation data.

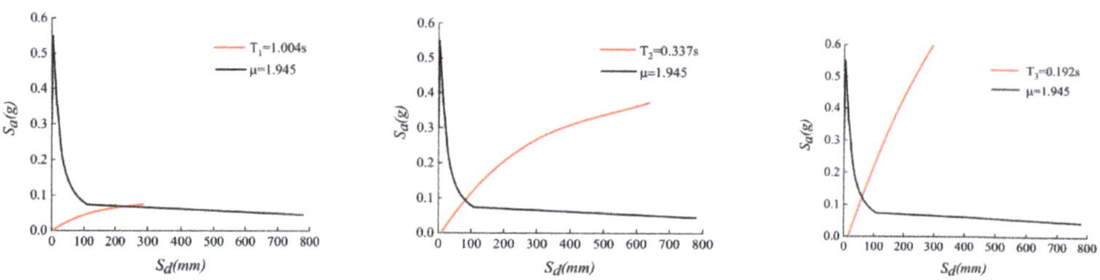

Figure 14. Performance point of model D1-12-0.9 under design earthquake.

(3) Calculation of Performance Factors for Case D1-12-0.9.

In the case of minor seismic excitations, specific structural parameters are determined as follows: the elastic design point (V_d) is computed as 927.83 kN, and the associated displacement (Δ_d) is found to be 37.01 mm. Furthermore, the initial stiffness of the structure (K_e) is established at 24.552 kN/mm, while the yield shear force (V_y) is calculated to be 4238.76 kN. Additionally, the displacement demand (Δ) is determined to be 340.5 mm.

By applying the formulas denoted as (7) to (10), the ensuing performance factors for case D1-12-0.9 are meticulously computed and derived.

$$R = \frac{V_e}{V_d} = \frac{K_e\Delta}{V_d} = \frac{24.552 \times 340.5}{927.83} = 9.010 \tag{17}$$

$$C_d = \frac{\Delta}{\Delta_d} = \frac{340.5}{37.01} = 9.20 \tag{18}$$

$$R_\mu = \frac{V_e}{V_y} = \frac{24.552 \times 340.5}{927.83} = 1.972 \tag{19}$$

$$R_\Omega = \frac{V_y}{V_d} = \frac{4238.76}{927.83} = 4.568 \tag{20}$$

4.3. Summary of Calculation Results

4.3.1. IDA Fitting Curves for Each Case

In accordance with the IDA analysis process outlined in the preceding section, the fitting performance curves for all cases subjected to multiple seismic waves are presented in Figures 15–20, and information on the seismic records are as Appendix A. These figures also incorporate the pushover curves for both uniformly distributed and inverted triangular loading modes. Notably, for the majority of cases, the initial stiffness of the structure, as determined from the IDA fitting curves, falls within the range defined by the inverted triangular loading and uniformly distributed loading modes. Furthermore, the structural capacity derived from the IDA fitting curves shows closer alignment with the results obtained under uniformly distributed loading conditions. This observation highlights the importance of employing the IDA approach to comprehensively assess structural behavior under diverse seismic scenarios, facilitating a more accurate representation of structural capacity.

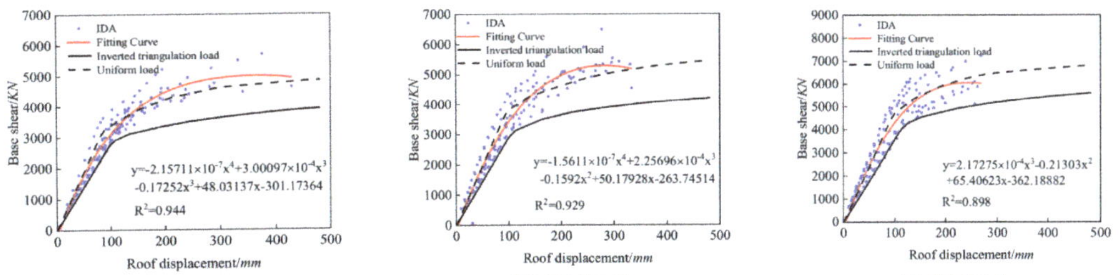

(a) D1-8-0.9 (b) D1-8-1.0 (c) D1-8-1.1

Figure 15. Structure–performance curve of the 8 stories (D1).

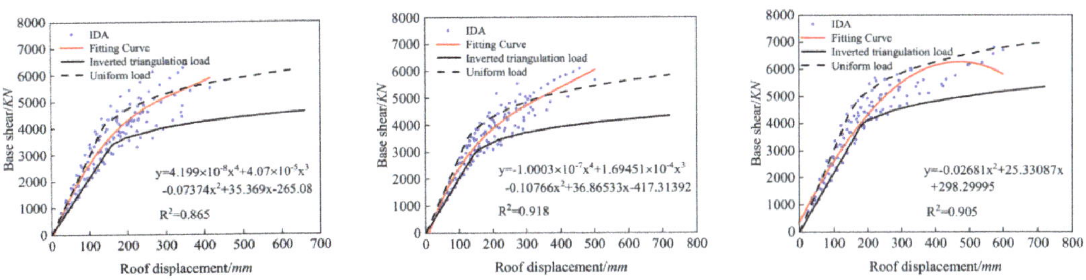

(a) D1-12-0.9 (b) D1-12-1.0 (c) D1-12-1.1

Figure 16. Structure–performance curve of the 12 stories (D1).

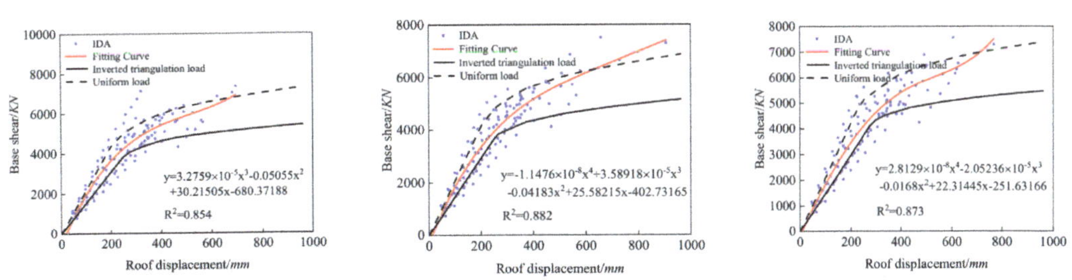

(a) D1-16-0.9 (b) D1-16-1.0 (c) D1-16-1.1

Figure 17. Structure–performance curve of the 16 stories (D1).

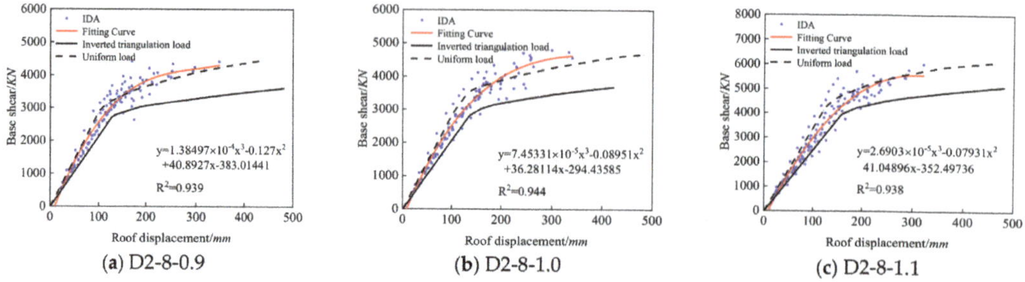

Figure 18. Structure–performance curve of the 8 stories (D2).

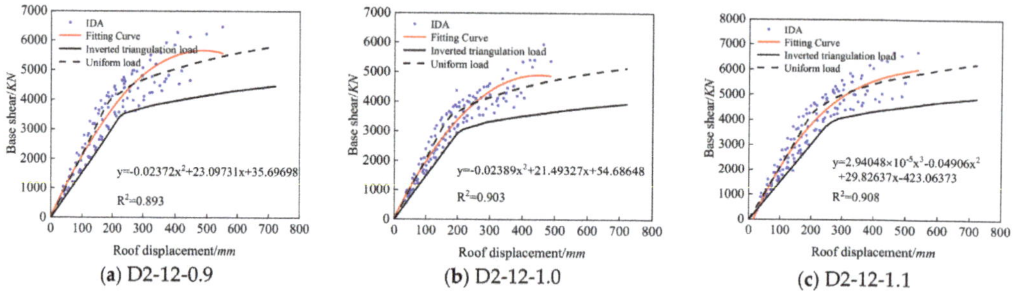

Figure 19. Structure–performance curve of the 12 stories (D2).

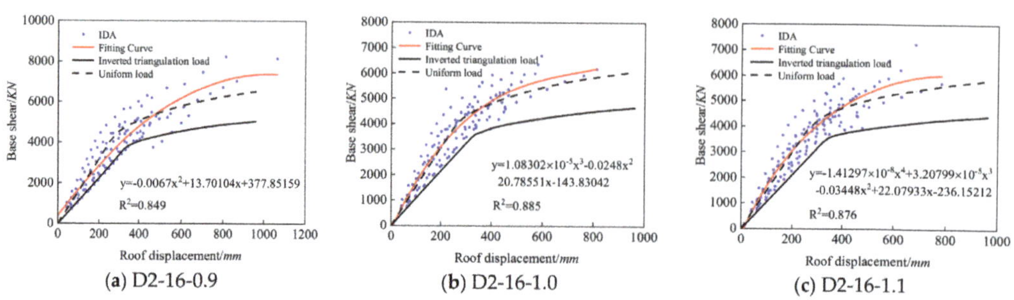

Figure 20. Structure–performance curve of the 16 stories (D2).

4.3.2. Initial Stiffness and Significant Yielding Point for Each Case

Derived from the IDA analysis, coupled with the obtained performance curves for each case, the initial stiffness and significant yielding point of the structures are meticulously determined. A comprehensive summary of these specific results is provided in Table 6.

Upon examining the table, a clear trend emerges: as the number of structural stories increases, the initial stiffness (K_e) of the structure gradually decreases, while the ultimate bearing capacity (V_m) shows an incremental rise. Additionally, it is noteworthy that for cases with the same number of stories, the Q690 composite case demonstrates lower initial stiffness and ultimate bearing capacity compared to the Q460 composite case. This observation highlights the significant influence of structural configuration and material properties on the dynamic response and ultimate performance of the analyzed structures.

Table 6. The initial stiffness and significant yield points.

Model	Δ_m (mm)	V_m (kN)	K_e (kN/mm)	A (kN·mm)	V_y (kN)	Δ_y (mm)
D1-8-0.9	430.667	4936.789	31.4	1,672,809	4459.219	141.981
D1-8-1.0	335.947	5195.302	34.4	1,316,090	4786.129	138.789
D1-8-1.1	270.211	6043.783	42.8	1,178,524	5608.73	130.975
D1-12-0.9	417.785	5893.627	24.2	1,601,000	4238.761	174.96
D1-12-1.0	505.353	6065.311	22.4	1,968,500	3703.977	164.862
D1-12-1.1	603.736	5819.289	24.0	2,830,006	5937.389	247.095
D1-16-0.9	697.921	6921.277	16.4	3,098,830	4932.992	299.898
D1-16-1.0	909.117	7410.585	16.1	4,432,884	4738.037	294.005
D1-16-1.1	772.672	7518.736	16.5	3,603,942	4398.911	266.075
D2-8-0.9	346.949	4301.339	25.6	1,062,025	3530.434	137.904
D2-8-1.0	339.558	4622.684	24.7	1,071,201	3747.746	151.393
D2-8-1.1	320.256	5543.01	28.6	1,194,575	4834.807	168.574
D2-12-0.9	549.343	5566.875	20.3	2,194,523	4834.114	237.954
D2-12-1.0	485.217	4859.031	19.2	1,646,965	4011.697	207.921
D2-12-1.1	536.004	5997.204	20.0	2,146,260	4557.528	227.582
D2-16-0.9	1062.703	7370.32	14.1	5,457,170	5673.852	399.947
D2-16-1.0	811.687	6180.02	15.3	3,484,873	4766.574	309.944
D2-16-1.1	781.336	6001.567	14.8	3,238,867	4754.159	320.933

4.3.3. Seismic Performance Points for Each Case

Utilizing the capacity spectrum method, capacity spectrum curves for the first three modes of each case are systematically plotted on the same coordinate system alongside the demand spectrum curve representative of moderate seismic excitations. Through an iterative computational process, the ductility demand (μ) and the seismic performance point for moderate seismic excitations are determined for each individual structure. These specific results are meticulously illustrated in Figures 21–26.

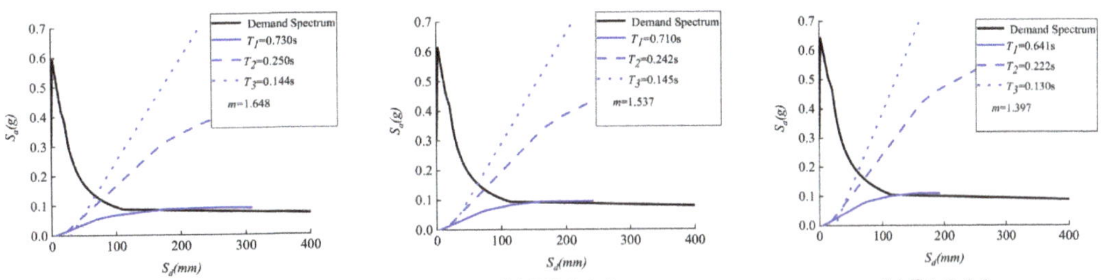

(a) D1-8-0.9 (b) D1-8-1.0 (c) D1-8-1.1

Figure 21. Seismic performance points of the 8 stories (D1).

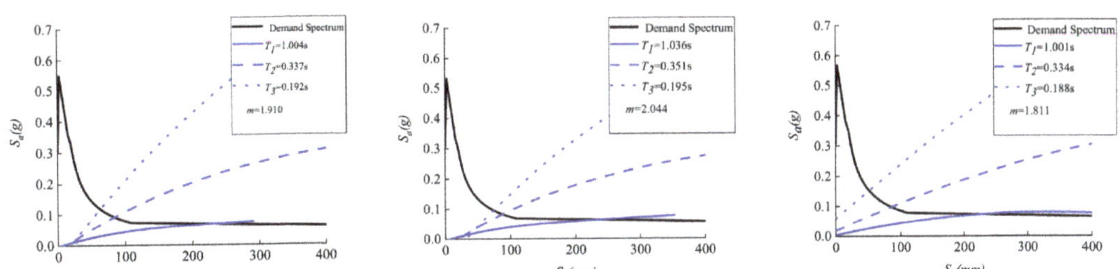

(a) D1-12-0.9 (b) D1-12-1.0 (c) D1-12-1.1

Figure 22. Seismic performance points of the 12 stories (D1).

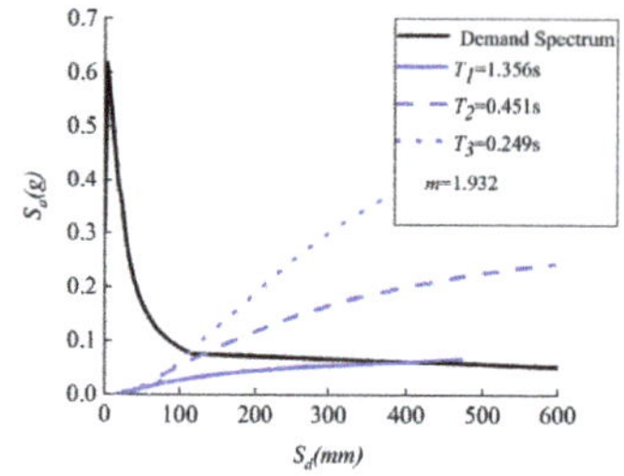

(**a**) D1-16-0.9

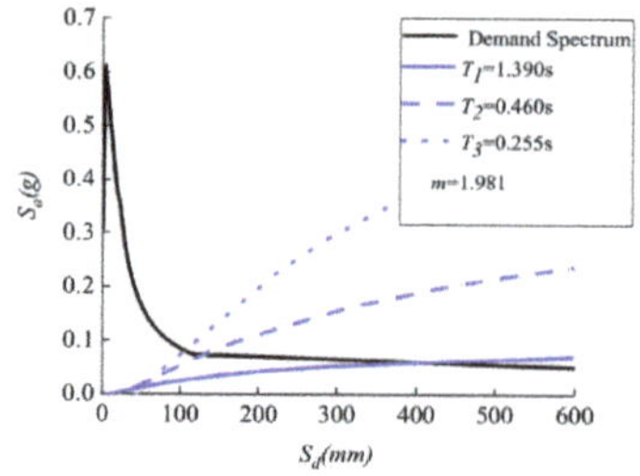

(**b**) D1-16-1.0

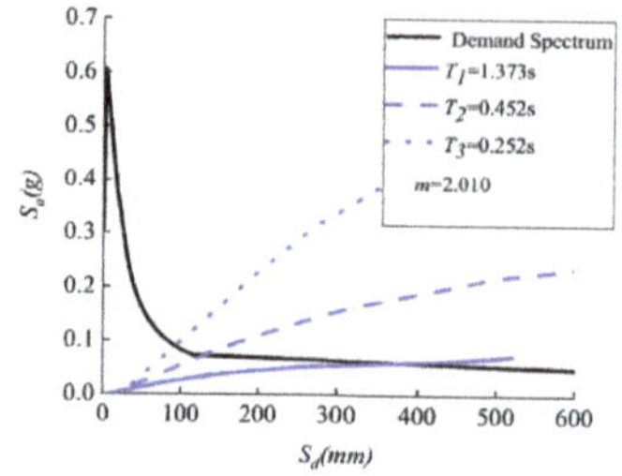

(**c**) D1-16-1.1

Figure 23. Seismic performance points of the 16 stories (D1).

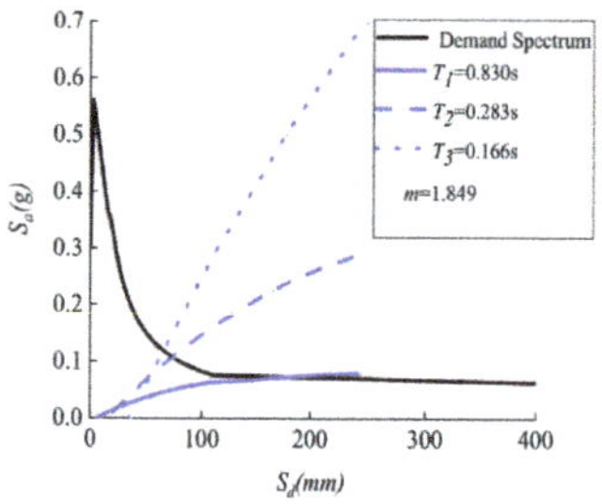

(**a**) D2-8-0.9

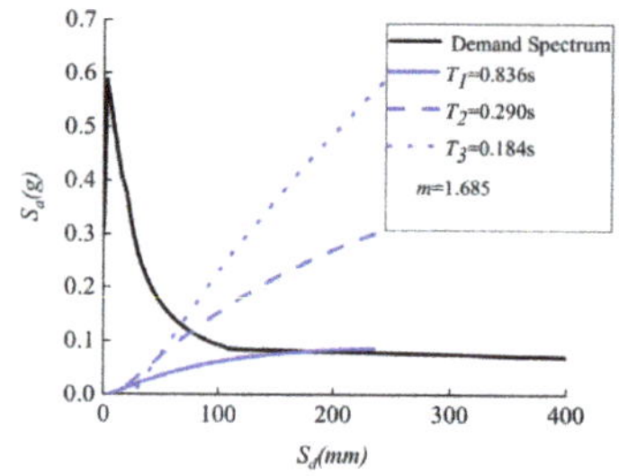

(**b**) D2-8-1.0

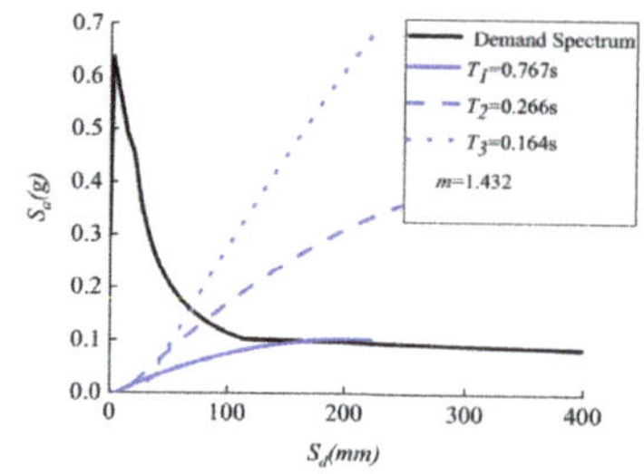

(**c**) D2-8-1.1

Figure 24. Seismic performance points of the 8 stories (D2).

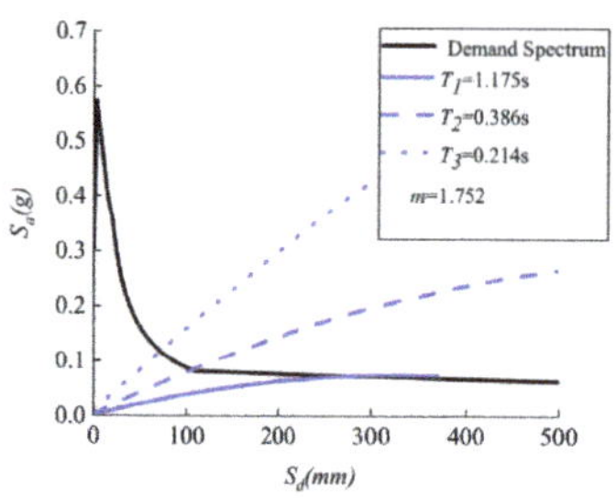

(**a**) D2-12-0.9

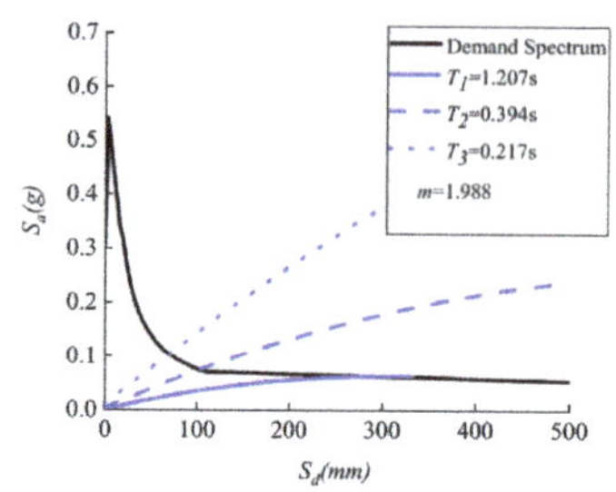

(**b**) D2-12-1.0

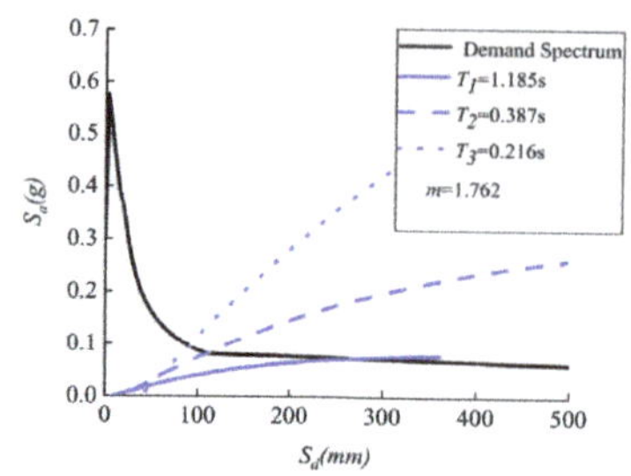

(**c**) D2-12-1.1

Figure 25. Seismic performance points of the 12 stories (D2).

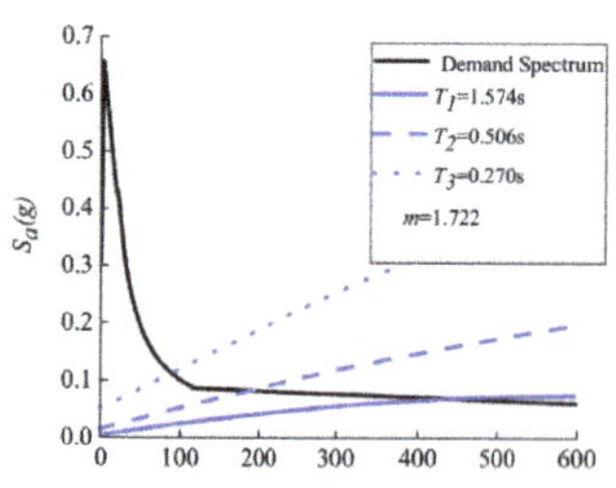

(**a**) D2-16-0.9

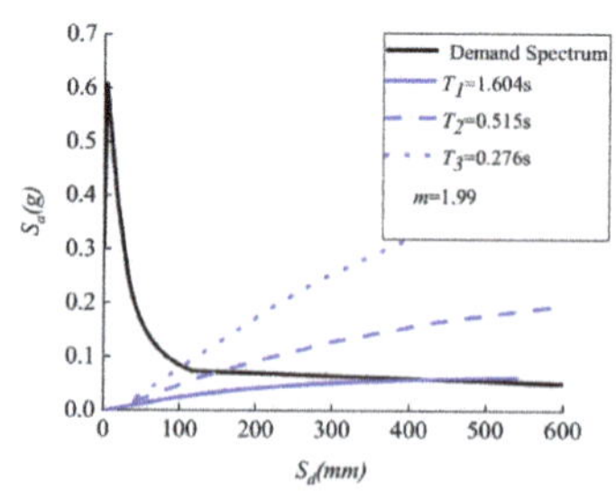

(**b**) D2-16-1.0

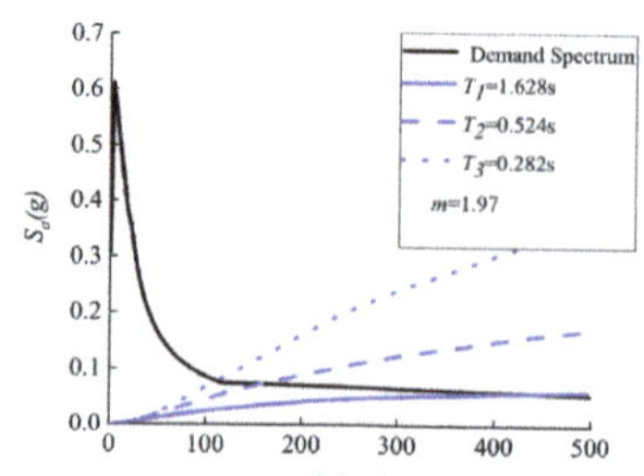

(**c**) D2-16-1.1

Figure 26. Seismic performance points of the 16 stories (D2).

4.3.4. Displacement Demand (Δ) for Each Case under Moderate Seismic Level

Drawing from the spectral displacements corresponding to the performance points derived from the demand spectrum and capacity spectrum for moderate seismic excitations, the vertex displacements Δ_i ($i = 1, 2, 3$) are systematically computed using the formula (12). Subsequently, the structural displacement demand Δ is determined by taking the square root of the sum of squares of the vertex displacements. A comprehensive summary of these specific results is presented in Table 7.

Table 7. Seismic displacement requirements in each example.

Model	S_{d1}	S_{d2}	S_{d3}	Δ_1	Δ_2	Δ_3	Δ
D1-8-0.9	164.557	74.853	65.984	229.523	41.942	20.219	234.198
D1-8-1.0	149.164	71.572	61.810	208.367	41.749	20.645	213.509
D1-8-1.1	125.95	68.88	58.96	177.191	40.671	18.649	182.753
D1-12-0.9	228.648	82.159	61.373	329.293	53.463	25.596	334.586
D1-12-1.0	231.879	85.481	72.583	332.494	53.135	27.044	337.797
D1-12-1.1	223.11	82.941	48.777	321.919	53.903	17.998	326.897
D1-16-0.9	388.069	125.779	108.095	571.235	88.159	44.572	579.714
D1-16-1.0	390.299	124.764	101.019	574.533	88.834	43.621	582.995
D1-16-1.1	355.598	124.532	87.967	525.528	89.948	40.725	534.723
D2-8-0.9	176.417	72.814	65.689	250.446	44.911	18.997	255.149
D2-8-1.0	173.917	75.126	66.941	249.411	50.554	22.602	255.485
D2-8-1.1	162.906	79.087	67.906	234.392	53.416	21.334	241.346
D2-12-0.9	276.216	104.061	70.611	409.138	74.309	27.791	416.759
D2-12-1.0	278.392	100.371	68.594	407.008	67.193	25.017	413.275
D2-12-1.1	264.846	108.341	89.743	392.209	77.868	34.257	401.329
D2-16-0.9	421.119	176.422	79.601	676.545	130.288	32.495	689.742
D2-16-1.0	405.683	143.051	97.358	607.085	106.063	41.380	617.669
D2-16-1.1	415.609	149.441	105.557	621.239	109.786	43.475	632.361

4.3.5. Elastic Base Shear (V_e) for Each Model under Moderate Seismic Level

The calculation of elastic base shear (V_e) for each case, under conditions of moderate seismic excitations, is predicated on the initial stiffness (K_e) of the structure and the displacement demand (Δ). Detailed outcomes of these calculations are meticulously presented in Table 8.

Table 8. The seismic elastic base shear in each example.

Model	**D1-8-0.9**	**D1-8-1.0**	**D1-8-1.1**	**D1-12-0.9**	**D1-12-1.0**	**D1-12-1.1**
V_e (KN)	7355.49	7362.83	7825.97	8105.99	7589.34	7854.94
Model	**D1-16-0.9**	**D1-16-1.0**	**D1-16-1.1**	**D2-8-0.9**	**D2-8-1.0**	**D2-8-1.1**
V_e (KN)	9535.63	9395.24	8840.37	6531.95	6324.52	6921.96
Model	**D2-12-0.9**	**D2-12-1.0**	**D2-12-1.1**	**D2-16-0.9**	**D2-16-1.0**	**D2-16-1.1**
V_e (KN)	8466.58	7973.85	8036.94	9785.02	9499.01	9367.52

4.3.6. Summary of Performance Factors for Each Case

Drawing upon the calculated results derived from the IDA analysis method elucidated earlier, comprehensive performance factors for each individual case can be ascertained. A comprehensive compilation of these specific results is succinctly summarized in Table 9.

Table 9. Summary table of performance coefficients of each calculation example.

Model	D1-8-0.9	D1-8-1.0	D1-8-1.1	D1-12-0.9	D1-12-1.0	D1-12-1.1
R	9.229	9.147	8.815	8.736	8.378	8.398
C_d	9.325	8.775	8.261	9.040	9.09	8.895
R_μ	1.649	1.538	1.395	1.912	2.048	1.322
R_Ω	5.595	5.946	6.317	4.568	4.089	6.348
Model	**D1-16-0.9**	**D1-16-1.0**	**D1-16-1.1**	**D2-8-0.9**	**D2-8-1.0**	**D2-8-1.1**
R	8.766	8.838	8.162	9.055	8.866	9.079
C_d	9.784	9.596	8.872	8.699	8.564	8.864
R_μ	1.933	1.982	2.009	1.850	1.687	1.431
R_Ω	4.535	4.457	4.061	4.894	5.254	6.341
Model	**D2-12-0.9**	**D2-12-1.0**	**D2-12-1.1**	**D2-16-0.9**	**D2-16-1.0**	**D2-16-1.1**
R	10.049	9.660	9.551	9.916	9.830	9.831
C_d	9.281	9.067	8.865	9.780	8.570	8.640
R_μ	1.751	1.987	1.763	1.724	1.992	1.970
R_Ω	5.737	4.86	5.416	5.749	4.933	4.989

5. The Impact of Design Parameters on R and C_d

Focusing on the number of floors (N), link length (e), and alterations in steel strength of non-dissipative frames, this section examines their influence on critical structural indicators such as the response modification factor (R), displacement amplification factor (C_d), and structural overstrength factor (R_Ω), all derived through Incremental Dynamic Analysis (IDA) methods. Compared to the data from the pushover analysis, it is noteworthy that the structure influence coefficient and displacement amplification factor obtained through IDA analysis are relatively smaller. This disparity can be attributed to the higher elastic stiffness of the fitted performance curve in IDA analysis, which generally surpasses the initial stiffness derived from the distributed lateral force adjustment method used in the pushover analysis. Additionally, the discrete nature of the IDA analysis method introduces certain distinctions in the structural performance factors obtained, highlighting the importance of considering the specific analytical method employed when evaluating the impact of design parameters on key structural response metrics.

5.1. Impact of Design Parameters on Response Modification Factor, R

(1) Influence of Story Number N

Figure 27 reveals that the response modification factor, as obtained from IDA analysis, does not exhibit a discernible regular pattern with respect to the story number. Specifically, as the story number increases from 8 to 12, there is a decrease in the R value for the Q460 combination case, while conversely, there is an upward trend in the R value for the Q690 combination case. However, the alteration in the R value is relatively minor when the story number increases from 12 to 16.

(2) Influence of Link Length e

In Figure 27, it is apparent that the response modification factor for most structural configurations tends to decline with an increase in the link length. Nevertheless, this decrease is not particularly substantial. Moreover, the correlation between the response modification factor and the link length is relatively modest, notably in the case of the 16-story configuration with the Q460 model and the 8-story configuration with the Q690 model.

(3) Influence of Variation in Steel Strength

A comparison of the response modification factor between two distinct high-strength steel combinations reveals nuanced variations. Specifically, for the 8-story configuration with the Q690 steel models, the response modification factor is marginally lower than that of its counterpart with the Q460 models. Conversely, for the 12-story and 16-story configurations with equivalent link lengths, the response modification factor for the Q690 high-strength steel surpasses the results obtained for the Q460 high-strength steel.

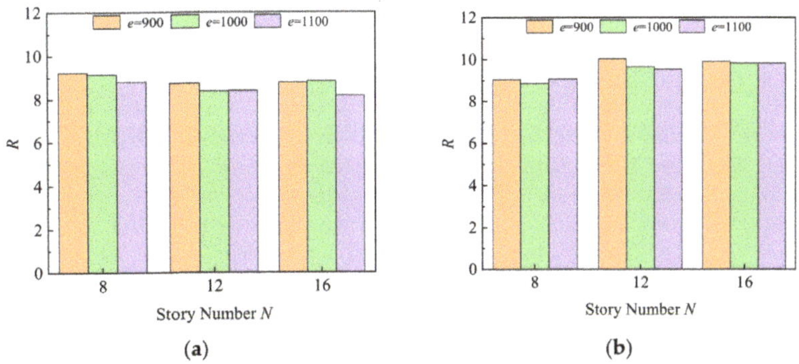

Figure 27. The influence of design parameters on R. (**a**) Example of Q460 high-strength steel combination calculation. (**b**) Example of Q690 high-strength steel combination calculation.

5.2. Impact of Design Parameters on Displacement Amplification Factor C_d

(1) Influence of Story Number N

Figure 28 reveals that the displacement amplification factor tends to increase with a rise in the story number, notably in the scenario with a link length of 900 mm. The most significant surge in the displacement amplification factor is observed during the transition from 12 stories to 16 stories. However, for other link lengths, the relationship between the displacement amplification factor and the story number is not as pronounced.

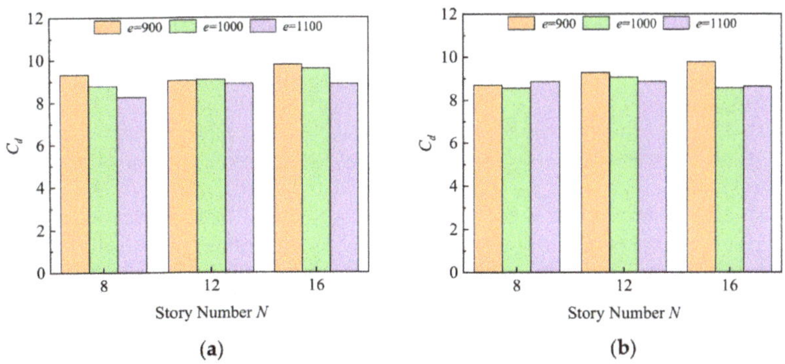

Figure 28. The influence of design parameters on C_d. (**a**) Example of Q460 high-strength steel combination calculation. (**b**) Example of Q690 high-strength steel combination calculation.

(2) Influence of Link Length e

From Figure 28, it becomes evident that the displacement amplification factor generally diminishes with an increase in the link length across various structural configurations. Notably, in the 12-story model with the Q460 high-strength steel and the 8-story model with the Q690 high-strength steel, the displacement amplification factor exhibits relatively stable changes with a weak correlation to the number of floors.

(3) Influence of Variation in Steel Strength

Figure 28 highlights that in specific cases, such as the 12-story configuration with link lengths of 1000 mm or 1100 mm, and the 16-story configuration with a link length of 900 mm, the disparities in displacement amplification factor C_d between the two steel strength combinations are marginal. However, in the remaining scenarios, the displacement amplification factor for the Q460 high-strength steel models slightly exceeds the results observed for the Q690 high-strength steel models.

5.3. *Impact of Design Parameters on Structure Overstrength Factor R_{Ω}*

(1) Influence of Story Number *N*

Figure 29 reveals noteworthy trends in the structure overstrength factor R_{Ω}. Specifically, in cases involving the Q460 high-strength steel combination, a substantial decline in R_{Ω} is evident as the number of stories increases from 8 to 12 for link lengths of 900 mm and 1000 mm. Moreover, in the scenario featuring a link length of 1100 mm, there is a more pronounced decrease in R_{Ω} when transitioning from 12 to 16 stories. In contrast, for the Q690 high-strength steel combination cases, the R_{Ω} values for the link length of 900 mm exhibit an increase as the number of stories rises, while other cases showcase varying degrees of decrease.

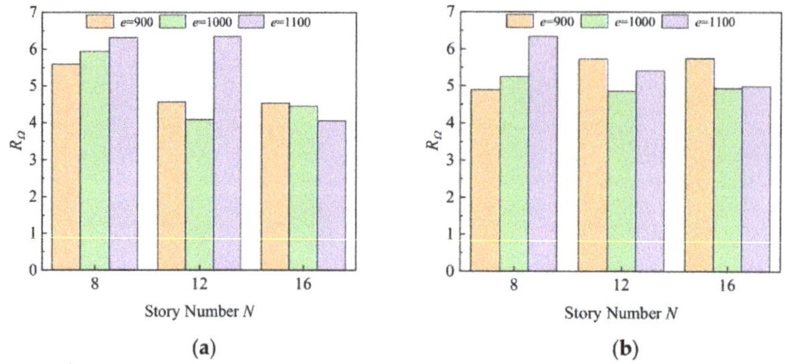

Figure 29. The influence of design parameters on R_{Ω}. (**a**) Example of Q460 high-strength steel combination calculation. (**b**) Example of Q690 high-strength steel combination calculation.

(2) Influence of Link Length *e*

Figure 29 illustrates that, for structures with 8 stories, the structure overstrength factor R_{Ω} typically increases with a growing number of stories. In the case of 12-story structures, there is a slight reduction in R_{Ω} as the link length extends from 900 mm to 1000 mm, while an increase is noted when the link length is further increased to 1100 mm. This increase is particularly pronounced in the Q460 high-strength steel combination cases. Meanwhile, in the context of 16-story structures, R_{Ω} exhibits a decreasing trend as the link length increases.

(3) Influence of Variation in Steel Strength

Figure 29 demonstrates that the influence of steel strength variation on the structure overstrength factor R_{Ω}, as gleaned from IDA analysis, lacks a distinct regular pattern. The average R_{Ω} values for the two different steel combinations, with the same number of stories, are as follows: 5.952, 5.496, 5.001, 5.337, 4.351, and 5.223.

6. Conclusions

In this study, the Incremental Dynamic Analysis (IDA) method was utilized to calculate key structural metrics for 18 structural models. These metrics include the response modification factor (*R*), displacement amplification factor (C_d), and structure overstrength factor (R_{Ω}). The analysis focused on the impact of three critical design parameters: the number of floors, the link lengths, and variations in steel combinations. The conclusions are as follows:

(1) The response modification factor (*R*) and structure overstrength factor (R_{Ω}) obtained through Incremental Dynamic Analysis (IDA) show only minor discrepancies compared to the results from pushover analysis. However, the displacement amplification factor (C_d) exhibits significant differences between the two methods. These differences are primarily due to variations in the initial stiffness of the structural performance curves derived from the two methods of analysis.

(2) The displacement amplification factor (C_d) derived from Incremental Dynamic Analysis (IDA) showed little variation with the number of stories, with no significant effect

observed. However, the impact of the link length (e) on the C_d value exhibits irregularity in IDA analysis. Moreover, for various high-strength steel composite models, the C_d value obtained from IDA analysis for the Q690 high-strength steel composite model shows inconsistent results.

(3) The influence of the story number on the response modification factor (R) is minimal, according to IDA analysis. Similarly, a clear relationship between the R value and the link length is not evident in the 16-story cases, which is consistent with findings from the pushover analysis. In the other cases, a general trend is observed where the R value decreases as the link length increases. Notably, for the 12-story cases, both methods of analysis yield congruent results, demonstrating that the structure influence coefficient is higher for the Q690 high-strength steel combination compared to the Q460 high-strength steel combination.

(4) For cases with a link length of 900 mm, the displacement amplification factor (C_d) from IDA analysis shows an increasing trend as the story number rises, though this effect is less pronounced in other cases. Comparing the two methods of analysis, pushover analysis demonstrates a more substantial increase in C_d with an increase in the story number. Additionally, the agreement between the two methods on the influence of link length and different high-strength steel combinations on C_d is relatively limited.

(5) For cases with a link length of 1100 mm, both methods of analysis show a significant decrease in the structure overstrength factor (R_Ω) as the number of floors increases. This decrease is particularly pronounced. Additionally, in the 8-story and 12-story cases, an increase in link length results in an improvement in R_Ω, displaying a consistent trend in both methods. Notably, pushover analysis yields a higher structure overstrength factor for the Q690 high-strength steel combination compared to the Q460 high-strength steel combination, while the trend obtained from IDA analysis is less pronounced.

These findings provide valuable insights into the impact of critical design parameters on key structural response metrics, clarifying the complexities involved in assessing structural behavior under seismic loading conditions.

Author Contributions: Methodology, Y.M.; Software, J.Y.; Validation, Y.M.; Investigation, J.Y.; Data curation, J.Y.; Writing—original draft, Y.M.; Writing—review & editing, X.M. All authors have read and agreed to the published version of the manuscript.

Funding: This study was supported by the Ningxia Natural Science Foundation (Grant No. 2021AAC03189, Grant No. 2023AAC03317) and the National Natural Science Foundation of China (Grant No. 52368011). The financial support is greatly appreciated.

Data Availability Statement: Data are contained within the article.

Conflicts of Interest: The authors declare no conflict of interest.

Appendix A

Table A1. Seismic records for model D1-8-0.9.

Seismic Wave Number	Seismic Wave	Seismic Wave Station	PGA/g	Earthquake Degree
RSN9	Borrego	El Centro Array #9	0.066	6.5
RSN161	Imperial Valley-06	Brawley Airport	0.163	6.53
RSN764	Loma Prieta	Gilroy—Historic Bldg.	0.285	6.93
RSN879	Landers	SCE 24 Lucerne	0.727	7.28
RSN1489	Chi-Chi Taiwan	TCU049	0.279	7.62
RSN1602	Duzce Turkey	Bolu	0.739	7.14
RSN3746	Cape Mendocino	Centerville Beach Naval Fac	0.318	7.01
RSN4040	Bam Iran	Bam	0.808	6.6
RSN6927	Darfield New Zealand	LINC	0.461	7
RSN6960	Darfield New Zealand	Riccarton High School	0.19	7

Table A2. Seismic records for models D1-8-1.0, D1-8-1.1.

Seismic Wave Number	Seismic Wave	Seismic Wave Station	PGA/g	Earthquake Degree
RSN161	Imperial Valley-06	Brawley Airport	0.163	6.53
RSN173	Imperial Valley-06	El Centro Array #10	0.11	6.53
RSN764	Loma Prieta	Gilroy—Historic Bldg.	0.285	6.93
RSN802	Loma Prieta	Saratoga—Aloha Ave	0.513	6.93
RSN879	Landers	SCE 24 Lucerne	0.727	7.28
RSN1085	Northridge-01	Sylmar—Converter Sta East	0.853	6.69
RSN1489	Chi-Chi Taiwan	TCU049	0.279	7.62
RSN1602	Duzce Turkey	Bolu	0.739	7.14
RSN3746	Cape Mendocino	Centerville Beach Naval Fac	0.318	7.01
RSN4040	Bam Iran	Bam	0.808	6.6

Table A3. Seismic records for model D1-12-0.9.

Seismic Wave Number	Seismic Wave	Seismic Wave Station	PGA/g	Earthquake Degree
RSN163	Imperial Valley	Calipatria Fire Station	0.129	6.53
RSN176	Imperial Valley	El Centro Array #13	0.049	6.53
RSN184	Imperial Valley-06	El Centro Differential Array	0.353	6.53
RSN802	Loma Prieta	Saratoga—Aloha Ave	0.513	6.93
RSN879	Landers	SCE 24 Lucerne	0.727	7.28
RSN1085	Northridge-01	Sylmar—Converter Sta East	0.853	6.69
RSN1602	Duzce Turkey	Bolu	0.739	7.14
RSN4228	Niigata Japan	NIGH11	0.599	6.63
RSN6927	Darfield New Zealand	LINC	0.461	7
RSN6933	Darfield New Zealand	MAYC	0.068	7

Table A4. Seismic records for model D1-12-1.0.

Seismic Wave Number	Seismic Wave	Seismic Wave Station	PGA/g	Earthquake Degree
RSN171	Imperial Valley-06	El Centro—Meloland Geot.Array	0.317	6.53
RSN184	Imperial Valley-06	El Centro Differential Array	0.353	6.53
RSN802	Loma Prieta	Saratoga—Aloha Ave	0.513	6.93
RSN1085	Northridge-01	Sylmar—Converter Sta East	0.853	6.69
RSN1148	Kocaeli Turkey	KOERI 99999 Arcelik	0.22	7.51
RSN1493	Chi-Chi Taiwan	TCU053	0.229	7.62
RSN1602	Duzce Turkey	Bolu	0.739	7.14
RSN4228	Niigata Japan	NIGH11	0.599	6.63
RSN5783	Wate Japan	Semine Kurihara City	0.142	6.9
RSN6927	Darfield New Zealand	LINC	0.461	7

Table A5. Seismic records for model D1-12-1.1.

Seismic Wave Number	Seismic Wave	Seismic Wave Station	PGA/g	Earthquake Degree
RSN171	Imperial Valley-06	El Centro—Meloland Geot.Array	0.317	6.53
RSN184	Imperial Valley-06	El Centro Differential Array	0.353	6.53
RSN802	Loma Prieta	Saratoga—Aloha Ave	0.513	6.93
RSN1004	Northridge-01	LA—Sepulveda VA Hospital	0.753	6.69
RSN1085	Northridge-01	Sylmar—Converter Sta East	0.853	6.69
RSN1119	Kobe_ Japan	Takarazuka	0.697	6.9
RSN1165	Kocaeli, Turkey	ERD 99999 Izmit	0.22	7.51
RSN1493	Chi-Chi Taiwan	TCU053	0.229	7.62
RSN1602	Duzce Turkey	Bolu	0.739	7.14
RSN3746	Cape Mendocino	Centerville Beach Naval Fac	0.318	7.01

Table A6. Seismic records for models D1-16-0.9, D1-16-1.0, D1-16-1.1.

Seismic Wave Number	Seismic Wave	Seismic Wave Station	PGA/g	Earthquake Degree
RSN171	Imperial Valley-06	El Centro—Meloland Geot.Array	0.317	6.53
RSN173	Imperial Valley-06	El Centro Array #10	0.11	6.53
RSN184	Imperial Valley-06	El Centro Differential Array	0.353	6.53
RSN185	Imperial Valley-06	Holtville Post Office	0.258	6.53
RSN767	Loma Prieta	Gilroy Array #3	0.342	6.93
RSN879	Landers	SCE 24 Lucerne	0.727	7.28
RSN1004	Northridge-01	LA—Sepulveda VA Hospital	0.753	6.69
RSN1085	Northridge-01	Sylmar—Converter Sta East	0.853	6.69
RSN1489	Chi-Chi Taiwan	TCU049	0.279	7.62
RSN1493	Chi-Chi Taiwan	TCU053	0.229	7.62

Table A7. Seismic records for models D2-8-0.9, D2-8-1.0, D2-8-1.1.

Seismic Wave Number	Seismic Wave	Seismic Wave Station	PGA/g	Earthquake Degree
RSN161	Imperial Valley-06	Brawley Airport	0.163	6.53
RSN764	Loma Prieta	Gilroy—Historic Bldg.	0.285	6.93
RSN802	Loma Prieta	Saratoga—Aloha Ave	0.513	6.93
RSN879	Landers	SCE 24 Lucerne	0.727	7.28
RSN1489	Chi-Chi Taiwan	TCU049	0.279	7.62
RSN1602	Duzce Turkey	Bolu	0.739	7.14
RSN3746	Cape Mendocino	Centerville Beach Naval Fac	0.318	7.01
RSN4040	Bam Iran	Bam	0.808	6.6
RSN6927	Darfield New Zealand	LINC	0.461	7
RSN6960	Darfield New Zealand	Riccarton High School	0.19	7

Table A8. Seismic records for models D2-12-0.9, D2-12-1.0, D2-12-1.1.

Seismic Wave Number	Seismic Wave	Seismic Wave Station	PGA/g	Earthquake Degree
RSN171	Imperial Valley-06	El Centro—Meloland Geot.Array	0.317	6.53
RSN173	Imperial Valley-06	El Centro Array #10	0.11	6.53
RSN802	Loma Prieta	Saratoga—Aloha Ave	0.513	6.93
RSN879	Landers	SCE 24 Lucerne	0.727	7.28
RSN1086	Northridge-01	Sylmar—Olive View Med FF	0.605	6.69
RSN1119	Kobe_ Japan	Takarazuka	0.697	6.9
RSN1165	Kocaeli, Turkey	ERD 99999 Izmit	0.22	7.51
RSN1602	Duzce Turkey	Bolu	0.739	7.14
RSN4228	Niigata Japan	NIGH11	0.599	6.63
RSN6927	Darfield New Zealand	LINC	0.461	7

Table A9. Seismic records for model D2-16-0.9.

Seismic Wave Number	Seismic Wave	Seismic Wave Station	PGA/g	Earthquake Degree
RSN173	Imperial Valley-06	El Centro Array #10	0.11	6.53
RSN184	Imperial Valley-06	El Centro Differential Array	0.353	6.53
RSN185	Imperial Valley-06	Holtville Post Office	0.258	6.53
RSN767	Loma Prieta	Gilroy Array #3	0.342	6.93
RSN802	Loma Prieta	Saratoga—Aloha Ave	0.513	6.93
RSN879	Landers	SCE 24 Lucerne	0.727	7.28
RSN1004	Northridge-01	LA—Sepulveda VA Hospital	0.753	6.69
RSN1602	Duzce Turkey	Bolu	0.739	7.14
RSN4228	Niigata Japan	NIGH11	0.599	6.63
RSN6897	Darfield New Zealand	DSLC	0.257	7

Table A10. Seismic records for models D2-16-1.0, D2-16-1.1.

Seismic Wave Number	Seismic Wave	Seismic Wave Station	PGA/g	Earthquake Degree
RSN173	Imperial Valley-06	El Centro Array #10	0.11	6.53
RSN185	Imperial Valley-06	Holtville Post Office	0.258	6.53
RSN828	Cape Mendocino	Petrolia	0.591	7.01
RSN879	Landers	SCE 24 Lucerne	0.727	7.28
RSN1086	Northridge-01	Sylmar—Olive View Med FF	0.605	6.69
RSN1493	Chi-Chi Taiwan	TCU053	0.229	7.62
RSN1602	Duzce Turkey	Bolu	0.739	7.14
RSN4228	Niigata Japan	NIGH11	0.599	6.63
RSN6897	Darfield New Zealand	DSLC	0.257	7
RSN6906	Darfield New Zealand	GDLC	0.765	7

References

1. *TJ11-78;* Code for Seismic Design of Industrial and Civil Buildings. China Architecture Industry Press: Beijing, China, 1978. (In Chinese)
2. *GBJ 11-89;* Code for Seismic Design of Buildings. China Architecture Industry Press: Beijing, China, 1989. (In Chinese)
3. *CECS 160:2004;* General Rule for Performance-Based Seismic Design of Buildings. China Plan Press: Beijing, China, 2004. (In Chinese)
4. International Council of Building Officials. *1997 Uniform Building Code;* International of Building Officials: Whittier, CA, USA, 1997.
5. *ASCE7-05;* Minimum Design Loads for Building and Other Structures. American Society of Civil Engineers: Reston, VA, USA, 2006.
6. Building Seismic Safety Council (BSSC). NEHRP Recommended Provisions for Seismic Regulations for New Buildings and Other Structures. In *National Earthquake Hazard Reduction Program (NEHRP);* Building Seismic Safety Council (BSSC): Washington, DC, USA, 2000.
7. Uang, C.M. Establishing R (or R_w) and C_d factors for building seismic provisions. *J. Struct. Eng.* **1991**, *117*, 19–28. [CrossRef]
8. Su, M.Z.; Liu, Y.L. Study on structural behavior factor of high-strength steel composite Y-type eccentrically braced frame based on incremental dynamic analysis. *J. Xi'an Univ. Archit. Technol. (Nat. Sci. Ed.)* **2014**, *46*, 635–642.
9. Xu, J.; Wang, S.L.; Mohamed, H.S. Seismic performance of shear energy dissipation beams in D-shaped eccentrically braced steel frames. *J. Constr. Steel Res.* **2021**, *180*, 106584. [CrossRef]
10. Li, S.; Wang, Z.; Guo, H. Seismic performance of high strength steel frames with variable eccentric braces based on PBSD method. *Earthq. Struct.* **2020**, *18*, 527–542.
11. Mohammadgholipour, A.; Billah, M.A. Performance-based plastic design and seismic fragility assessment for chevron braced steel frames considering aftershock effects. *Soil Dyn. Earthq. Eng.* **2024**, *178*, 108440. [CrossRef]
12. Kalapodis, N.A.; Muho, E.V.; Papagiannopoulos, G.A. Integration of peak seismic floor velocities and accelerations into the performance-based design of steel structures. *Soil Dyn. Earthq. Eng.* **2022**, *154*, 107160. [CrossRef]
13. *GB50011-2010;* Code for seismic design of buildings. China Architecture Industry Press: Beijing, China, 2010. (In Chinese)
14. Zheng, L.; Qin, C. Research on the structural influencing coefficient of steel plate shear wall. *Ind. Constr.* **2020**, *50*, 154–161.
15. Cheng, Q.Q.; Su, M.Z. Response modification factor of high-strength steel fabricated framed-tube with replaceable shear links. *Eng. Mech.* **2020**, *37*, 145–158.
16. Mussa, M.; Mahdi, Z. Determination of the Response Modification Factors of Buckling Restrained Braced Frames. *Procedia Eng.* **2013**, *54*, 222–231.
17. Yousef, H.; Al, Q.; Besan, Y.A. Evaluating seismic response modification factor of steel frames with different bracing systems. *Int. J. Earthq. Impact Eng.* **2018**, *2*, 203–225.
18. Li, S.; Liang, W.G. Response modification factor and displacement amplification factor of K-shaped eccentrically braced high-strength steel frames. *J. Asian Archit. Build. Eng.* **2022**, *21*, 2399–2425. [CrossRef]
19. *SAP2000;* Integrated Software for Structural Analysis & Design, Version 14.2.4—Analysis Reference Manual. Computers and Structures, Inc. (CSI): Berkeley, CA, USA, 2010.
20. Li, S.; Tian, J.B. Response on the seismic behavior of Y-type eccentrically braced frames combined with high strength steels based on PBSD method. *Ind. Constr.* **2018**, *48*, 140–150.
21. *GB50009-2012;* Load Code for Design of Building Structures. China Architecture Industry Press: Beijing, China, 2012. (In Chinese)
22. Li, S.; Gao, Y.T. Story shear distribution of high strength steel frame with D-eccentric braces based on multi-seismic-hazard levels. *Structures* **2023**, *48*, 616–641.
23. Richards, P.W. *Cyclic Stability and Capacity Design of Steel Eccentrically Braced Frames;* University of California: San Diego, CA, USA, 2004.

24. Rides, J.M.; Popov, E.P. Inelastic Link Element for EBF Seismic Analysis. *J. Struct. Eng.* **1994**, *120*, 441–463. [CrossRef]
25. Ramadan, T.; Ghobarah, A. Analytical model for shear-link behavior. *J. Struct. Eng.* **1995**, *121*, 1574–1580. [CrossRef]

MDPI AG
Grosspeteranlage 5
4052 Basel
Switzerland
Tel.: +41 61 683 77 34

Buildings Editorial Office
E-mail: buildings@mdpi.com
www.mdpi.com/journal/buildings

www.ingramcontent.com/pod-product-compliance
Lightning Source LLC
Chambersburg PA
CBHW061817210926
51850CB00006B/730
9783725852093